Horticulture Plants Stress Physiology

Horticulture Plants Stress Physiology

Guest Editor

Hakim Manghwar

MDPI

Basel • Beijing • Wuhan • Barcelona • Belgrade • Novi Sad • Cluj • Manchester

Guest Editor
Hakim Manghwar
Lushan Botanical Garden
Jiangxi Province and Chinese
Academy of Sciences
Jiujiang
China

Editorial Office
MDPI AG
Grosspeteranlage 5
4052 Basel, Switzerland

This is a reprint of the Special Issue, published open access by the journal *Horticulturae* (ISSN 2311-7524), freely accessible at: https://www.mdpi.com/journal/horticulturae/special_issues/B85450K0JB.

For citation purposes, cite each article independently as indicated on the article page online and as indicated below:

Lastname, A.A.; Lastname, B.B. Article Title. *Journal Name* **Year**, *Volume Number*, Page Range.

ISBN 978-3-7258-6752-3 (Hbk)
ISBN 978-3-7258-6753-0 (PDF)
https://doi.org/10.3390/books978-3-7258-6753-0

Contents

About the Editor

Hakim Manghwar

Hakim Manghwar is an Associate Researcher at the Lushan Botanical Garden, Chinese Academy of Sciences, China, and an approved Master's supervisor at the School of Life Sciences, Nanchang University, Nanchang, China. He received his PhD in Crop Breeding and Genetics from Huazhong Agricultural University, Wuhan, China. His research interests focus on plant stress physiology, plant molecular biology, functional genomics, genome editing, and crop resistance to biotic and abiotic stresses. Dr. Manghwar has published extensively in high-impact journals, including *Trends in Plant Science*, *Advanced Science*, and *Plant Biotechnology Journal*, with over 4400 citations and an h-index of 30. He actively serves as an editor and reviewer for multiple international journals and was listed among the Top 2% of Scientists Worldwide (2025) by Stanford/Elsevier.

horticulturae

Editorial

Horticulture Plants' Stress Physiology

Hakim Manghwar

Lushan Botanical Garden, Jiangxi Province and Chinese Academy of Sciences, Jiujiang 332000, China; hakim@lsbg.cn

The growth and productivity of horticultural crops have been significantly impacted by human activities and global climate change [1]. In order to effectively meet the demands of a rapidly growing global population, which is projected to surpass 3 billion, it is anticipated that agricultural production must significantly increase by up to 70% by the middle of this century [2]. Abiotic stress conditions include drought, salinity, heat, sodic alkaline, heavy metal exposure, etc. [3–5], and biotic stress conditions, including the presence of bacteria, fungi, viruses, nematodes, and insects, significantly impact the productivity and vigor of horticultural crops [4,6–9]. These biotic and abiotic stresses impact seed germination in horticultural crops as well as their reproduction, quality, growth, and yield [4]. Consequently, it is important to investigate the physiological, molecular, and biochemical mechanisms at work in horticultural crops to determine the impact of these stresses on their growth and identify mitigation strategies and potential resistance mechanisms [4]. In addition, plants have evolved complex adaptive mechanisms that enhance their resilience to adverse conditions [10]. A clear understanding of these biotic and abiotic stresses and their interaction with the physiological processes of horticultural crops is essential for improving these crops' resistance to stress.

This Editorial refers to the Special Issue "Horticulture Plants' Stress Physiology", which focuses on exploring novel opportunities and addressing the challenges in enhancing the performance of horticultural plants under diverse biotic and abiotic stresses and the molecular mechanisms that they use to cope with these stresses. This Special Issue attracted a significant number of submissions, all of which underwent a thorough peer-review process. A total of thirteen papers were accepted for publication in this Special Issue, comprising twelve original research articles and one review. The final contributions are listed below.

List of Contributions:

1. Stefanakis, M.K.; Giannakoula, A.E.; Ouzounidou, G.; Papaioannou, C.; Lianopoulou, V.; Philotheou-Panou, E. The Effect of Salinity and Drought on the Essential Oil Yield and Quality of Various Plant Species of the Lamiaceae Family (*Mentha spicata* L., *Origanum dictamnus* L., *Origanum onites* L.). *Horticulturae* **2024**, *10*, 265.
2. He, Y.; Yadav, V.; Bai, S.; Wu, J.; Zhou, X.; Zhang, W.; Han, S.; Wang, M.; Zeng, B.; Wu, X.; et al. Performance Evaluation of New Table Grape Varieties under High Light Intensity Conditions Based on the Photosynthetic and Chlorophyll Fluorescence Characteristics. *Horticulturae* **2023**, *9*, 1035.
3. Parri, S., Romi, M., Hoshika, Y., Giovannelli, A., Dias, M.C., Piritore, F.C., Cai, G. and Cantini, C. Morpho-physiological responses of three Italian olive tree (*Olea europaea* L.) cultivars to drought stress. *Horticulturae* **2023**, *9*, 830.
4. Jing, J.; Liu, M.; Yin, B.; Liang, B.; Li, Z.; Zhang, X.; Xu, J.; Zhou, S. Effects of 10 Dwarfing Interstocks on Cold Resistance of 'Tianhong 2'Apple. *Horticulturae* **2023**, *9*, 827.
5. Ahmed, S.; Wan Azizan WA, S.; Akhond MA, Y.; Juraimi, A.S.; Ismail, S.I.; Ahmed, R.; Md Hatta, M.A. Optimization of In Vitro Regeneration Protocol of Tomato cv. MT1 for Genetic Transformation. *Horticulturae* **2023**, *9*, 800.

6. Wu, J.; Abudureheman, R.; Zhong, H.; Yadav, V.; Zhang, C.; Ma, Y.; Liu, X.; Zhang, F.; Zha, Q.; Wang, X. The impact of high temperatures in the field on leaf tissue structure in different grape cultivars. *Horticulturae* **2023**, *9*, 731.
7. Fu, B.; Tian, Y. Combined Study of Transcriptome and Metabolome Reveals Involvement of Metabolites and Candidate Genes in Flavonoid Biosynthesis in *Prunus avium* L. *Horticulturae* **2023**, *9*, 463.
8. Park, B.M.; Jeong, H.B.; Yang, E.Y.; Kim, M.K.; Kim, J.W.; Chae, W.; Lee, O.J.; Kim, S.G.; Kim, S. Differential responses of cherry tomatoes (*Solanum lycopersicum*) to long-term heat stress. *Horticulturae* **2023**, *9*, 343.
9. Taher, D.; Nofal, E.; Hegazi, M.; El-Gaied, M.A.; El-Ramady, H.; Solberg S, Ø. Response of warm season turf grasses to combined cold and salinity stress under foliar applying organic and inorganic amendments. *Horticulturae* **2023**, *9*, 49.
10. Bantis, F.; Koukounaras, A. Ascophyllum nodosum and Silicon-Based Biostimulants differentially affect the physiology and growth of watermelon transplants under abiotic stress factors: The case of drought. *Horticulturae* **2022**, *8*, 1177.
11. Reyad NE, H.A.; Azoz, S.N.; Ali, A.M.; Sayed, E.G. Mitigation of Powdery Mildew Disease by integrating Biocontrol Agents and Shikimic acid with modulation of antioxidant defense system, anatomical characterization, and improvement of Squash Plant Productivity. *Horticulturae* **2022**, *8*, 1145.
12. Yang, L.; Li, P.; Qiu, L.; Ahmad, S.; Wang, J.; Zheng, T. Identification and comparative analysis of the rosaceae RCI2 gene family and characterization of the cold stress response in Prunus mume. *Horticulturae* **2022**, *8*, 997.
13. Ahammed, G.J.; Li, X. Melatonin-induced detoxification of organic pollutants and alleviation of phytotoxicity in selected horticultural crops. *Horticulturae* **2022**, *8*, 1142.

References

1. Varshney, R.K.; Singh, V.K.; Kumar, A.; Powell, W.; Sorrells, M.E. Can genomics deliver climate-change ready crops? *Curr. Opin. Plant Biol.* **2018**, *45*, 205–211. [CrossRef]
2. Francini, A.; Sebastiani, L. Abiotic stress effects on performance of horticultural crops. *Horticulturae* **2019**, *5*, 67. [CrossRef]
3. Lal, M.K.; Tiwari, R.K.; Gahlaut, V.; Mangal, V.; Kumar, A.; Singh, M.P.; Paul, V.; Kumar, S.; Singh, B.; Zinta, G. Physiological and molecular insights on wheat responses to heat stress. *Plant Cell Rep.* **2021**, *41*, 501–518. [CrossRef] [PubMed]
4. Lal, M.K.; Tiwari, R.K.; Altaf, M.A.; Kumar, A.; Kumar, R. Abiotic and biotic stress in horticultural crops: Insight into recent advances in the underlying tolerance mechanism. *Front. Plant Sci.* **2023**, *14*, 1212982. [CrossRef] [PubMed]
5. Manghwar, H.; Hussain, A. Mechanism of tobacco osmotin gene in plant responses to biotic and abiotic stress tolerance: A brief history. *Biocell* **2022**, *46*, 623. [CrossRef]
6. Kumar, R.; Tiwari, R.K.; Jeevalatha, A.; Siddappa, S.; Shah, M.A.; Sharma, S.; Sagar, V.; Kumar, M.; Chakrabarti, S.K. Potato apical leaf curl disease: Current status and perspectives on a disease caused by tomato leaf curl new delhi virus. *J. Plant Dis. Prot.* **2021**, *128*, 897–911. [CrossRef]
7. Lal, M.K.; Tiwari, R.K.; Kumar, R.; Naga, K.C.; Kumar, A.; Singh, B.; Raigond, P.; Dutt, S.; Chourasia, K.N.; Kumar, D. Effect of potato apical leaf curl disease on glycemic index and resistant starch of potato (*Solanum tuberosum* L.) tubers. *Food Chem.* **2021**, *359*, 129939. [CrossRef] [PubMed]
8. Gul, S.; Hussain, A.; Ali, Q.; Alam, I.; Alshegaihi, R.M.; Meng, Q.; Zaman, W.; Manghwar, H.; Munis, M.F.H. Hydropriming and osmotic priming induce resistance against aspergillus niger in wheat (*Triticum aestivum* L.) by activating β-1, 3-glucanase, chitinase, and thaumatin-like protein genes. *Life* **2022**, *12*, 2061. [CrossRef] [PubMed]
9. Manghwar, H.; Hussain, A.; Ali, Q.; Saleem, M.H.; Abualreesh, M.H.; Alatawi, A.; Ali, S.; Munis, M.F.H. Disease severity, resistance analysis, and expression profiling of pathogenesis-related protein genes after the inoculation of fusarium equiseti in wheat. *Agronomy* **2021**, *11*, 2124. [CrossRef]
10. Manzoor, M.A.; Xu, Y.; Xu, J.; Wang, Y.; Sun, W.; Liu, X.; Wang, L.; Wang, J.; Liu, R.; Whiting, M.D. Fruit crop abiotic stress management: A comprehensive review of plant hormones mediated responses. *Fruit Res.* **2023**, *3*, 30. [CrossRef]

horticulturae

MDPI

Article

Performance Evaluation of New Table Grape Varieties under High Light Intensity Conditions Based on the Photosynthetic and Chlorophyll Fluorescence Characteristics

Yawen He [1,2,†], Vivek Yadav [1,†], Shijian Bai [3], Jiuyun Wu [4], Xiaoming Zhou [1], Wen Zhang [1], Shouan Han [1], Min Wang [1], Bin Zeng [2], Xinyu Wu [1], Haixia Zhong [1,*] and Fuchun Zhang [1,*]

1 The State Key Laboratory of Genetic Improvement and Germplasm Innovation of Crop Resistance in Arid Desert Regions (Preparation), Key Laboratory of Genome Research and Genetic Improvement of Xinjiang Characteristic Fruits and Vegetables, Institute of Horticulture Crops, Xinjiang Academy of Agricultural Sciences, Urumqi 830091, China; vivekyadav@nwafu.edu.cn (V.Y.); hanshouan@163.com (S.H.)
2 College of Horticulture, Xinjiang Agricultural University, Urumqi 830091, China
3 Research Institute of Grape and Melon Fruits in Xinjiang Uygur Autonomous Region, Turpan 838200, China
4 Turpan Research Institute of Agricultural Sciences, Xinjiang Academy of Agricultural Science, Turpan 830000, China
* Correspondence: zhonghaixia1@xaas.ac.cn (H.Z.); zhangfc@xaas.ac.cn (F.Z.); Tel.: +86-18-119-152-951 (H.Z.); +86-15-299-198-295 (F.Z.)
† These authors contributed equally to this work.

Abstract: The evaluation of photosynthetic characteristics of plants is important for the success rate of germplasm introduction. To select grape varieties with higher adaptability and trait performance, this experiment is aimed at evaluating and comparing the photosynthetic indices, chlorophyll fluorescence parameters, photosynthetic pigment content, and leaf characteristics of five Chinese hybrid varieties. The results showed that under high light intensity stress, the leaf growth of 'Ruidu Cuixia' was most affected and its specific leaf weight was the lowest, while 'Jing Hongbao' had the highest chlorophyll content. The maximum net photosynthetic rate (P_{nmax}), maximum light quantum yield (F_v/F_m), and apparent quantum efficiency (AQE) were different among varieties. It was reported that the 'Ruidu Zaohong' variety had the highest P_{nmax}. 'Ruidu Wuheyi' was found to have the highest F_v/F_m, while the highest AQE was recorded for 'Ruidu Cuixia', with intercellular CO_2 concentration (C_i) and stomatal conductance (gs) at 292.56 μmol·mol^{-1}, 766.56 mmol·m^{-2}·s^{-1}, and 66.8 μmol·m^{-2}·s^{-1}, respectively. The indices of ABS/CS_m, TR_o/CS_m, and DI_o/CS_m were significantly different among varieties, and these indices of 'Ruidu Zaohong' were the highest. P_n was positively correlated with C_i and T_r, gs were positively correlated with F_v and TR_o/CS_m. The specific leaf area was negatively correlated with F_v/F_m and Φ_{DIo}. The results of the principal component analysis and TOPSIS comprehensive evaluation showed that 'Jing Hongbao' and 'Ruidu Cuixia' performed best. Overall, the measurement of the photosynthetic characteristics of the plants during the growing period provided valuable data for the varietal introduction strategies. The better photosynthetic performance of 'Jing Hongbao' and 'Ruidu Cuixia' indicates more adaptability to the long day, high light intensity, and the high-temperature climate of Xinjiang.

Keywords: grapes; grape hybrid varieties; adaptation; photochromism; fluorescence

1. Introduction

A natural climate of large temperature differences between day and night, long sunshine hours, and a dry climate [1] have always been advantageous for quality grape production. Traditional local varieties, owing to their geographical advantages, dominate most of the grape market and are the best-selling fruit products of the season in grape-growing areas. China is the largest producer of table and fresh grapes in the world (OIV 2022; https://www.oiv.int/what-we-do/country-report?oiv, accessed on 6 September

2023), and the Xinjiang region holds the top position in China. Grapes grown in Xinjiang are of good quality due to the unique climatic conditions in the region. However, the Xinjiang grape industry relies excessively on local varieties, resulting in a monotonous product structure that no longer meets the development needs of the grape industries.

To enrich the table grape variety resources in the region and enhance the efficiency and competitiveness of the Xinjiang grape industry, the Institute of Horticultural Crops at the Xinjiang Academy of Agricultural Sciences introduced several superior new grape varieties in 2019 for regional trial observation. The aim was to offer wider varieties for production. The current study hypothesized that the performance of newly introduced table grape varieties, when exposed to high light intensity conditions, will demonstrate significant differences in their photosynthetic and chlorophyll fluorescence characteristics, suggesting that certain varieties will exhibit superior adaptability and resilience to elevated light levels compared to others. The assessment of the success of introduced species depends significantly on the adaptive capacity of the introduced plants, including their ability to adapt to the local environment through seasonal rhythmic growth and development patterns, high production yield, and other relevant ecological and economic factors [2,3].

Photosynthesis is an important indicator of plant growth and production [4] and consists of components such as photosynthetic pigments, electron transport systems, and photosystems, each of which can potentially be affected by abiotic stresses [5]. Therefore, the study of photosynthesis performance in plants can reveal their growth potential [6], and it can be used as a basis for judging the success of plant variety introduction. Du et al. [7] showed that carbon metabolism was severely impaired under low nitrogen stress, leading to a decrease in the CO_2 assimilation rate, which accumulates in the cells and affects the overall photosynthetic rate. This phenomenon is mainly caused by stomatal and non-stomatal factors. In recent years, chlorophyll fluorescence detection techniques have been widely used to monitor the photosynthetic capacity of plants under different growth conditions, such as drought stress [8], salt stress [9], nitrogen stress [10], and high-temperature stress [11], etc. The results of Kromdijk et al. [12] showed that the q_P and NPQ of plants were always fluctuating under different stress conditions and became the standard to measure the inhibition of the electron transport chain. According to the results of Zhao et al. [13], the study of fluorescence kinetics is helpful in understanding the light-capturing ability of photosynthetic pigments and their tolerance to high-photon flux density in real-time, and to judge the photosynthetic capacity of plants under the current growth environment.

Fluorescence characteristics are extensively used in many studies related to plant physiology and photochemistry [14]. Chlorophyll fluorescence studies can also detect gross photosynthesis in large areas. In photosystem II (PSII), three pathways—chlorophyll fluorescence, photochemical reactions, and non-photochemical quenching (NPQ)—dissipate all of the light energy absorbed by the leaf. Some recent studies have demonstrated that stress conditions in plants can significantly influence photosynthetic physiology. Hazrati et al. [15] identified that both light intensity and water stress have a drastic impact on phytochemistry and fluorescence in *Aloe vera* plants. In a separate study, it was discovered that heat stress has a pronounced impact on the chlorophyll fluorescence properties of Rhododendron leaves [16]. Some studies including peony plants revealed that high temperature stress directly influences chlorophyll fluorescence induction kinetics [17]. The direct impact of heat stress on plant fluorescence activity suggests its potential use as an indicator of heat stress [18]. Therefore, leaf photosynthesis measurements can be used as an indicator of plant adaptability to environmental changes and as a criterion for predicting plant domestication potential, and provide a scientific basis for enriching table grape variety resources in Xinjiang [19,20].

In this study, five Chinese own hybrid varieties, namely 'Ruidu Xiangyu', 'Ruidu Cuixia', 'Ruidu Zaohong', 'Ruidu Wuheyi', and 'Jing Hongbao', were used as indicators of plant adaptation. This study aimed to evaluate the physiological parameters of photosynthetic characteristics, chlorophyll fluorescence, chlorophyll content, and leaf appearance of these five new Chinese own hybrid varieties. The objective was to assess their adaptability to the climate in Xinjiang and provide a reference for the introduction of suitable newer grape varieties.

2. Materials and Methods

2.1. Experimental Site Overview

The study was carried out at the grape research base (87°28′ E, 45°56′ N) of the Urumqi Anningqu Experimental Field, Xinjiang Academy of Agricultural Sciences, Urumqi, China. The base is located on the northern slope of the Tianshan Mountains on the southern margin of the Junggar Basin. The average altitude is 600~800 m, and the terrain is gentle. The area is under the typical temperate range of arid and semi-arid continental climates. For this experimental field, the mean annual temperature was recorded at 7.13 °C, the accumulated temperature of ≥10 °C was 3000~3500 °C, and the annual sunshine hours were 2500~3000 h.

2.2. Experimental Plant Materials

Five Chinese hybrid grape varieties, 'Ruidu Xiangyu', 'Ruidu Cuixia', 'Ruidu Zaohong', 'Ruidu Wuheyi', and 'Jing Hongbao', were used as experimental materials (Table 1). The introduced varieties were planted in 2019. The plant rows were oriented north to south, with vine spacing of 1 by 3.5 m. The Y-shaped tree planting was adopted. The soil of the vineyard is sandy loam. Recommended vineyard practices, including canopy and disease management, were followed during the growing season. Normal soil fertilizer and drip system was installed for water management.

Table 1. Introduction of five new table grape varieties.

Varieties	Species	Parent	Breeding Units	Breeding Year
Ruidu Xiangyu	Eurasian	Jingxiu × Xiangfei	Institute of Forestry and Fruit Science, Beijing Academy of Agriculture and Forestry Science	In December 2007, it was approved by Beijing Forest Variety Examination and Approval Committee
Ruidu Cuixia	Eurasian	Jingxiu × Xiangfei	Institute of Forestry and Fruit Science, Beijing Academy of Agriculture and Forestry Science	In December 2007, it was approved by Beijing Forest Variety Examination and Approval Committee
Ruidu Zaohong	Eurasian	Jingxiu × Xiangfei	Institute of Forestry and Fruit Science, Beijing Academy of Agriculture and Forestry Science	In December 2014, it was approved by Beijing Forest Variety Examination and Approval Committee
Ruidu Wuheyi	Eurasian	Xiangfei × Hongbaoshi seedless	Institute of Forestry and Fruit Science, Beijing Academy of Agriculture and Forestry Science	In 2009, it was approved by the Beijing Forest Variety Examination and Approval Committee
Jinghongbao	Eurasian	Guibao × Wuhebai Jixin	Fruit research institute of Shanxi Academy of Agricultural Sciences	In 2012, it was approved by Shanxi Provincial Crop Variety Examination and Approval Committee

2.3. Test Equipment and Test Reagents

CIRAS-3 PP Systems photosynthetic analyzer (Amesbury, MA, USA; JUNIOR-PAM fluorometer (Heinz Walz GmbH, Effeltrich, Germany). Anhydrous ethanol procured from Tianjin Kaitong Chemical Reagent Co., LTD, Tianjin, China.

2.4. Test Methods

2.4.1. Photo-Response Curve Determination

P_n-PAR response curve related measurements were recorded at 10:30 and 12:30 (UTC +08.00, Beijing Time) on sunny days. Three disease-free plants with moderate vigor were selected from each variety. From the fourth to fifth nodes of the new shoots, three leaves were chosen. These leaves had good leaf color, similar size dimensions, and were free from diseases and insect pests [21]. The CIRAS-3 PP Systems photosynthetic analyzer (Amesbury, MA, USA) was used to measure the light response indices of the leaves [22,23]. A PLC3 universal leaf cuvette light source leaf chamber was utilized, and 10 gradients ranging from 0 to 2500 μmol·m^{-2}·s^{-1} (2500, 2000, 1500, 1000, 750, 500, 300, 150, 75, 0 μmol·m^{-2}·s^{-1}) were set by the light source. The data were automatically recorded, with each gradient being stable for 90 s. The net photosynthetic rate (P_n), stomatal conductance (gs), transpiration rate (T_r), intercellular CO_2 volume fraction (C_i), and water use efficiency (WUE) were measured using CIRAS-3 portable photosynthesis system (PP Systems, Amsbury, MA, USA). The readings were automatically recorded by CIRAS-3 after a certain interval. Few parameters like relative humidity (60%), CO_2 concentration (380 μmol mol^{-1}), and leaf temperature (28 °C) were maintained using an automatic control device on the instrument. Red-blue light (90%: 10%) was provided inbuilt LED light unit in the CIRAS-3. Photosynthesis-light response simulations were conducted using the leaf drift model [24], and the model fitting equation was employed.

$$P_n = \alpha \frac{1 - \beta I}{1 + \gamma I} I - R_d$$

Note: α is initial quantum efficiency; I symbolize photosynthetically active radiation (μmol·m^{-2}·s^{-1}); R_d is dark respiration rate (μmol·m^{-2}·s^{-1}); β is photoinhibition coefficient (m^2·s·μmol^{-1}); γ is light saturation coefficient (m^2·s·μmol^{-1}).

2.4.2. Measurement of Chlorophyll Fluorescence Parameters

The chlorophyll fluorescence parameters of varieties were measured with the same leaves that were used for photosynthetic index measurements. The JUNIOR-PAM fluorometer was used, with a 30-min dark acclimatization period before the measurements. Based on studies by Strasser et al. [25] and Tsimilli Michael [26], the following parameters were defined and calculated: actual light energy conversion efficiency (Φ_{PSII}), non-fluorescence quenching (Y_{NPQ}), photosynthetic electron transport rate (ETR), maximum light quantum yield (F_v/F_m), as well as other indicators. These parameters include the maximum yield of primary photochemical reactions (Φ_{Po}), heat dissipation per unit area (Φ_{Do}), light energy captured per unit reaction center (RC) (TR_o/CS_m), and heat dissipation per unit area (DI_o/CS_m).

2.4.3. Measurement of Leaf Appearance Traits

Twenty mature leaves were randomly selected from both, the sunny and shaded sides of the grape trellis for all three plants. These leaves were measured for photosynthetic indicators. The leaf area was measured using a leaf area meter, and the leaf weight was determined using digital electronic balance. The specific leaf weight (calculated as the single leaf weight divided by the leaf area) and specific leaf area (calculated as the leaf area divided by the single leaf weight) were calculated.

2.4.4. Measurement of Chlorophyll Content

The chlorophyll content of each of the three plants was measured using the photosynthetic index determination method. Five leaves with similar leaf color, size, and exposure to sunlight were selected. The leaves were ground into a powder, with a weight of 0.2 g selected for analysis. They were then mixed with 80% acetone and kept in darkness for 12 h until the sample turned white. Afterward, the mixture was filtered, and absorbance values were measured at wavelengths of 645 nm, 663 nm, and 470 nm. These values were recorded, and the contents of Chl a (chlorophyll a), Chl b (chlorophyll b), and carotenoids were determined, following Arnon's method [27].

2.4.5. TOPSIS Evaluation Method

The TOPSIS integrated evaluation method was used to synthesize the chlorophyll content, chlorophyll fluorescence parameters, and photosynthetic characteristic parameters of the leaves of the five varieties to comprehensively evaluate the photosynthetic strength of the five varieties.

In step 1, the indicators were homogenized to avoid affecting the description of the results.

Step 2, normalization of the data.

$$Y_{ij} = \frac{X}{\sqrt{\sum_{j=1}^{m} X_j^2}}, \ (j = 1, 2, \ldots m)$$

In the formula, j represents a certain evaluation indicator, and m represents the number of evaluation indicators.

Step 3: Calculate the distance between positive and negative ideal solutions (D±) and the relative closeness degree (C):

$$D^+ = \sqrt{\sum_{j=1}^{m} W_j \left(A_j^+ - Y_{ij}\right)^2}, \ (j = 1, 2, \ldots m)$$

$$D^- = \sqrt{\sum_{j=1}^{m} W_j \left(A_j^- - Y_{ij}\right)^2}, \ (j = 1, 2, \ldots m)$$

In the formula, j represents an evaluation index, m represents the number of evaluation indexes, W_j represents the weight value of the jth index, A_j^+ represents the optimal scheme data of the jth index, A_j^- represents the worst scheme data of the jth index and Y^{ij} represents the corresponding data of a certain evaluation object i for the jth indicator.

$$C_j = \frac{D^-}{D^- + D^+}$$

In the formula, the value of C_j ranges from 0 to 1. The larger C_j is, the stronger the photosynthetic capacity of the jth new variety is, and the closer the variety's adaptability to the climate in Xinjiang is to the optimal level.

2.5. Data Processing and Statistical Analysis

All the data were collated in at least three replications and tabulated using Microsoft Excel 2010, and the results were statistically analyzed by analysis of variances tests (one-way ANOVA). We used SPSS 25.0 (SPSS Inc., Chicago, IL, USA) to perform Pearson correlation analysis and principal component analysis, and photo-response curves were fitted and plotted using Origin 2019.

3. Results

3.1. Chlorophyll Content and Leaf Appearance Traits

There were significant differences in leaf characteristics among the five new table grape varieties ($p < 0.05$) (Figure 1). The results showed that the leaf area and single-leaf weight of 'Ruidu Zaohong' were the highest among the five varieties (Figure 1A,B). The specific leaf area of 'Ruidu Cuixia' was 8.99% higher than that of 'Ruidu Zaohong' (Figure 1D). As shown in Figure 1C, there was no significant difference in specific leaf weight among the five new table grape varieties.

The chlorophyll content of higher plants affects the metabolic rate of the plant and is an index for judging plant health and local adaptation. The results showed that the chlorophyll content of the five new varieties differed significantly ($p < 0.05$) (Figure 2). 'Jing Hongbao' had the highest chlorophyll a, chlorophyll b, and total chlorophyll content, which were 29.36%, 139.02%, and 59.33% higher than those of 'Ruidu Zaohong', respectively.

Interestingly, 'Jing Hongbao' had the lowest values of carotenoid content and chlorophyll a/b, 32.08% lower than 'Ruidu Cuixia' and 46.47% lower than 'Ruidu Zaohong'.

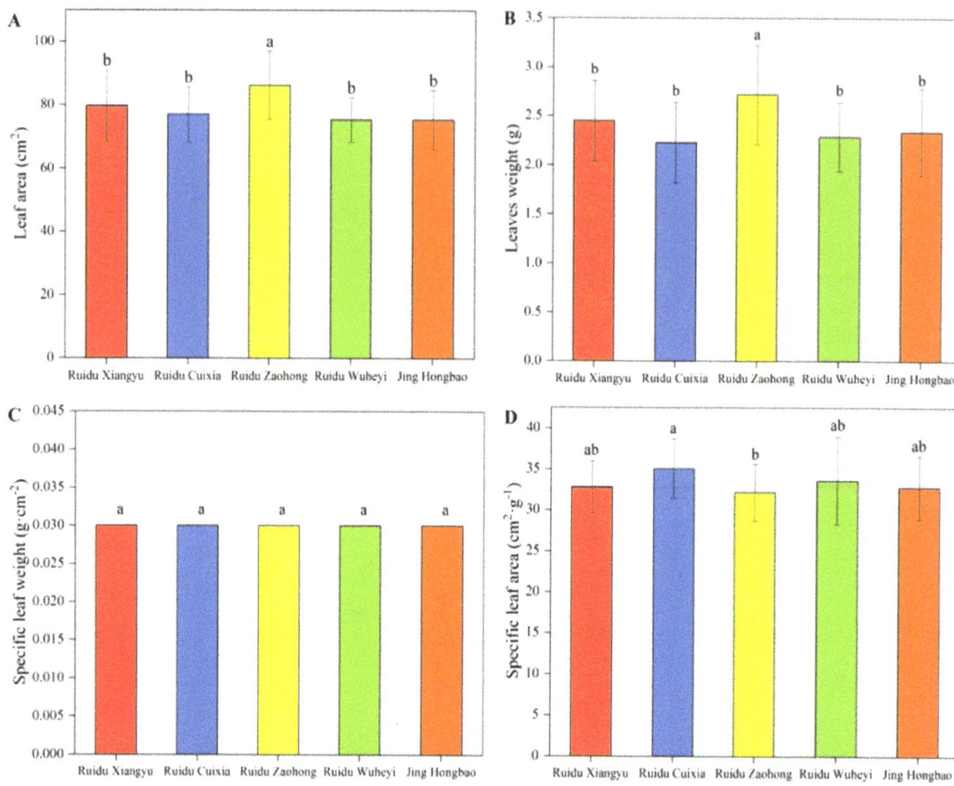

Figure 1. Comparison of leaf characteristic parameters of five new table grape varieties. Same letter in the same figure indicates that there is no significant difference. The data in the figure are mean ± standard deviation, and different lowercase letters indicate significant differences ($p < 0.05$). (**A**) Leaf area; (**B**) Leaf weight; (**C**) specific leaf weight; (**D**) specific leaf area.

3.2. Photosynthetic Parameters and Photo-Response Curve

Photosynthetic parameters for all the varieties were measured. From Figure 3, it can be determined that all five photosynthetic parameters of 'Ruidu Xiangyu' were lower than those of the other four varieties, with an average C_i of 194.67 μmol·mol^{-1}, which is 33.46% lower than that of 'Ruidu Cuixia' (Figure 3A). The gs is 145.56 mmol·m^{-2}·s^{-1}, which is 81.07% lower than that of 'Ruidu Cuixia' (Figure 3B). The P_n is 9.61 μmol m^{-2} s^{-1}, which is 52.14% lower than that of 'Ruidu Zaohong' (Figure 3C). The T_r is 5.88 mmol·mol^{-1}, which is 49.18% lower than that of 'Jing Hongbao' (Figure 3D).

Finally, the average WUE is 1.64 mmol·m^{-2}·s^{-1}, which is 15.46% lower than that of 'Ruidu Zaohong' (Figure 3E).

Fitting curves of the light response of five new table grape varieties were also observed. The fitting curve presented in Figure 4 showed that the light response curve of all five varieties shows a similar trend. With the increase in photosynthesis active radiation, the net photosynthesis rate gradually increases. After reaching the saturation light intensity, it stabilizes or slightly decreases. There are significant differences in the net photosynthesis rate of the five varieties under high light intensity.

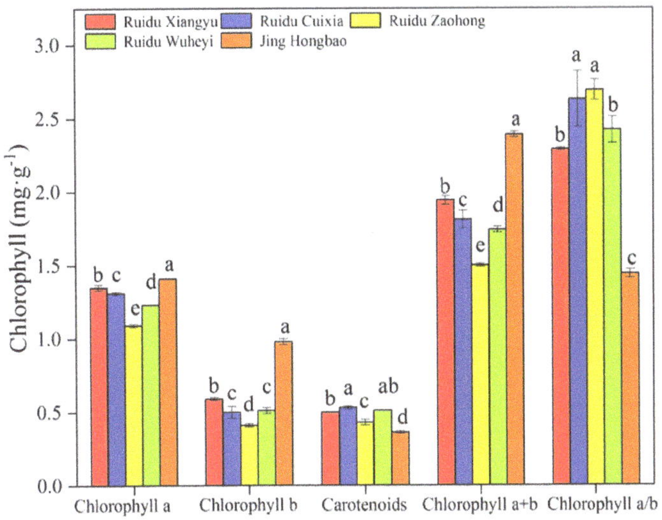

Figure 2. Comparison of chlorophyll content in varieties. The error bar indicates the standard deviation obtained from three biological replicates. Same letter in the same figure indicates that there is no significant difference. The data in the figure are mean ± standard deviation, and different lowercase letters indicate significant differences ($p < 0.05$).

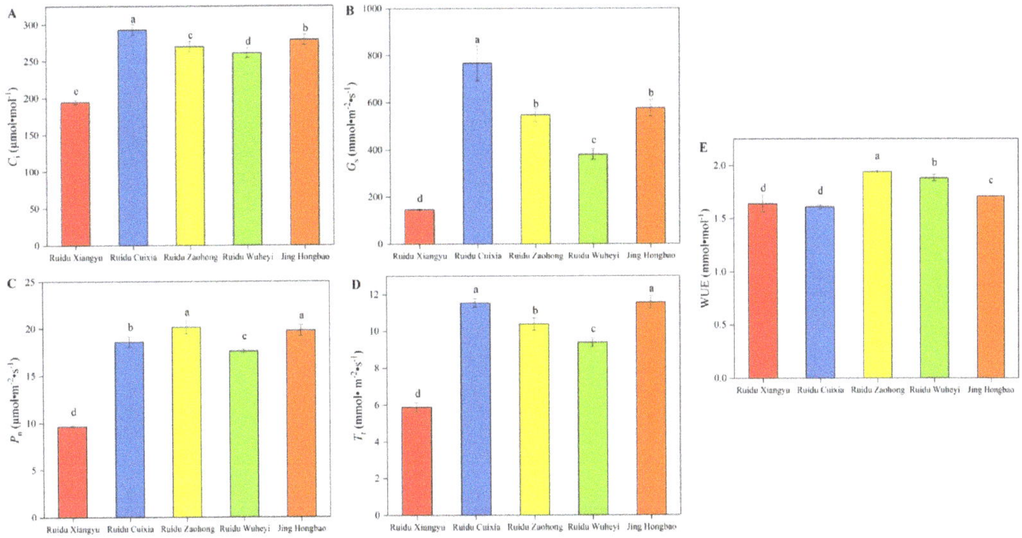

Figure 3. Comparison of photosynthetic parameters of five new table grape varieties. C_i (**A**), gs (**B**), P_n (**C**), T_r (**D**), and WUE (**E**) in the figure showed the difference in photosynthetic parameters of the five new varieties. Same letter in the same figure indicates that there is no significant difference. The data in the figure are mean ± standard deviation, and different lowercase letters indicate significant difference ($p < 0.05$).

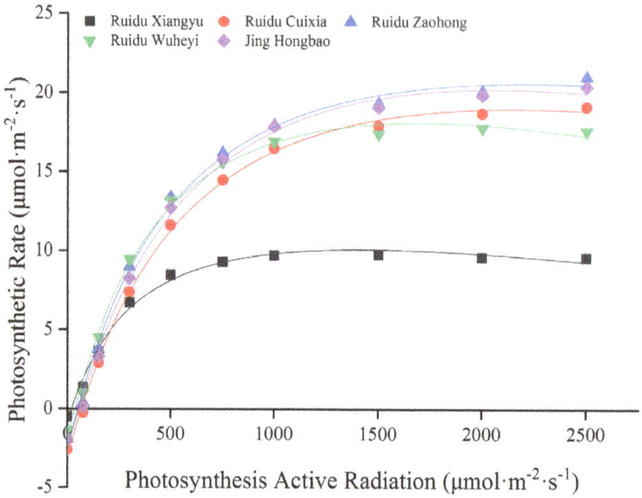

Figure 4. Fitting curves of light response of five new table grape varieties.

Based on the measured parameters of the P_n-PAR light response curve (Table 2), it can be observed that the photosynthetic characteristics of 'Ruidu Xiangyu' are the lowest among the five varieties. The mean value of its apparent quantum efficiency is 0.0298, which is 14.83% lower than that of 'Ruidu Wuheyi'. Its dark respiration rate is 0.85 μmol·m^{-2}·s^{-1}, which is 72.84% lower than that of 'Ruidu Cuixia'. Its light saturation intensity is 1437.1 μmol·m^{-2}·s^{-1}, which is 39.55% lower than that of 'Ruidu Zaohong'. Its light compensation point is 20 μmol·m^{-2}·s^{-1}, which is 70.06% lower than that of 'Ruidu Cuixia'. Finally, its maximum net photosynthesis rate is 9.8 μmol·m^{-2}·s^{-1}, which is 51.96% lower than that of 'Jing Hongbao'.

Table 2. Comparison of characteristic parameters of response curves of five new table grape varieties P_n-*PAR*.

Varieties	Right Angle Hyperbolic Modified Model	Apparent Quantum Efficiency	Adjust R-Square	Dark Respiration Rate/ (μmol·m^{-2}·s^{-1})	Light Saturation Point/ (μmol·m^{-2}·s^{-1})	Light Compensation Point/(μmol·m^{-2}·s^{-1})	Maximum Net Photosynthetic Rate/ (μmol·m^{-2}·s^{-1})
Ruidu Xiangyu	$y = 0.04488x\frac{1-0.00012x}{1+0.00273x} - 0.85047$	0.0298	0.991	0.85	1437.1	20	9.8
Ruidu Cuixia	$y = 0.05172x\frac{1-0.00008x}{1+0.00147x} - 3.12694$	0.0371	0.998	3.13	2290.5	66.8	19.2
Ruidu Zaohong	$y = 0.05921x\frac{1-0.00007x}{1+0.00170x} - 2.69752$	0.0384	0.995	2.70	2377.4	49.6	21.0
Ruidu Wuheyi	$y = 0.05966x\frac{1-0.00011x}{1+0.00182x} - 2.02591$	0.0391	0.995	2.03	1734.3	36.4	17.8
Jing Hongbao	$y = 0.05297x\frac{1-0.00009x}{1+0.00139x} - 2.70940$	0.0356	0.996	2.71	2171.3	55.4	20.4

3.3. Chlorophyll Fluorescence Parameters

The dynamic parameters of chlorophyll fluorescence were mathematically analyzed for these five different grape varieties, aiming to characterize the structural and electron transfer performance of their photosynthetic apparatus. The analysis results reflect the photosynthetic performance of these grape varieties. Representative chlorophyll fluorescence parameters were summarized, revealing significant differences in the regulation ability of chlorophyll fluorescence in response to light intensity among the different varieties (Figure 5).

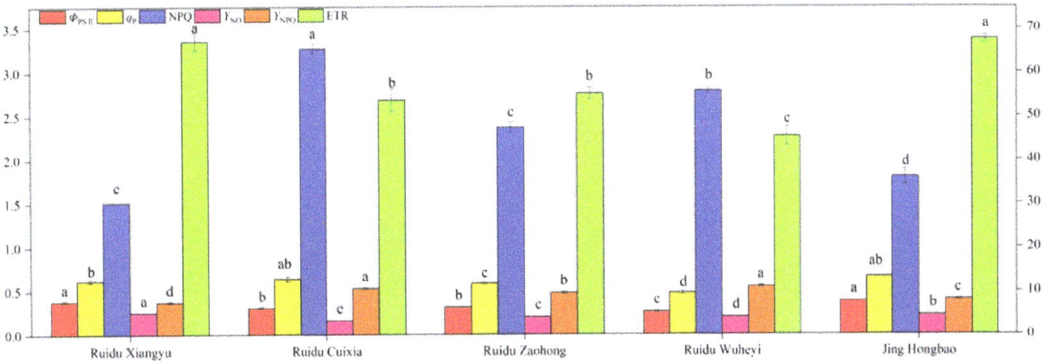

Figure 5. Significant differences in chlorophyll fluorescence parameters among five new table grape varieties. ETR in the figure is the right coordinate axis degree, and other indicators are the left coordinate axis degree. Same letter in the same figure indicates that there is no significant difference. The data in the figure are mean ± standard deviation, and different lowercase letters indicate significant difference ($p < 0.05$).

Analysis of chlorophyll fluorescence parameters revealed that the highest and lowest values of Φ_{PSII}, ETR, and q_P were observed in 'Jing Hongbao' and 'Ruidu Wuheyi', respectively (Figure 5). The mean values of Φ_{PSII}, ETR, and q_P for 'Jing Hongbao' were 0.38, 67.6, and 0.66, respectively, representing higher of 46.15%, 49.00%, and 37.50% compared to 'Ruidu Wuheyi'. Conversely, 'Ruidu Wuheyi' exhibited the highest mean value of Y_{NPQ} (0.55), while 'Ruidu Xiangyu' had the lowest mean value (0.37). In Figure 6, 'Ruidu Wuheyi' displayed the highest values of F_v/F_m and Φ_{Po}, while 'Ruidu Zaohong' had the lowest values. Additionally, 'Ruidu Zaohong' demonstrated superior performance in parameters such as Φ_{Do} and ABS/CS_m. The lowest values of ABS/CS_m and DI_o/CS_m were observed in 'Ruidu Wuheyi', whereas 'Ruidu Xiangyu' had the lowest value of TR_o/CS_m.

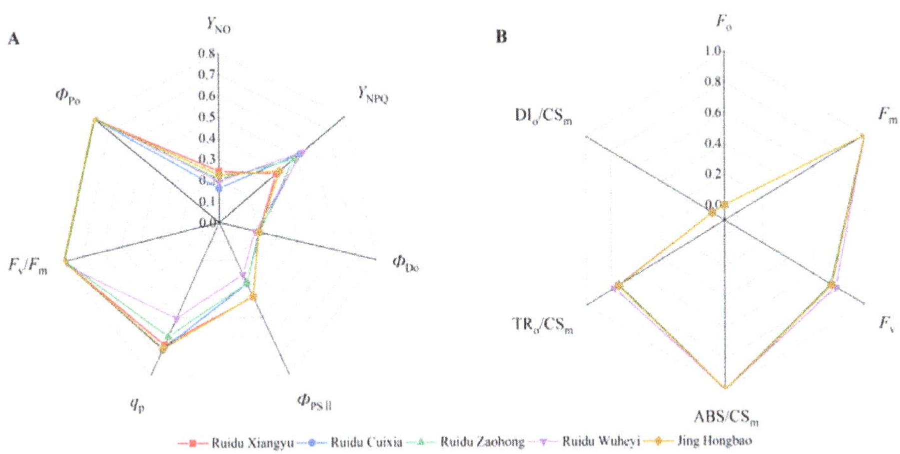

Figure 6. Comparison of fluorescence characteristic parameters of five new table grape varieties. **(A)** Φ_{Do}, Y_{NO}, Y_{NPQ}, Φ_{PO}, q_p, F_v/F_m, and Φ_{PSII}; **(B)** ABS/CS_m, DI_o/CS_m, TR_o/CS_m, F_v, and F_m.

3.4. Correlation Analysis and Hierarchical Cluster Analysis

The light adaptation ability of the five varieties was analyzed by hierarchical cluster analysis, and the results are presented in Figure 7A. From Figure 7A, it can be observed that the five varieties were divided into two categories based on clades, and their prominent characteristics. The first category includes 'Jing Hongbao', 'Ruidu Cuixia', and 'Ruidu Zaohong', which have the highest chlorophyll content and chlorophyll fluorescence parameters and lower light saturation intensity.

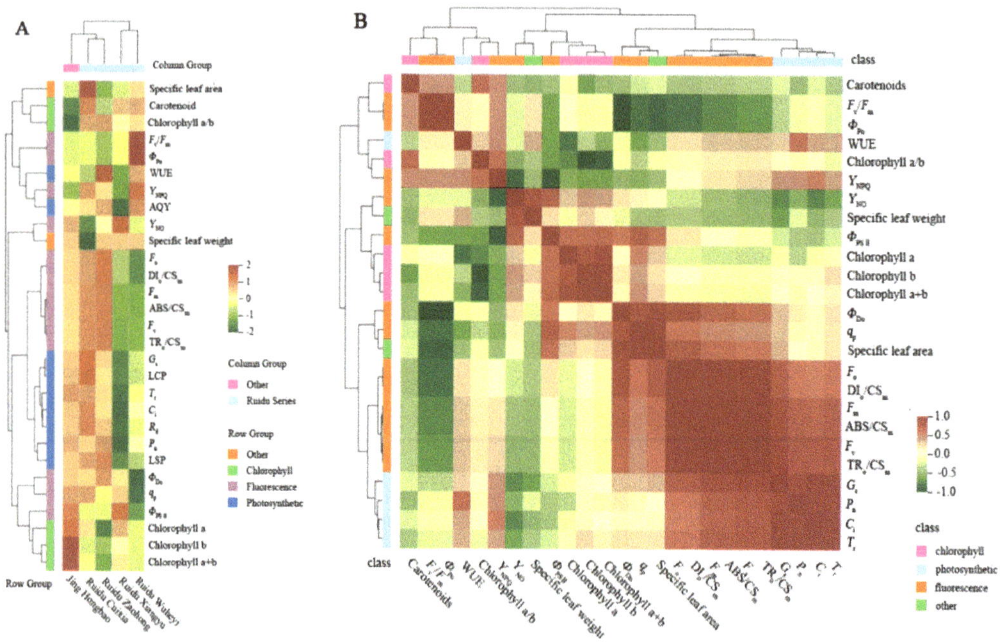

Figure 7. Correlation and hierarchical clustering analysis of chlorophyll content, chlorophyll fluorescence, and photosynthetic characteristic parameters of five new table grape varieties. (**A**) represents hierarchical cluster analysis, and (**B**) represents correlation analysis. Different colors of red and green in Panel B indicate a significant correlation at the 0.05 level (two-tailed). Red indicates a high positive correlation, green indicates a high negative correlation, the redder the color, the higher the positive correlation between different indicators, the greener the color, and the negative phase between different indicators.

The second category includes 'Ruidu Xiangyu' and 'Ruidu Wuheyi', which have lower chlorophyll content and chlorophyll fluorescence parameter values but higher apparent quantum efficiency values.

Pearson correlation analysis was performed on the 25 photosynthetic phenotypic indices of the five varieties (Figure 7B). From Figure 7B, it can be observed that P_n is significantly positively correlated with C_i and T_r ($p < 0.05$), and their correlation coefficients are all greater than 0.92. gs is significantly positively correlated with F_v and TR_o/CS_m, and its correlation coefficient is 0.885. Specific leaf area is significantly positively correlated with Φ_{Do} and q_P, and negatively correlated with Φ_{Po} and F_v/F_m ($p < 0.05$). Their correlation coefficients are all greater than 0.92.

Further correlation analysis results clearly showed that there is a good correlation between photosynthetic characteristics and chlorophyll fluorescence parameters, indicating that the evaluation of plant photosynthetic capacity needs to comprehensively consider both photosynthetic and chlorophyll fluorescence indicators.

3.5. Principal Component Analysis

Principal component analysis was performed on the 25 photosynthetic and chlorophyll fluorescence indicators in the experiment. Four principal components with eigenvalues greater than 1 were extracted (Table 3), accounting for a cumulative contribution rate of 100% and effectively retaining most of the information from the original variables. These four principal components were used for a comprehensive analysis of the photochemical efficacy of the five varieties. The eigenvectors of the principal components were calculated based on the principal component loading matrix and eigenvalues.

Table 3. Principal component characteristic values, contribution rate, and cumulative contribution rate of five new table grape varieties.

Principal Component Number	Eigenvalue	Rate of Contribution/%	Accumulating Contribution Rate/%
1	10.9385	43.75%	43.75%
2	7.3320	29.33%	73.08%
3	3.5369	14.15%	87.23%
4	3.1926	12.77%	100.00%

By multiplying the obtained eigenvectors with the standardized data and considering the proportion of eigenvalues corresponding to the four principal components relative to the total sum of eigenvalues, weights were determined. These weights were then utilized to calculate the composite scores of the principal components.

The results (Table 4) revealed the relative photosynthetic abilities of the five varieties as follows: 'Jing Hongbao' > 'Ruidu Cuixia' > 'Ruidu Zaohong' > 'Ruidu Xiangyu' > 'Ruidu Wuheyi'. To allow for comparison of the indicator scores, the scores of the four principal components were multiplied and summed with the squared variances' percentages of the extracted loadings. These values were subsequently divided by the cumulative percentages. The results (Table 5) showcased the varying weightings of the 25 indicators in leaf photosynthesis, with chlorophyll b, T_r, chlorophyll a + b, q_P, and gs ranking among the top 5.

Table 4. Comprehensive principal component scores of five new table grape varieties.

Varieties	F1	F2	F3	F4	F	Rank
Ruidu Xiangyu	−3.6070	2.3626	−0.6990	−1.8853	−1.2248	4
Ruidu Cuixia	2.9872	−1.5209	1.9314	−1.8010	0.9041	2
Ruidu Zaohong	2.8615	−0.7777	−2.8697	0.3004	0.6561	3
Ruidu Wuheyi	−3.4800	−3.2592	0.4389	1.3790	−2.2403	5
Jinghongbao	1.2383	3.1953	1.1984	2.0069	1.9048	1

Table 5. Scores of photosynthetic indexes of five new table grape varieties.

Indexes	F1	F2	F3	F4	F	Rank
Chlorophyll a	−0.0683	0.2336	0.3939	−0.0271	0.0673	16
Chlorophyll b	−0.0022	0.2730	0.2428	0.2771	0.1105	1
Carotenoids	−0.1108	−0.2200	0.1283	−0.3760	−0.1063	25
Chlorophyll a + b	−0.0272	0.2729	0.3128	0.1776	0.1003	3
Chlorophyll a/b	0.0285	−0.2691	−0.1971	−0.3176	−0.1002	24
C_i	0.2296	−0.1361	0.2000	0.2138	0.0863	12
gs	0.2706	−0.0866	0.1834	0.0891	0.0968	5
P_n	0.2299	−0.1105	0.0299	0.3205	0.0843	13
T_r	0.2424	−0.0692	0.1815	0.2534	0.1069	2
WUE	0.0835	−0.1826	−0.3069	0.3293	−0.0134	19

Table 5. Cont.

Indexes	F1	F2	F3	F4	F	Rank
F_v/F_m	−0.2076	−0.2165	0.1546	0.1772	−0.0816	22
Φ_{Po}	−0.2076	−0.2165	0.1546	0.1772	−0.0816	23
Φ_{Do}	0.2076	0.2165	−0.1546	−0.1772	0.0816	14
Φ_{PSII}	−0.0033	0.3635	0.0179	−0.0962	0.0708	15
q_P	0.1554	0.2587	0.1377	−0.2358	0.0988	4
Y_{NO}	−0.1478	0.2805	−0.2165	0.0768	−0.0023	18
Y_{NPQ}	0.0621	−0.3569	0.0759	0.0332	−0.0465	21
F_o	0.2910	0.0678	−0.0871	−0.0637	0.0942	6
F_m	0.3011	0.0068	−0.0422	−0.0205	0.0930	8
F_v	0.3017	−0.0156	−0.0254	−0.0046	0.0916	10
ABS/CS_m	0.3011	0.0068	−0.0422	−0.0205	0.0930	9
Tr_o/CS_m	0.3017	−0.0156	−0.0254	−0.0046	0.0916	11
DI_o/CS_m	0.2910	0.0678	−0.0871	−0.0637	0.0942	7
Specific leaf area	0.0419	−0.1743	0.4141	−0.2195	−0.0019	17
Specific leaf weight	−0.1528	0.1154	−0.3097	0.3106	−0.0273	20

3.6. Comprehensive Evaluation of Photosynthetic Capacity

The results (Tables 6 and 7) show that the photosynthetic capacity of the five new varieties of table grapes from strongest to weakest are: 'Jing Hongbao' > 'Ruidu Cuixia' > 'Ruidu Zhaohong' > 'Ruidu Wuheyi' > 'Ruidu Xiangyu'.

Table 6. Positive and negative ideal solutions of photosynthetic indexes of five new table grape varieties.

Indexes	Positive Ideal Solution A^+	Negative Ideal Solution A^-	Indexes	Positive Ideal Solution A^+	Negative Ideal Solution A^-
Chlorophyll a	0.492	0.38	Φ_{Po}	0.457	0.443
Chlorophyll b	0.695	0.291	Φ_{Do}	0.465	0.409
Carotenoids	0.506	0.344	Φ_{PSII}	0.513	0.356
Chlorophyll a + b	0.563	0.353	q_P	0.482	0.364
Chlorophyll a/b	0.513	0.276	Y_{NO}	0.523	0.349
C_i	0.499	0.341	Y_{NPQ}	0.518	0.355
gs	0.63	0.135	F_o	0.508	0.357
P_n	0.521	0.237	F_m	0.492	0.391
T_r	0.521	0.269	F_v	0.487	0.397
WUE	0.503	0.397	ABS/CS_m	0.492	0.391
Specific leaf area	0.462	0.424	Tr_o/CS_m	0.487	0.397
Specific leaf weight	0.453	0.424	DI_o/CS_m	0.508	0.357
F_v/F_m	0.457	0.443			

Table 7. TOPSIS evaluation and calculation results of five newly introduced table grape varieties.

Varieties	Positive Ideal Solution Distance D^+	Negative Ideal Solution Distance D^-	Degree of Relative Proximity C	Sorting Result
Ruidu Xiangyu	0.779	0.389	0.333	5
Ruidu Cuixia	0.434	0.769	0.639	2
Ruidu Zaohong	0.53	0.663	0.556	3
Ruidu Wuheyi	0.624	0.464	0.426	4
Jing Hongbao	0.367	0.78	0.68	1

4. Discussion

Photosynthetic pigments are an important part of the photosynthesis mechanism. Under the long-term strong light irradiation in Xinjiang during the daytime, the leaf pigment contents of the five varieties were quite different. The chlorophyll a and chlorophyll b contents of 'Ruidu Zaohong' were significantly lower than those of the other four varieties. The chlorophyll a + b content of 'Jing Hongbao' reached 2.39 mg·g^{-1}, with chlorophyll a at

1.41 mg·g^{-1} and chlorophyll b at 0.98 mg·g^{-1}, respectively. Its growth potential was better than that of the 'Ruidu' series.

The results of this experiment were consistent with Yan's findings in 2021. High temperature and strong light stress reduced the pigment content of the leaves, damaged the chloroplast structure in them, and inhibited photosynthesis. However, the chlorophyll a and b contents of 'Jing Hongbao' were higher, indicating that the plants can initiate self-protection mechanisms to meet their own growth needs under stress conditions [28].

Photosynthesis serves as an important indicator for testing the sensitivity of plants to environmental stress [29]. In the high temperature and high light intensity conditions of Xinjiang, the average P_n of 'Ruidu Zaohong' was as high as 20.14 μmol·m^{-2}·s^{-1}, whereas 'Ruidu Xiangyu' only reached 9.64 μmol·m^{-2}·s^{-1}. Moreover, 'Ruidu Cuixia' exhibited mean gs and C_i values that were 50.29% and 426.63% higher than those of 'Ruidu Xiangyu', respectively.

These findings suggest that higher temperatures and stronger light may lead to stomatal closure in 'Ruidu Xiangyu' leaves, thereby impacting the gas exchange rate. Consequently, this closure causes a decrease in gs and C_i, inhibiting photosynthetic efficiency by reducing photosynthetic assimilation substances and water loss, which aligns with the observations made by Tang et al. [30].

Plants possess a light radiation signal regulation system [31] that modulates both stomatal and non-stomatal factors based on the effective light radiation intensity. This regulation system influences various photosynthetic parameters in leaves. Research findings indicate that leaf damage due to excessive light results in reduced chlorophyll content, T_r value, and P_n value, ultimately diminishing photosynthetic capacity. This outcome is similar to the findings reported by Negi et al. [32]. Additionally, correlation analysis demonstrates a significant positive relationship between P_n, C_i, and T_r. Consequently, it is speculated that stomatal factors play a pivotal role in a plant's growth potential. Different grape varieties employ diverse mechanisms to coordinate their photosynthesis, utilizing CO_2 absorption, water uptake, and inorganic ion transport to adapt to their specific growth environments.

Chlorophyll fluorescence is a commonly used, non-destructive method for detecting plant physiological characteristics and stress traits, which further helps in increasing our understanding of the behavior of plants in their natural environments [33]. Previous research results have shown that the instantaneous fluorescence signal of PSII is primarily caused by the oxidation-reduction reaction of plastoquinone A (QA) [34]. QA represents the reduction state of the photosynthetic electron transport chain and is manifested as photochemical energy conversion and thermal dissipation [35,36]. F_v/F_m can estimate the maximum quantum yield of QA reduction, representing the potential efficiency of plant PSII [37].

Under non-stress conditions, the normal range of F_v/F_m for plant leaves is between 0.80 and 0.85. When under environmental stress, the F_v/F_m value will significantly decrease. The results of this study show that only 'Ruidu Wuheyi' F_v/F_m is greater than 0.81, indicating that the light duration and intensity in Xinjiang are suitable for the growth needs of this variety. The F_v/F_m of the other 4 varieties was slightly lower than 0.80, indicating that the plants were under environmental stress, which is speculated to be related to the reversible inactivation or downregulation of PSII caused by high light and heat [38]. F_v is related to the photo-acclimation state of the dark-adapted reaction center [39]. Q_P represents the proportion of the PSII reaction center capturing the excitation energy, TR_o/CS_m represents the light energy captured per unit area, Φ_{Po} represents the maximum quantum yield of the primary photochemical reaction, and Φ_{Do} represents the quantum yield of energy dissipation.

In this study, 'Ruidu Zhaohong' had the highest TR_o/CS_m and Φ_{Do} values, indicating that QA was affected by high light and heat stress and could not effectively transmit electrons to the next level Quinone receptor [40], resulting in severe energy loss.

In addition, the correlation analysis shows that gs is significantly positively correlated with F_v and TR_o/CS_m; specific leaf area is significantly positively correlated with Φ_{Do}

and q_P, and significantly negatively correlated with Φ_{Po} and F_v/F_m, which is inconsistent with the results of previous studies [9,41]. The reasons may be due to differences in the experimental plant varieties, changes in the growth environment, and uncertainties in the correlations between various photosynthetic parameters, and further research is needed to determine the specific reasons.

Photosynthesis in plants is controlled by several factors, such as environmental factors, growth and developmental stages, and nutritional status, which can lead to differences in response to plant traits. In this study, five Chinese own hybrid varieties grapes were selected and cultivated under normal water and fertilizer management. Their photosynthetic capacity was evaluated in relation to their phenotypic, physiological, and biochemical indicators. The varieties were ranked according to the combined scores of principal component analysis and TOPSIS.

The results showed that 'Jing Hongbao', 'Ruidu Cuixia', and 'Ruidu Zaohong' ranked in the top three positions in both evaluation methods. However, further research is needed to determine the photosynthetic capacity between 'Ruidu Xiangyu' and 'Ruidu Wuheyi', as they ranked in the bottom two positions.

Moreover, since grapes are a berry plant, the adaptability of grapes to the regional environment of Xinjiang still needs to be evaluated in terms of fruit quality and internal tissue structure, among other factors.

This study only analyzed the photosynthetic characteristics of the leaves, and further research will be conducted on the physiological characteristics of the fruits.

5. Conclusions

There were some differences in leaf photosynthetic performance, photosynthetic pigment content, and chlorophyll fluorescence parameters between the five hybrid varieties under cultivation conditions in Xinjiang. Overall, 'Jing Hongbao' and 'Ruidu Cuixia' exhibited a stronger ability to accumulate organic matter content through photosynthesis, and their utilization efficiency in a strong light environment was significantly higher than that of 'Ruidu Zaohong', 'Ruidu Xiangyu', and 'Ruidu Wuheyi', which demonstrated a strong resistance to strong light intensity stress. Based on the comprehensive evaluation results of photosynthetic traits for each variety using principal component analysis and TOPSIS analysis, it can be preliminarily concluded that 'Jing Hongbao' and 'Ruidu Cuixia' displayed stronger adaptability to the climate in Xinjiang and are suitable for cultivation in the region.

Author Contributions: F.Z., X.W. and H.Z. conceived and designed the experiments. S.B., J.W., X.Z. and W.Z. provide technical guidance. Y.H. and F.Z. performed the experiments. Y.H. and V.Y. wrote the manuscript. Y.H. analyzed the data. S.H., M.W., B.Z. and X.W. revised the manuscript and provided technical support. All authors have read and agreed to the published version of the manuscript.

Funding: Supported by the Agricultural Science and Technology Innovation long-term support project, Xinjiang Academy of Agricultural Sciences (xjnkywdzc-2022001-9), Key research and development project of autonomous region (2022B02045-1-1), China Agriculture Research System of MOF and MARA (CARS-29-30), Xinjiang Uygur Autonomous Region Tianshan Talents Training Program-Young top-notch scientific and technological talents (2022TSYCJC0036) and Xinjiang Uygur Autonomous Region Innovation Environment Construction Special Project (PT2314).

Data Availability Statement: Data are contained within the article.

Acknowledgments: Horticultural Crops Institute of Xinjiang Academy of Agricultural Sciences and various regional botanical gardens for sampling support during the progress of this project.

Conflicts of Interest: The authors declare no conflict of interest.

References

1. He, B.B.; Sheng, Y.; Cao, W.; Wu, J.C. Characteristics of Climate Change in Northern Xinjiang in 1961–2017, China. *Chin. Geogr. Sci.* **2019**, *30*, 249–265. [CrossRef]
2. Ivashchenko, I.B.A.; Rakhmetov, D.; Fishchenko, V. Biochemical peculiarities of Glebionis coronaria (*Asteraceae*) introduced in Central Polissya of Ukraine. *Plant Fungal Res.* **2019**, *2*, 32–39. [CrossRef]
3. Ivashchenko, I.; Kotyuk, L.; Bakalova, A. Morphology and productivity of tarragon (*Artemisia dracunculus* L.) in Central Polissya (Ukraine). *Ukr. J. Ecol.* **2020**, *10*, 48–55.
4. Tsai, Y.C.; Chen, K.C.; Cheng, T.S.; Lee, C.; Lin, S.H.; Tung, C.W. Chlorophyll fluorescence analysis in diverse rice varieties reveals the positive correlation between the seedlings salt tolerance and photosynthetic efficiency. *BMC Plant Biol.* **2019**, *19*, 403. [CrossRef] [PubMed]
5. Ashraf, M.; Harris, P.J.C. Photosynthesis under stressful environments: An overview. *Photosynthetica* **2013**, *51*, 163–190. [CrossRef]
6. Wei, S.S.; Wang, X.Y.; Shi, D.; Li, Y.H.; Zhang, J.W.; Liu, P.; Zhao, B.; Dong, S. The mechanisms of low nitrogen induced weakened photosynthesis in summer maize (*Zea mays* L.) under field conditions. *Plant Physiol. Biochem.* **2016**, *105*, 118–128. [CrossRef] [PubMed]
7. Du, X.H.; Peng, F.R.; Jiang, J.; Tan, P.P.; Wu, Z.Z.; Liang, Y.W.; Zhong, Z.K. Inorganic nitrogen fertilizers induce changes in ammonium assimilation and gas exchange in *Camellia sinensis* L. *Turk. J. Agric. For.* **2015**, *39*, 28–38. [CrossRef]
8. Mathur, S.; Tomar, R.S.; Jajoo, A. Arbuscular Mycorrhizal fungi (AMF) protects photosynthetic apparatus of wheat under drought stress. *Photosynth. Res.* **2019**, *139*, 227–238. [CrossRef]
9. Wang, Y.; Jie, W.; Peng, X.; Hua, X.; Yan, X.; Zhou, Z.; Lin, J. Physiological adaptive strategies of oil seed crop Ricinus communis early seedlings (*cotyledon* vs. *true leaf*) under salt and alkali stresses: From the growth, photosynthesis and chlorophyll fluorescence. *Front. Plant Sci.* **2019**, *9*, 1939. [CrossRef]
10. Tantray, A.Y.; Bashir, S.S.; Ahmad, A. Low nitrogen stress regulates chlorophyll fluorescence in coordination with photosynthesis and Rubisco efficiency of rice. *Physiol. Mol. Biol. Plants* **2020**, *26*, 83–94. [CrossRef]
11. Yan, Z.; Ma, T.; Guo, S.; Liu, R.; Li, M. Leaf anatomy, photosynthesis and chlorophyll fluorescence of lettuce as influenced by arbuscular mycorrhizal fungi under high temperature stress. *Sci. Hortic.* **2021**, *280*, 109933. [CrossRef]
12. Kromdijk, J.; Głowacka, K.; Leonelli, L.; Gabilly, S.T.; Iwai, M.; Niyogi, K.K.; Long, S.P. Improving photosynthesis and crop productivity by accelerating recovery from photoprotection. *Science* **2016**, *354*, 857–861. [CrossRef]
13. Zhao, L.S.; Li, K.; Wang, Q.M.; Song, X.Y.; Su, H.N.; Xie, B.B.; Zhang, X.Y.; Huang, F.; Chen, X.L.; Zhou, B.C. Nitrogen Starvation Impacts the Photosynthetic Performance of *Porphyridium cruentum* as Revealed by Chlorophyll a Fluorescence. *Sci. Rep.* **2017**, *7*, 8542. [CrossRef] [PubMed]
14. Bark, K.M.; Forcé, R.K. Fluorescence properties of fluoranthene as a function of temperature and environment. *Spectrochim. Acta Part A Mol. Spectrosc.* **1993**, *49*, 1605–1611. [CrossRef]
15. Hazrati, S.; Tahmasebi-Sarvestani, Z.; Modarres-Sanavy, S.A.M.; Mokhtassi-Bidgoli, A.; Nicola, S. Effects of water stress and light intensity on chlorophyll fluorescence parameters and pigments of *Aloe vera* L. *Plant Physiol. Biochem.* **2016**, *106*, 141–148. [CrossRef] [PubMed]
16. Liu, W.; Yuan, Y.; Wang, W.; Xie, Q.; Chen, Q. Effects of heat stress on chlorophyll fluorescence characteristics of Rhododendron leaves. *Jiangsu Agric. Sci.* **2019**, *47*, 144–148.
17. Li, Z.; Ji, W.; Hong, E.; Fan, Z.; Lin, B.; Xia, X.; Chen, X.; Zhu, X. Study on Heat Resistance of Peony Using Photosynthetic Indexes and Rapid Fluorescence Kinetics. *Horticulturae* **2023**, *9*, 100. [CrossRef]
18. Van der Westhuizen, M.M.; Oosterhuis, D.M.; Berner, J.M.; Boogaers, N. Chlorophyll a fluorescence as an indicator of heat stress in cotton (*Gossypium hirsutum* L.). *S. Afr. J. Plant Soil* **2020**, *37*, 116–119. [CrossRef]
19. Carlson, J.E.; Adams, C.A.; Holsinger, K.E. Intraspecific variation in stomatal traits, leaf traits and physiology reflects adaptation along aridity gradients in a South African shrub. *Ann. Bot.* **2016**, *117*, 195–207. [CrossRef]
20. Liu, Q.Q.; Huang, Z.J.; Wang, Z.N.; Chen, Y.F.; Wen, Z.M.; Liu, B.; Tigabu, M. Responses of leaf morphology, NSCs contents and C:N:P stoichiometry of Cunninghamia lanceolata and Schima superba to shading. *BMC Plant Biol.* **2020**, *20*, 354. [CrossRef]
21. Wu, J.A.R.; Zhong, H.; Yadav, V.; Zhang, C.; Ma, Y.; Liu, X.; Zhang, F.; Zha, Q.; Wang, X. The Impact of High Temperatures in the Field on Leaf Tissue Structure in Different Grape Cultivars. *Horticulturae* **2023**, *9*, 731. [CrossRef]
22. Altaf, M.A.; Hao, Y.; Shu, H.; Jin, W.; Chen, C.; Li, L.; Zhang, Y.; Mumtaz, M.A.; Fu, H.; Cheng, S.; et al. Melatonin mitigates cold-induced damage to pepper seedlings by promoting redox homeostasis and regulating antiox-idant profiling. *Hortic. Plant J.* **2023**, *9*, 1–13.
23. Altaf, A.; Nawaz, F.; Majeed, S.; Ahsan, M.; Ahmad, K.S.; Akhtar, G.; Shehzad, M.A.; Rashad Javeed, H.M.; Farman, M. Foliar Humic Acid and Salicylic Acid Application Stimulates Physiological Responses and Antioxidant Systems to Im-prove Maize Yield under Water Limitations. *JSFA Rep.* **2023**, *3*, 119–128. [CrossRef]
24. Ye, Z.P.; Li, J.S. Comparative investigation light response of photosynthesis on non-rectangular hyperbola model and modified model of rectangular hyperbola. *J. Jinggangshan Univ.* **2010**, *31*, 38–44.
25. Strasser, R.J.; Srivastava, A.; Tsimilli-Michael, M. The fluorescence transient as a tool to characterize and screen photosynthetic samples. *Probing Photosynth. Mech. Regul. Adapt.* **2000**, *25*, 445–483.
26. Tsimilli-Michael, M. Revisiting JIP-test: An educative review on concepts, assumptions, approximations, definitions and terminology. *Photosynthetica* **2019**, *57*, 90–107.

27. Arnon, D.I. Copper enzymes in isolated chloroplasts. polyphenoloxidase in beta Vulgaris. *Plant Physiol.* **1949**, *24*, 1–15. [CrossRef]
28. Ma, X.H.; Song, L.L.; Yu, W.W.; Hu, Y.Y.; Liu, Y.; Wu, J.S.; Ying, Y.Q. Growth, physiological, and biochemical responses of Camptothecaacuminata seedlings to different light environments. *Front. Plant Sci.* **2015**, *6*, 321. [CrossRef]
29. Chaves, M.M.; Flexas, J.; Pinheiro, C. Photosynthesis under drought and salt stress: Regulation mechanisms from whole plant to cell. *Ann. Bot.* **2009**, *103*, 551–560. [CrossRef]
30. Tang, H.; Hu, Y.Y.; Yu, W.W.; Song, L.L.; Wu, J.S. Growth, photosynthetic and physiological responses of Torreya grandis seedlings to varied light environments. *Trees* **2015**, *29*, 1011–1022. [CrossRef]
31. Huang, Z.; Liu, Q.; An, B.; Wu, X.; Sun, L.; Wu, P.; Liu, B.; Ma, X. Effects of Planting Density on Morphological and Photosynthetic Characteristics of Leaves in Different Positions on Cunninghamia lanceolata Saplings. *Forests* **2021**, *12*, 853. [CrossRef]
32. Negi, S.; Barry, A.N.; Friedland, N.; Sudasinghe, N.; Subramanian, S.; Pieris, S.; Holguin, F.O.; Dungan, B.; Schaub, T.; Sayre, R. Impact of nitrogen limitation on biomass, photosynthesis, and lipid accumulation in Chlorella sorokiniana. *J. Appl. Phycol.* **2016**, *28*, 803–812. [CrossRef]
33. Jiang, Y.; Ding, X.; Zhang, D.; Deng, Q.; Yu, C.L.; Zhou, S.; Hui, D. Soil salinity increases the tolerance of excessive sulfur fumigation stress in tomato plants. *Environ. Exp. Bot.* **2017**, *133*, 70–77. [CrossRef]
34. Hichem, H.; Naceur, A.E.; Mounir, D. Effects of salt stress on photosynthesis, PSII photochemistry and thermal energy dissipation in leaves of two corn (*Zea mays* L.) varieties. *Photosynthetica* **2009**, *47*, 517–526. [CrossRef]
35. Lin, Z.H.; Zhong, Q.S.; Chen, C.S.; Ruan, Q.C.; Chen, Z.H.; You, X.M. Carbon dioxide assimilation and photosynthetic electron transport of tea leaves under nitrogen deficiency. *Bot. Stud.* **2016**, *57*, 37. [CrossRef]
36. Lin, Z.H.; Chen, L.S.; Chen, R.B.; Zhang, F.Z.; Jiang, H.X.; Tang, N. CO_2 assimilation, ribulose-1,5-bisphosphate carboxylase/oxygenase, carbohydrates and photosynthetic electron transport probed by the JIP-test, of tea leaves in response to phosphorus supply. *BMC Plant Biol.* **2009**, *9*, 43. [CrossRef]
37. Murchie, E.H.; Lawson, T. Chlorophyll fluorescence analysis: A guide to good practice and understanding some new applications. *J. Exp. Bot.* **2013**, *64*, 3983–3998. [CrossRef] [PubMed]
38. Hanachi, S.; Van Labeke, M.C.; Mehouachi, T. Application of chlorophyll fluorescence to screen eggplant (*Solanum melongena* L.) cultivars for salt tolerance. *Photosynthetica* **2014**, *52*, 57–62. [CrossRef]
39. Hanelt, D. *Photosynthesis Assessed by Chlorophyll Fluorescence*; Elsevier: Amsterdam, The Netherlands, 2018.
40. Mathur, S.; Allakhverdiev, S.I.; Jajoo, A. Analysis of high temperature stress on the dynamics of antenna size and reducing side heterogeneity of Photosystem II in wheat leaves (*Triticum aestivum*). *Biochim. Biophys. Acta (BBA) Bioenerg.* **2011**, *1807*, 22–29. [CrossRef]
41. Martins, S.C.; Galmés, J.; Molins, A.; DaMatta, F.M. Improving the estimation of mesophyll conductance to CO_2: On the role of electron transport rate correction and respiration. *J. Exp. Bot.* **2013**, *64*, 3285–3298. [CrossRef]

horticulturae

Article

The Impact of High Temperatures in the Field on Leaf Tissue Structure in Different Grape Cultivars

Jiuyun Wu [1,2,†], Riziwangguli Abudureheman [1,2], Haixia Zhong [2,3,†], Vivek Yadav [2,3], Chuan Zhang [2,3], Yaning Ma [1,2], Xueyan Liu [1,2], Fuchun Zhang [2,3], Qian Zha [2,4,*] and Xiping Wang [1,2,5,*]

1. Turpan Research Institute of Agricultural Sciences, Xinjiang Academy of Agricultural Sciences, Turpan 838000, China; kobewjy@163.com (J.W.)
2. Xinjiang Grape Engineering Technology Research Center, Turpan 838000, China
3. The State Key Laboratory of Genetic Improvement and Germplasm Innovation of Crop Resistance in Arid Desert Regions (Preparation), Key Laboratory of Genome Research and Genetic Improvement of Xinjiang Characteristic Fruits and Vegetables, Institute of Horticultural Crops, Xinjiang Academy of Agricultural Sciences, Urumqi 830091, China
4. Research Institute of Forestry and Pomology, Shanghai Academy of Agricultural Science, Shanghai 201403, China
5. Colleges of Horticulture, Northwest A&F University, Xianyang 712100, China

* Correspondence: zhaqian@saas.sh.cn (Q.Z.); wangxiping@nwsuaf.edu.cn (X.W.); Tel.: +86-21-62208171 (Q.Z.); +86-29-87082129 (X.W.)

† These authors contributed equally to this work.

Abstract: Global warming will significantly affect grapevine growth and development. To analyze the effects of high temperature on the leaf tissue structure of grapevines in the field, 19 representative cultivars were selected from the grapevine germplasm resources garden in Turpan Research Institute of Agricultural Sciences, XAAS. Twelve tissue structure indexes of grapevine leaves, including the thickness of the upper epidermis (TUE), the thickness of palisade tissue (TPT), leaf vein (LV), the thickness of spongy tissue (TST), the thickness of the lower epidermis (TLE), stoma (St), guard cell (GC), cuticle (Cu), leaf tissue compactness (CTR) and leaf tissue porosity (SR), were measured during the natural high-temperature period in Turpan. The results showed significant differences in the leaf tissue structure of the 19 grapevine cultivars under natural high temperature. Based on the comprehensive comparative analysis of the leaf phenotype in the field, we identified that the leaves of some cultivars, including 'Zaoxia Wuhe', 'Centennial Seedless' and 'Kyoho' showed strong heat tolerance, whereas grapevine cultivars 'Golden Finger', 'Shine Muscat', 'Flame Seedless', 'Bixiang Wuhe' and 'Thompson Seedless' showed sensitivity to high temperature. We further evaluated the heat tolerance of different grapevine cultivars by principal component analysis and the optimal segmentation clustering of ordered samples. These findings provide a theoretical basis for adopting appropriate cultivation management measures to reduce the effect of high temperatures and offer fundamental knowledge for future breeding strategies for heat-tolerant grapevine varieties.

Keywords: grapevine; high temperature; heat tolerance; leaf anatomical structure

1. Introduction

High temperature is a major factor in the threat to global plant production and distribution. Each degree Celsius of the average growing season temperature may reduce crop production and plant distribution. Grapevine (*Vitis* L.) has to face the variety of biological and abiotic stresses that are inevitable during its growth and development. High temperature is one of the primary abiotic stress factors that restrict the yield and quality of grapes [1,2]. The most obvious sign of a plant response to heat stress is the change in the tissue structure of grapevine leaves. When evaluating how well various grapevine genetic resources can withstand heat, this method is frequently applied as one of the evaluation indicators of the extent of high temperature damage to the plant [3,4]. Leaves, as the primary organs of land plants, play a crucial role in photosynthesis. Their traits are closely linked to the physiological functions of plants [5]. Several aspects of leaf morphology have been definitively recognized as functional, displaying clear associations with abiotic stress. Leaf thickness, which represents the distance between the upper (adaxial) and lower (abaxial) leaf surfaces, has been shown to correlate with environmental factors such as water availability, temperature and light levels [6]. Additionally, studies indicate that plant diversity and the capacity of plants to adapt to arid conditions often lead to thicker leaves [7,8]. It is important to differentiate between leaf thickness within the context of typical leaf morphology, characterized by clear adaxial/abaxial flattening, and extremely thick leaves known as succulent leaves, which often exhibit a more radial structure. On an organismal level, thicker leaves present a trade-off between rapid growth and tolerance to drought and heat stress [9]. High temperatures not only damage the tissue structure of grapevine leaves but also hinder photosynthesis and nutrient metabolism. Furthermore, they inhibit fruit metabolism and the synthesis of aroma-related compounds, significantly impacting the commercial and market value of grapes [10,11]. When high temperature damage exceeds its ability to regulate adversity, serious heat damage symptoms and even plant death may occur [12,13]. The Turpan region in Xinjiang has great potential to produce high quality grape because of its suitable growing conditions. Abundant sunlight and optimum temperature during grape-growing seasons provide the opportunity to produce high quality grapes in this region. It is an important grape production region in China, covering an area of 38,025 hectares with a yield of 1,211,509 tons per year. However, the unique geography of the Turpan region means that a temperature of beyond 40 °C spans more than 35 days per year. This is likely to induce water loss, the wilting of grapevine leaves, damage to the cell structure, and a destruction of other factors that cause photosynthesis, as well as complicating the way grapes use nutrients. Furthermore, limiting the synthesis and transportation of photosynthetic products will lead to a significant decrease in grape yield and quality. With the increasing frequency and duration of extremely high temperature in the world [14], global grape-producing regions involving Turpan will face more severe challenges from high temperature stress [15]. Therefore, improving plant heat tolerance will be an important research field in response to global warming [16]. Numerous studies have been performed in the past by simulating high temperatures under artificial conditions. However, grapevines often face more complex environments during the field growth and development process, such as the intensity and occurrence time of high temperatures in the field fluctuating and changing. In addition, the high humidity or dry environment often accompanied by high temperature will also experience substantial differences. Therefore, it is difficult to fully and truly reflect the heat tolerance of grapevines in the natural environment when only simulating high-temperature climate indoors [17–19]. Therefore, under the unique arid and heat conditions in Turpan, research was carried out with a number of important grape varieties. The present research includes field phenotype observation, tissue structure analysis of different grapevine varieties, and a comprehensive evaluation of leaf heat tolerance to screen and identify the grapevine germplasm with different high-temperature tolerance types. The study's results will provide an in-depth understanding of the growth of different grapevines under field conditions and assist with information for breeding heat resistant varieties and cultivating stress resistance.

2. Materials and Methods

2.1. Grapevine Resources and Cultivation Conditions

The representative grapevine varieties available in the grapevine resource garden of the Turpan Research Institute of Agricultural Sciences, Xinjiang Academy of Agricultural Sciences (XAAS), were used as the test materials. The grapevine garden is located at 89°11′ E, 42°56′ N, at an altitude of 0 m. In July, with an average air humidity of 37.20%, the area received a mere 3.4 mm of precipitation, while the humidity ranged from 18.68% to 58.72%. The average photosynthetic radiation measured 163.52 W/m^2, and the average light intensity was 46.21 Klux. The detailed information regarding the grapevine varieties used in the current experiment with species characteristics and the parental sources of 19 grapevine varieties are presented in Table 1. In the trial field, a V-shaped leaf curtain was used as a trellis, and the distance between plants and rows was maintained at 1.2 m to 2.5 m. The garden soil is sandy loam with a pH of 8.04, and an organic content of approximately 12.8 g/kg. The area benefits from favorable water conservancy conditions, and fertilization is carried out using a water and fertilizer all-in-one machine. The plants were planted in a south-to-north direction, and they were all 6 years old, moderate and well-handled.

Table 1. The tested grapevine cultivars and parental origins.

No.	Cultivar	Species	Parental Origin (Female/Male)
1	Golden Finger	*V. vinifera* × *V. labrusca*	Manicure Finger × Seneca
2	Zhengyan Wuhe	*V. vinifera*	Jingxiu × Bronx Seedless
3	Flame Seedless	*V. vinifera*	Unknown
4	Jumeigui	*V. vinifera* × *V. labrusca*	Shenyang Meigui × Kyoho
5	Kyoho	*V. vinifera* × *V. labrusca*	Ishiharawase × Centennial
6	Cardinal	*V. vinifera*	Flame Tokay × Ribier
7	Bixiang Wuhe	*V. vinifera*	Zhengzhou Zaoyu × Pearlof Csaba
8	Qingfeng	*V. vinifera* × *V. labrusca*	Jingxiu × Bronx Seedless
9	Jintian Meigui	*V. vinifera*	Muscat Hamburg × Red Globe
10	Centennial Seedless	*V. vinifera*	Gold × Q25-6
11	Thompson Seedless	*V. vinifera*	Unknown
12	Summer Black	*V. vinifera* × *V. labrusca*	Kyoho × Thompson Seedless
13	Xinyu	*V. vinifera*	Red Globe × Rizamat
14	Shine Muscat	*V. vinifera* × *V. labrusca*	Akitsu21 × Hakunan
15	Zhengmei	*V. vinifera*	Manicure Finger × Zhengzhou Zaohong
16	Zitian Wuhe	*V. vinifera*	Niunai × Autumroyal
17	Zuijinxiang	*V. vinifera* × *V. labrusca*	7601 × Kyoho
18	Zaoxia Wuhe	*V. vinifera* × *V. labrusca*	Summer Black Mutation
19	Brilliant Seedless	*V. vinifera*	Red Globe × Centennial Seedless

The numbers represent the name of grapevine varieties. 1: 'Golden Finger'; 2: 'Zhengyan Wuhe'; 3: 'Flame Seedless'; 4: 'Jumeigui'; 5: 'Kyoho'; 6: 'Cardinal'; 7: 'Bixiang Wuhe'; 8: 'Qingfeng'; 9: 'Jintian Meigui'; 10: 'Centennial Seedless'; 11: 'Thompson Seedless'; 12: 'Summer Black'; 13: 'Xinyu'; 14: 'Shine Muscat'; 15: 'Zhengmei'; 16: 'Zitian Wuhe'; 17: 'Zuijinxiang'; 18: 'Zaoxia Wuhe'; 19: 'Brilliant Seedless'.

2.2. Temperature Measurement Tested

The experiment was conducted in Turpan, Xinjiang Uygur Autonomous Region, in 2022 during the high-temperature period (July). A MicroLite USB Temperature data Logger (Fourier Systems, Fourtec-Fourier Technologies., Ltd., San Francisco, CA, USA) was used to monitor the temperature data at a point in the grapevine resource garden. The temperature was measured and recorded once an hour throughout the entire high-temperature period.

2.3. Leaf Blade Morphology Observation

The plant growth status was observed in the field on 27 July 2022, and the grapevine leaves were photographed immediately using a digital compact camera (Canon-G15, Canon Corporation, Tokyo, Japan) by placing the leaves against a fixed black background in a photography chamber (Zhejiang standard photography equipment Corporation., Ltd., Ningbo, Zhejiang, China). Each variety selected functional leaves with essentially the same growth at the 5 to 7th nodes of the middle branch of the vine for photo observation.

2.4. Sampling for Leaf Structure Observation

The leaf sampling for microscopy was performed in July 2022. For detailed structure observation, healthy grapevine plants of the representative population were selected. Mature, healthy, without-disease leaves from the 5th and 7th nodes of middle branches with uniform shape and size were collected during a period of naturally high temperature. A clean and sharp pair of scissors was used to cut the leaves from the plants, preventing any damage to the leaf tissue. Each sample was labeled with basic information and transferred to a clean and dry container for transportation to the laboratory. The leaves were wiped to remove any dust with damp clothes and 0.5~1.0 cm pieces of tissue were cut intercostal between the veins. The leaves were fixed in FAA fixative solution, and a continuous paraffin section of 10 μm was made. The leaf samples were dyed with toluidine blue, soaked in xylene for 5 minutes, and then sealed and dried for later use. The fine tissue structures were observed with the Nikon Digital SightDS-L1 digital microscope camera system (Nikon Instruments Inc., Tokyo, Japan). The thickness of the leaves, upper epidermis, lower epidermis, palisade tissue, spongy tissue, cuticle, stomata and guard cells were evaluated using Image J, ver.1.47 (July 2013, National Institutes of Health, Bethesda, Maryland, USA). In total, 30 visual areas of each sample were measured and used to calculate the leaf tissue compactness (CTR) and leaf tissue porosity (SR) according to the below formulas [19,20]:

$$\text{CTR} = (\text{palisade tissue thickness}/\text{leaf thickness}) \times 100\%$$

$$\text{SR} = (\text{spongy tissue thickness}/\text{leaf thickness}) \times 100\%$$

2.5. Data Analysis

The test data were analyzed by variance, and the Duncan method was used for multiple comparisons and the significance of the difference was tested. $p < 0.05$. was the significant level of the difference, and $p < 0.01$ was the extremely significant level of the difference. GraphPad Prism ver.9.0 (October 2020, Dotmatics Corporation, Boston, MA, USA) and IBM SPSS Statistics (Statistical Product and Service Solutions) program, ver. 19.0 (August 2010, IBM Corporation, New York, NY, USA) were used for correlation analysis and cluster analysis. Data in tables and figures show means and standard errors representing 30 technical replications.

3. Results

3.1. Temperature Dynamics in the Field

The average temperature of the research field in July 2022 was 33.62 °C. The daily average high temperature was 39.78 °C, and the average low temperature was 27.41 °C. The highest and lowest temperatures for the month were reported 44.28 °C and 22.10 °C, respectively. Thirty days into the month, temperatures exceeded 35 °C, including 16 days with temperatures between 35 °C and 40 °C and 14 days with temperatures over 40 °C (Figure 1). The samples were collected on 28 July 2022, between 15:00 and 17:00.

3.2. Phenotypic Observation of Leaves

Under the natural high-temperature conditions in the field, only the 'Centennial Seedless', 'Brilliant Seedless', 'Jumeigui' and 'Zhengmei' possessed no obvious heat damage symptoms on leaves (Figure 2). The other varieties showed different degrees of heat damage symptoms on leaves. Among them, the leaves of 'Thompson Seedless', 'Shine Muscat' and 'Golden Finger' became yellowish green. The leaves of 'Jumcigui', 'Kyoho', 'Summer Black' and 'Zaoxia Wuhe' turned to dark green, and other varieties remained green. The leaves of 'Golden Finger', 'Zhengyan Wuhe', 'Flame Seedless', 'Bixiang Wuhe' and 'Shine Muscat' showed curly leaf margins. Except for the varieties 'Zhengyan Wuhe', 'Jumeigui', 'Centennial Seedless', 'Zhengmei' and 'Brilliant Seedless', the leaf margins of other varieties were all dry to varying degrees, and the leaf margins of 'Golden Finger', 'Zuijinxiang' and 'Shine Muscat' were more obviously dry. The leaves of 'Golden Finger', 'Flame Seedless', 'Thompson Seedless', 'Bixiang Wuhe' and 'Shine Muscat' were accompanied by spots. The results of the morphological observation showed that 'Thompson Seedless', 'Golden Finger', 'Flame Seedless', 'Bixiang Wuhe' and 'Shine Muscat' had heat damage symptoms that were more evident, and based on the preliminarily observation, these were of a high-temperature sensitive variety. The varieties 'Centennial Seedless', 'Brilliant Seedless', 'Zhengmei', and 'Jumeigui' were found to be high-temperature-tolerant varieties.

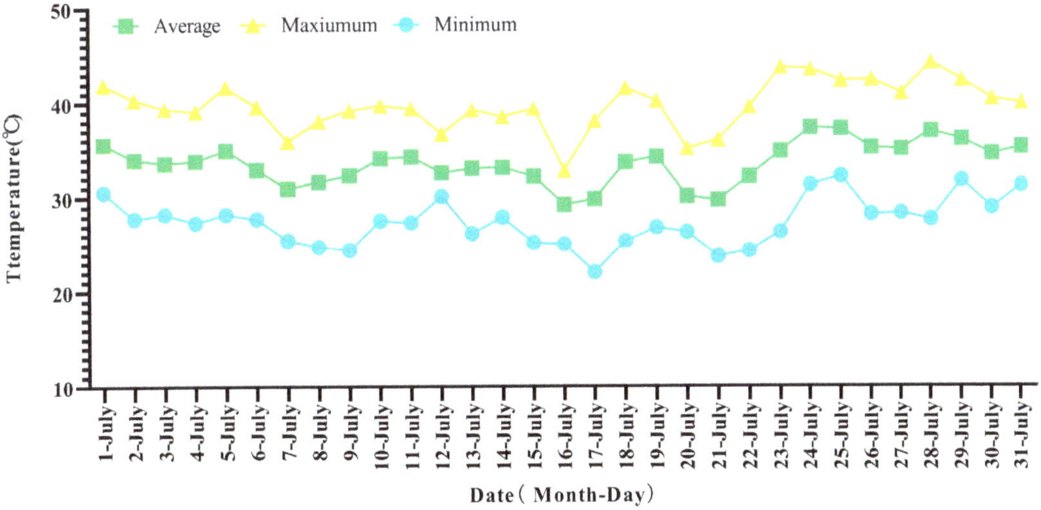

Figure 1. The air temperature of the viticultural region of XAAS in July, 2022, in Turpan.

3.3. Observation of the Cell Structure of Grapevine Leaves

The observations from the anatomical diagram of grapevine leaves show that the leaf structure is primarily composed of the thickness of upper epidermis (TUE), thickness of palisade tissue (TPT), leaf vein (LV), thickness of spongy tissue (TST), thickness of lower epidermis (TLE), stoma (St), guard cell (GC), cuticle (Cu), etc. (Figure 3). The thickness of leaf (TL) in 19 grapevine varieties ranged from 85.00~168.26 μm and the average TL was 114.34 ± 12.20 μm. The upper and lower epidermal cells were composed of long oval monolayer cells, and the length of the TUE ranged from 9.88~32.57 μm with an average length of 19.73 ± 6.92 μm. The TLE was observed on palisade tissue, and the shapes of upper and lower epidermal cells were found to be similar (Figure 3). The length of upper and lower epidermal cells varied from to 10.19–24.31 μm with an average length of 17.49 ± 4.27 μm.

The mesophyll tissue is mainly composed of palisade tissue and spongy tissue. The palisade tissue cells are shaped in the form of long columns and are arranged neatly and closely in the mesophyll, with a length of 19.36–57.66 μm; the average length is about 32.68 ± 5.44 μm. Spongy tissue is composed of oval cells located between the palisade tissue and lower epidermis. They are loosely and irregularly arranged in the mesophyll. The cell length varies from 24.65 to 11.77 μm, and the average length is about 18.74 ± 3.88 μm. There were obvious traces of cambium and relatively developed xylem in the leaves. The cambium is composed of several layers of parenchyma cells, and the xylem is composed of five to seven layers of radial and orderly arranged cells. There are vessels in the xylem, and the vascular bundles are uniformly and closely arranged in the veins. The lengths of the veins were in the range of 9.69–34.52 μm; the average length is about 19.48 ± 4.13 μm (Figure 3). The phloem cells are relatively small and thin, arranged in an orderly manner.

Figure 2. The leaf phenotypes of different grapevine cultivars after exposure to high temperatures in the field. 1: 'Golden Finger'; 2: 'ZhengyanWuhe'; 3: 'Flame Seedless'; 4: 'Jumeigui'; 5: 'Kyoho'; 6: 'Cardinal'; 7: 'BixiangWuhe'; 8: 'Qingfeng'; 9: 'JintianMeigui'; 10: 'Centennial Seedless'; 11: 'Thompson Seedless'; 12: 'Summer Black'; 13: 'Xinyu'; 14: 'Shine Muscat'; 15: 'Zhengmei'; 16: 'ZitianWuhe'; 17: 'Zuijinxiang'; 18: 'ZaoxiaWuhe'; 19: 'Brilliant Seedless'.

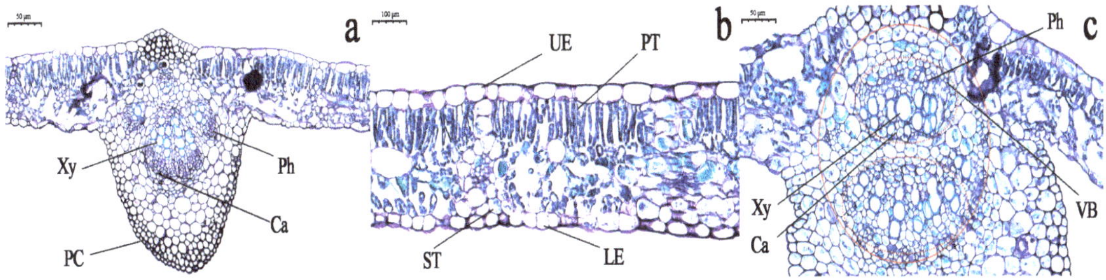

Figure 3. Anatomical structures of grapevine leaves. Anatomical structure of grapevine leaves and response to heat stress. (**a**–**c**): Different views of leaf cross-section; AS: abaxial surface; Ca: cambium; CU: cuticle; Ep: epidermis; LE: lower epidermis; LV: lateral vein; Me: mesophyll; MV: main vein; PC: parenchymal cells; PT: palisade tissue; ST: spongy tissue; VB: vascular bundles; Ve: vein; VS: ventral surface; Xy: xylem. Scale bars: (**a**,**c**), 50 μm; (**b**), 100 μm.

3.4. Leaf Thickness and Epidermal Cells

The thickness of leaves (TLs) is one of the indicators to evaluate the heat tolerance of plants. Under high temperature environment, the TLs slow down transpiration to reduce water loss and alleviate heat damage. Depending on their thickness, epidermal cells can affect the heat tolerance of grapevines to a certain extent. There are certain statistical differences in leaf thickness and epidermal cells between the 19 grapevine varieties (Table 2). The TLs of 'Kyoho' and 'Zaoxia Wuhe' are greater, which are 168.26 μm and 161.29 μm, respectively. The minimum TL of 'Zhengmei' (TL = 85.00 μm) is significantly lower than those of other varieties ($p < 0.05$) except for 'Shine Muscat'. Comparing TUE and TLE, TUE arranges more orderly and is slightly larger than TLE, but there is no significant difference in cell size ($p < 0.05$). Under the naturally high-temperature conditions in Turpan, the TUE of 'Summer Black' and 'Centennial Seedless' are significantly higher than other varieties ($p < 0.05$), which are 32.57 μm and 30.19 μm, respectively. 'Flame Seedless' presents the minimum in 19 grapevine varieties (TUE = 9.88 μm); it is significantly lower than other varieties, except for 'Qingfeng', 'Cardinal' and 'Centennial Seedless' ($p < 0.01$). The thicknesses of TLE and TUE resemble each other. The TLE of 'Zitian Wuhe', 'Zaoxia Wuhe', 'Zuijinxiang' and 'Centennial Seedless' are significantly greater than that of other varieties, except of 'Centennial Seedless' ($p < 0.01$), with 'Zitian Wuhe' having a maximum TLE of 24.31 μm, and 'Golden Finger' having a significantly lower minimum TLE than other varieties (10.19 μm; $p < 0.01$) (Figure 3).

3.5. Palisade Tissue, Spongy Tissue and Palisade Tissue/Spongy Tissue

The structural characteristics of mesophyll cells are directly related to photosynthesis, and their degree of differentiation can indirectly indicate the heat tolerance of plants, while the larger the palisade tissue/spongy tissue (P/S), the stronger the heat tolerance. The palisade tissue is significantly the largest in 'Cardinal' and 'Kyoho', at 57.66 μm and 54.63 μm, respectively (Table 2; $p < 0.01$). However, it is significantly the smallest in 'Thompson Seedless' ($p < 0.01$). The significantly largest spongy tissue is that of 'Centennial Seedless' (24.65 μm), whereas 'Zhengyan Wuhe' and 'Kyoho' have the significantly smallest one ($p < 0.05$). The P/S ratio ranges from 1.12 to 3.14. Four varieties have a significantly higher P/S than 2.0, including 'Qingfeng', 'Kyoho', 'Cardinal' and 'Zaoxia Wuhe' ($p < 0.01$). Among them, 'Qingfeng' (P/S = 3.14) reaches the maximum, indicating that its heat tolerance is relatively strong, while 'Zhengyan Wuhe' has the minimum P/S of 1.12, indicating that its heat tolerance is relatively weak.

Table 2. Comparison of leaf anatomical structure indexes of different grapevine varieties.

No.	TUE (μm)	TPT (μm)	LV (μm)	TST (μm)	TLE (μm)	St (μm)	GC (μm)	Cu (μm)	TL (μm)	P/S
1	14.24 ± 2.28 gh	36.8 ± 7.64 cd	18.01 ± 3.62 e	20.12 ± 4.69 de	10.19 ± 3.18 h	7.34 ± 3.25 a	6.66 ± 2.46 defgh	2.89 ± 1.51 g	111.33 ± 7.69 e	1.91 ± 0.53 d
2	13.4 ± 3.32 hi	25.45 ± 2.84 f	18.4 ± 3.49 e	14.35 ± 3.33 f	16.96 ± 3.4 bcde	4.31 ± 1.74 fgh	6.06 ± 2.41 fgh	3.55 ± 1.03 defg	93.34 ± 9.78 hi	1.12 ± 0.31 h
3	9.88 ± 3.32 j	24.07 ± 4.01 fg	14.14 ± 3.27 fg	14.03 ± 4.07 f	17.91 ± 3.48 bcd	4.08 ± 2.06 ghi	6.79 ± 3.19 defgh	3.47 ± 1.23 defg	103.77 ± 14.34 f	1.84 ± 0.53 de
4	17.12 ± 3.5 fg	33.94 ± 6.14 d	34.52 ± 7.92 a	22.09 ± 4.25 bcd	18.86 ± 4.81 bc	5.57 ± 1.85 bcde	6.68 ± 1.94 defgh	3.15 ± 1.15 efg	114.44 ± 10.5 e	1.58 ± 0.37 defg
5	30.39 ± 37.78 cd	54.63 ± 8.27 a	31.72 ± 5.81 b	22.7 ± 3.82 abc	13.45 ± 3.22 fg	5.47 ± 3.24 cdef	7.86 ± 2.8 bcd	2.94 ± 1.14 g	161.29 ± 11.03 b	2.47 ± 0.56 c
6	11.32 ± 3.52 hij	57.66 ± 12 a	26.94 ± 5.77 c	21.07 ± 4.1 bcd	17.86 ± 3.37 bcd	4.53 ± 2.35 efgh	6.47 ± 2.98 defgh	2.87 ± 1.26 g	149.98 ± 13.12 c	2.85 ± 0.95 b
7	16.37 ± 6.65 fg	19.36 ± 3.94 h	23.04 ± 4 22 d	15.29 ± 3.85 f	13.6 ± 3.34 fg	4.8 ± 2.5 defgh	5.5 ± 2.84 h	3.05 ± 1.02 fg	101.03 ± 21.9 fg	1.35 ± 0.47 gh
8	12.33 ± 3.74 hij	47.92 ± 7.94 b	23.26 ± 5 78 d	16.26 ± 4.12 f	16.3 ± 3.72 cde	5.98 ± 2.12 bc	5.66 ± 2.32 gh	1.34 ± 1.34 h	144.93 ± 13.14 cd	3.14 ± 0.91 a
9	22.08 ± 5.28 d	23.6 ± 2.18 fg	15.11 ± 2.86 f	21.38 ± 5.43 bcd	15.68 ± 5.04 def	4.89 ± 1.6 cdefgh	5.37 ± 1.82 h	3.02 ± 1.79 fg	92.89 ± 12.84 hi	1.17 ± 0.3 h
10	30.19 ± 9.08 a	39.13 ± 5.11 c	23.66 ± 5.52 d	24.65 ± 5.03 a	21.96 ± 5.61 a	5.41 ± 1.5 cdef	8.9 ± 2.22 ab	3.96 ± 1.33 cd	139.34 ± 11.73 d	1.65 ± 0.38 defg
11	10.55 ± 4.39 ij	20.87 ± 7.18 gh	12.08 ± 4.96 gh	11.77 ± 2.38 g	12.52 ± 3.49 g	4.71 ± 1.38 defgh	5.72 ± 2.08 gh	3.8 ± 1.15 de	92.2 ± 9.18 hi	1.85 ± 0.84 de
12	25.6 ± 7.27 bc	29.47 ± 4.13 e	18.92 ± 4 47 e	18.41 ± 4.15 e	19.22 ± 5.34 b	3.14 ± 1.25 i	8.39 ± 2.71 abc	4.67 ± 0.99 bc	100.61 ± 6.98 fg	1.67 ± 0.4 def
13	21.1 ± 5.69 de	22.51 ± 4.73 fgh	11.37 ± 2.52 hi	15.51 ± 2.94 f	16.44 ± 3.77 cde	4.57 ± 1.32 efgh	6.27 ± 1.88 efgh	4.6 ± 1.37 bc	95.54 ± 11.63 gh	1.5 ± 0.39 fg
14	19.04 ± 4.69 ef	24.43 ± 2.39 f	9.69 ± 2.15 i	16.33 ± 2.91 f	17.66 ± 4.39 bcd	3.95 ± 0.82 hi	7.1 ± 3.01 cdefg	4.96 ± 1.11 ab	87.87 ± 7.48 ij	1.54 ± 0.31 efg
15	17.37 ± 5.42 f	25.45 ± 2.58 f	12.39 ± 2.98 gh	15.39 ± 2.3 f	14.8 ± 3.49 efg	5.22 ± 2.03 cdefg	7.36 ± 2.24 cdef	4.03 ± 1.42 cd	85 ± 14.73 j	1.68 ± 0.25 def
16	26.84 ± 8.01 b	34.58 ± 5.02 d	15.47 ± 2.81 f	20.57 ± 3.79 cd	24.31 ± 8.19 a	5.8 ± 1.39 bcd	8.39 ± 2.47 abc	4.15 ± 1.37 cd	112.96 ± 17.16 e	1.74 ± 0.42 def
17	21.92 ± 6.62 de	29.32 ± 4.48 e	20.2 ± 3.75 e	22.15 ± 3.2 bcd	22.38 ± 4.65 a	5.64 ± 1.55 bcde	7.62 ± 1.68 bcde	4.15 ± 1.37 cd	112.96 ± 17.16 e	1.35 ± 0.26 gh
18	32.57 ± 5.46 a	47.05 ± 8.92 b	23 ± 3.59 d	21.08 ± 4.9 bcd	22.77 ± 4.34 a	6.64 ± 1.35 ab	9.51 ± 2.12 a	3.74 ± 1.57 def	168.26 ± 13.56 a	2.35 ± 0.75 c
19	22.54 ± 5.44 d	24.71 ± 3.94 f	18.18 ± 2.79 e	22.93 ± 4.46 ab	19.46 ± 4.37 b	4.65 ± 1.63 defgh	7.42 ± 2 cdef	5.5 ± 1.1 a	104.71 ± 7.92 f	1.89 ± 0.58 d

The average value of TUE, TPT, LV, TST, TLE, St, GC, Cu, TL and P/S are means ± S.E. Different lowercase letters in each column indicate significant difference at $p < 0.05$. TUE: thickness of upper epidermis; TPT: thickness of palisade tissue; LV: leaf vein; TST: thickness of spongy tissue; TLE: thickness of lower epidermis; St: stoma; GC: guard cell; Cu: cuticle; TL: thickness of leaves; P/S: palisade tissue/spongy tissue. 1: 'Golden Finger'; 2: 'Zhengyan Wuhe'; 3: 'Flame Seedless'; 4: 'Jumeigui'; 5: 'Kyoho'; 6: 'Cardinal'; 7: 'Bixiang Wuhe'; 8: 'Qingfeng'; 9: 'Jintian Meigui'; 10: 'Centennial Seedless'; 11: 'Thompson Seedless'; 12: 'Summer Black'; 13: 'Xinyu'; 14: 'Shine Muscat'; 15: 'Zhengmei'; 16: 'Zitian Wuhe'; 17: 'Zuijinxiang'; 18: 'Zaoxia Wuhe'; 19: 'Brilliant Seedless'.

3.6. Stomata, Guard Cells and Cuticle

Stomata (St) are the main pathways of photosynthetic gas exchange metabolism in plants. Stomatal opening is mainly regulated by guard cells (GC), which generally increases with the rise in temperature. However, high temperature (>30 °C) will lead to strong transpiration. At this temperature, guard cells close stomata to avoid water loss. Under the natural high-temperature conditions in Turpan, 'Golden Finger' (St = 7.34 μm) has the maximum stomata number on leaves, 'Zaoxia Wuhe' has the second highest number (St = 6.44 μm), and 'Summer Black' has the minimum (St = 3.14 μm) (Table 2). The guard cell dimension of 'Golden Finger' is significantly higher than that of other varieties. The guard cell dimension ranges from 5.37 μm to 9.51 μm in 19 grapevine varieties; four varieties bear guard cells longer than 8 μm, including 'Zaoxia Wuhe', 'Centennial Seedless', 'Summer Black' and 'Zitian Wuhe'. 'Jintian Meigui' has the significantly shortest guard cells at 5.37 μm ($p < 0.01$). The cuticle (Cu) is a membrane composed of impermeable adipose tissue, which can improve heat tolerance in plants by reducing transpiration and prevent water loss in high-temperature environments [21]. The cuticle of 'Zhengyan Wuhe' is significantly the thickest one (Cu = 5.50 μm) (Figure 2). 'Shine Muscat' ($p < 0.05$) and 'Qingfeng' have the significantly thinnest cuticle (1.34 μm) ($p < 0.01$).

3.7. Leaf Vein, CTR and SR

Leaf veins (LV) are composed of vascular bundles distributed in the mesophyll tissue and play a role in conduction and physical support. The veins of different grape varieties have different compactness and arrangement. Some have thicker vascular bundles. Under the natural high-temperature conditions in Turpan, the stomata of the leaves of 'Jumeigui' were significantly the largest (diameter 34.52 μm) of the 19 grapevine varieties, followed by 'Kyoho' (diameter: 31.72 μm), 'Cardinal' (diameter: 26.94 μm) and other varieties ($p < 0.05$). 'Shine Muscat' significantly has the smallest veins (diameter: 9.69 μm) ($p < 0.05$). The CTR value is the ratio of palisade tissue thickness to leaf thickness, indicating the compactness of tissue structure. CTR values are commonly used as an important indicator to describe the heat tolerance of plants [19]. The CTR values of 19 grapevine varieties lie between 20.30% and 38.63%. The CTR of 'Cardinal' is significantly the highest value (38.63%) ($p < 0.01$) (Figure 4). 'Bixiang Wuhe' has the lowest CTR (20.30%). SR is the ratio of spongy tissue thickness to leaf thickness, indicating the porosity of the tissue structure. The smaller the SR, the stronger the heat resistance [20]. 'Jintian Meigui' has the significantly highest SR (23.43%). 'Zhengyan Wuhe' and 'Qingfeng' have the significantly lowest SR (11.30%).

3.8. Correlation Analysis

Pearson correlation analysis of the 12 cell structure indicators was carried out on the leaves of different grapevine varieties. Results showed that there was a certain correlation between the indicators (Figure 5). The TUE was positively correlated with TST (0.650 **), TLE (0.524 **) and GC (0.781 **). The TPT was positively correlated with LV (0.670 **), TST (0.513 **) and TL (0.917 **), and negatively correlated with Cu (−0.499 *). LV was positively correlated with TST (0.565 **), TL (0.700 **) and CTR (0.472 *), and negatively correlated with Cu (−0.482 *). TST was positively correlated with GC (0.537 *) and TL (0.537 *). The TLE was also significantly positively correlated with GC (0.665 *), St and TL (0.481 *). The TL was significantly positively correlated with CTR (0.551 *). A high positive correlation was found with P/S (0.712 **) and a significantly negative correlation was found with Cu (−0.497 *), while SR was significantly positively correlated with Cu (0.495 *).

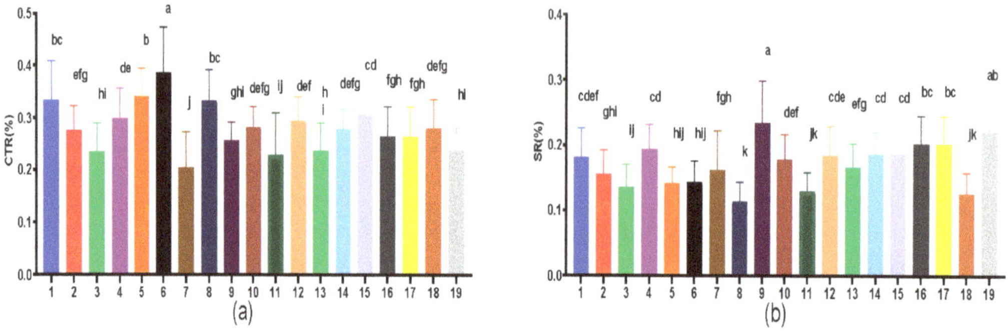

Figure 4. The CTR and SR values of grapevine leaf anatomy. Different letters indicate means which are significantly different at $p < 0.05$. (**a**) The CTR values of 19 grapevines. (**b**) The SR values of 19 grapevines. X-axis represent grape genotypes. 1: 'Golden Finger'; 2: 'Zhengyan Wuhe'; 3: 'Flame Seedless'; 4: 'Jumeigui'; 5: 'Kyoho'; 6: 'Cardinal'; 7: 'Bixiang Wuhe'; 8: 'Qingfeng'; 9: 'Jintian Meigui'; 10: 'Centennial Seedless'; 11: 'Thompson Seedless'; 12: 'Summer Black'; 13: 'Xinyu'; 14: 'Shine Muscat'; 15: 'Zhengmei'; 16: 'Zitian Wuhe'; 17: 'Zuijinxiang'; 18: 'Zaoxia Wuhe'; 19: 'Brilliant Seedless'.

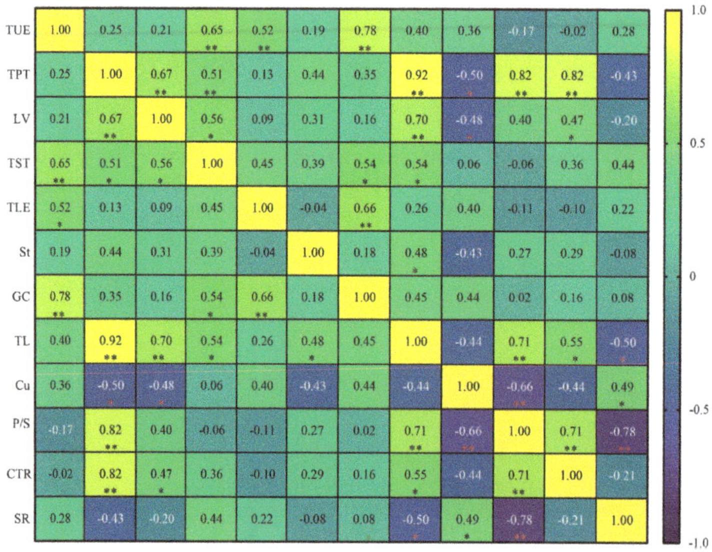

Figure 5. Correlation analysis diagram. ** shows highly significant correlation at 0.01 level and * indicates significant correlation at 0.05 level. CTR: Palisade tissue/leaf thickness ratio; Cu: cuticle; GC: guard cell; P/S: ratio of palisade tissue/spongy tissue; LV: leaf vein; SR: spongy tissue/thickness of leaf; St: stoma; TL: thickness of leaf; TLE: thickness of lower epidermis; TPT: thickness of palisade tissue; TST: thickness of spongy tissue; TUE: thickness of upper epidermis.

3.9. *Principal Component Analysis and Heat Tolerance Evaluation*

The principal component analysis (PCA) method can be used to obtain a clear picture of how different grapevine leaf cell structure components contribute to a particular trait [20,22]. The K value (>0.6) and p value (<0.01) obtained from KMO (Kaiser–Meyer–Olkin) and the Bartlett tests showed suitability of the principal component analysis [23]. Through a factor analysis of the 12 main indicators, the cumulative contribution rate of the two principal components is 69.41%, and the eigenvalue is greater than 1.0, which ultimately covers most of the information of each character indicator and can be used as the principal component of the comprehensive character indicator. The variance contribution rate of PC1 is 41.09%, which mainly represents the index information of TPT, TL, P/S, CTR, LV, etc., while the variance contribution rate of PC2 is 28.32%, which mainly represents the index information of the TUE, GC, TLE, TST, Cu, SR, etc. As shown in Figure 6, the selected 12 indicators only formed two clusters with good separation, most of which are located on PC1 (clusters 1 and 2), where cluster 1 is mainly located on PC1, including P/S, CTR, SR, TPT, TL, LV, St, etc., while cluster 2 is mainly located under PC2 and on PC1, including TST, TLE, GC, TUE, etc., while the cuticle is located under PC1 and PC2 (cluster 2).

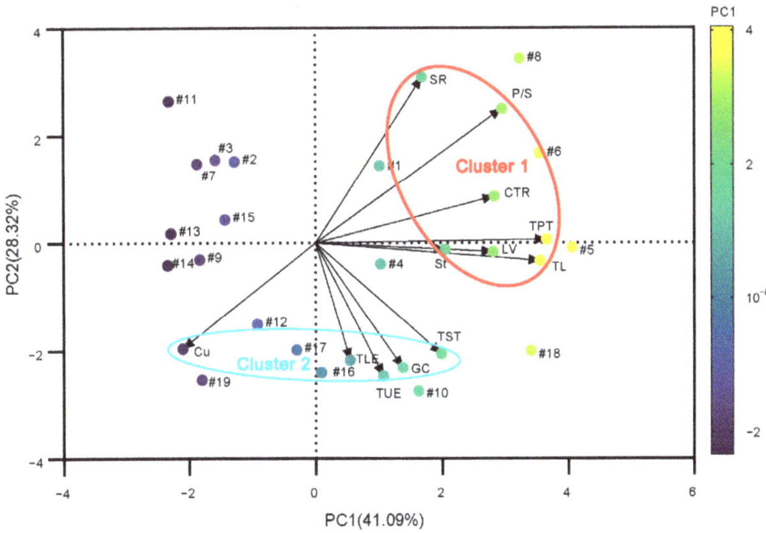

Figure 6. Principal component analysis of leaf cell structure components. Vectors indicate the direction and strength of each variable to the overall distribution. Colored symbols correspond to leaf structure components. Cluster 1: CTR, P/S, TPS, TL, LV, St, SR, and cluster 2: Cu, TLE, GC, TST, TUE.

Based on PCA, the results showed that the 'Kyoho', 'Cardinal', 'Zaoxia Wuhe', 'Qingfeng', 'Centennial Seedless', 'Jumeigui', 'Golden Finger' and 'Zitian Wuhe' have higher scores of PC1 in 19 grapevine varieties, while the 'Centennial Seedless', 'Brilliant seedless', 'Zitian Wuhe', 'Zaoxia Wuhe' and 'Zuijinxiang' have higher scores of PC2 in 19 grapevine varieties. The model was used to analyze and rank the comprehensive scores of leaf traits of 19 grapevine varieties. The comprehensive score of the principal components is ranked as follows (Table 3): 'Zaoxia Wuhe' > 'Kyoho' > 'Centennial Seedless' > 'Cardinal' > 'Zitian Wuhe' > 'Jumeigui' > 'Zuijinxiang' > 'Qingfeng' > 'Summer Black' > 'Brilliant seedless' > 'Golden Finger' > 'Jintian Meigui' > 'Zhengmei' > 'Shine Muscat' > 'Xinyu' > 'Zhengyan Wuhe' > 'Flame Seedless' > 'Bixiang Wuhe' > 'Thompson Seedless'.

Table 3. Principal component scores of cultivars and rank.

No.	Cultivars	F_1	$Rank_1$	F_2	$Rank_2$	F	Rank	Heat Tolerance
1	Golden Finger	0.43701	7	−0.76282	13	−0.052	11	Medium
2	Zhengyan Wuhe	−0.56488	11	−0.82448	15	−0.6747	16	Weak
3	Flame Seedless	−0.6953	13	−0.8409	16	−0.7592	17	Weak
4	Jumeigui	0.46852	6	0.19565	8	0.3595	6	Medium
5	Kyoho	1.81753	1	0.07831	10	1.1159	2	Strong
6	Cardinal	1.61936	2	−0.89711	17	0.5981	4	Medium
7	Bixiang Wuhe	−0.8295	15	−0.8212	14	−0.8311	18	Weak
8	Qingfeng	1.47415	4	−1.85897	19	0.1174	8	Medium
9	Jintian Meigui	−0.82192	14	0.13828	9	−0.4335	12	Weak
10	Centennial Seedless	0.69555	5	1.51558	1	1.0358	3	Strong
11	Thompson Seedless	−1.01656	17	−1.45255	18	−1.2014	19	Weak
12	Summer Black	−0.42692	10	0.78397	6	0.0666	9	Medium
13	Xinyu	−1.03539	18	−0.08435	11	−0.652	15	Weak
14	Shine Muscat	−1.05973	19	0.22746	7	−0.5387	14	Weak
15	Zhengmei	−0.65071	12	−0.23027	12	−0.4824	13	Weak
16	Zitian Wuhe	0.03903	8	1.29195	3	0.5526	5	Medium
17	Zuijinxiang	−0.13273	9	1.05875	5	0.3547	7	Medium
18	Zaoxia Wuhe	1.51481	3	1.12429	4	1.364	1	Strong
19	Brilliant Seedless	−0.83232	16	1.3584	2	0.0603	10	Medium

The F (factor) value is represented as the comprehensive principal component factor score, and rank is the classification of heat tolerance of different grapevine varieties by SPSS 19.0. The F_1 (factor 1) value is represented as principal component factor 1, and R_1 ($rank_1$) is represented as the rank of F_1 value in 19 grapevines. The F_2 (factor 2) value is represented as the principal component factor 2, and R_2 ($rank_2$) is represented as the rank of F_2 value in 19 grapevines.

In addition, the F value was further classified by using the ordered sample optimal segmentation clustering method to obtain the optimal segmentation error function and classification results of all varieties for heat tolerance (Table 4). With the increase in classification numbers, the error function tended to be stabled, and it was categorized into three grades. The F test shows that the difference between each grade is very significant ($p < 0.01$). Therefore, a three-level segmentation model was found suitable and all 19 varieties were screened and evaluated according to this standard. It was observed that 'Zaoxia Wuhe', 'Centennial Seedless' and 'Kyoho' were defined as heat-tolerant leaf varieties, and 'Golden Finger', 'Jintian Meigui', 'Zhengmei', 'Shine Muscat', 'Xinyu', 'Zhengyan Wuhe', 'Flame Seedless', 'Bixiang Wuhe' and 'Thompson Seedless' were defined as heat-sensitive varieties.

Table 4. Classification results of F value under different cluster numbers.

Cluster Number	Error Function	Optimal Segmentation Results
2	0.4129	1–11, 12–19
3	0.1355	1–3, 4–11, 12–19
4	0.0823	1–3, 4–7, 8–11, 12–19
5	0.0381	1–3, 4–7, 8–11, 12–18, 19
6	0.0225	1–3, 4–7, 8–11, 12–14, 15–18, 19
7	0.0141	1, 2–3, 4–7, 8–11, 12–14, 15–18, 19

The column number of "optimal segmentation results" indicates the rank of 19 cultivars, the same as Table 3.

4. Discussion

Temperature is one of the most important factors influencing the global grapevine distribution [15,24]. With the average temperatures continuously rising due to global warming, high temperature has become one of the most significant negative factors that inevitably limits grape yield and quality [3,25]. In order to adapt to the high-temperature environment, grapevine plants have also evolved their ecological habit of adapting to the high temperature stress by responding to high temperature stress in a timely manner. However, there are significant differences in heat tolerance among grapevine genotypes. Heat tolerance is a genetic characteristic of adaptation for plants to high temperature stress within a long-term period, and it is not only dependent on their internal physiological and biochemical structures, but also mainly on their genetic characteristics and morphological and organizational structures [4,19,26]. Wu et al., (2019) found that physiological and biochemical indexes are susceptible to environmental factors and show different changes [27,28]. Moreover, leaves are mostly composed of plastic tissues and organs which are established during plant evolution. Leaf shape and anatomy result from long-term evolution. Different adaptation types establish under different environmental conditions and are less relatively affected by transient environmental factors [29,30]. Therefore, an important evaluation index in the study of plant heat tolerance has often been used. Previous studies on the heat tolerance of the genera *Vitis* [19], *Jujube* [31], *Rhododendron* [20] and other crops, based on their leaf structures, suggested that the stability of the leaf cell structure is related to the heat tolerance of grapevines. There are many methods and indicators to evaluate plant heat tolerance based on cell structure, among which principal component analysis (PCA) can integrate its performance and simplify the selection process of various indicators, which is convenient for a comprehensive evaluation of the heat tolerance of various plants. The anatomical changes of leaves under high temperature and drought stress are relatively similar [32], so there are many applied studies on the evaluation of drought resistance and heat tolerance of fruit trees, flowers, vegetables, wheat, and other crops using the PCA method [33–35]. For instance, Guo et al. (2020) screened 10 relevant indexes as typical indicators for the comprehensive evaluation of drought resistance of 238 chestnut varieties (lines) after PCA analysis on the leaf anatomy of 238 varieties (lines) of Chinese chestnut [36]. Ding et al. (2022) evaluated the leaf cell structure of twenty-five grapevine rootstock cultivars based on PCA analysis and screened five rootstock varieties with strong drought resistance [22]. Qiu et al. (2022) conducted PCA analysis of seven rhododendron species based on leaf structure and found that the damage symptoms under artificial simulation of high temperature stress, and the conclusions reached by the evaluation and screening of field heat-tolerant varieties were ultimately consistent [20]. This previous approach revealed that it is convenient to use PCA analysis while screening and evaluating the heat tolerance of grapevines based on their cells. PCA was used in our study and the results showed that the 'Kyoho', 'Cardinal', 'Zaoxia Wuhe', 'Qingfeng', 'Centennial Seedless', 'Jumeigui', 'Golden Finger' and 'Zitian Wuhe' had higher scores of PC1 in 19 grapevine varieties. Therefore, the present study used PCA dimension reduction and cluster analysis to conduct a comprehensive evaluation on multiple indicators, allowing for the relative screening of heat-tolerant variety resources while also providing a reference for the evaluation and classification of grapevine heat tolerance.

The heat tolerance of plants is largely related to the cell structure of their leaves, especially the thickness of leaves, epidermal cells, palisade tissue, spongy tissue and cuticle, which have a significant impact on their heat tolerance [20,30]. This study found that under the high temperature (drought) stress conditions in the Turpan area, the cell and tissue structures of the leaves of 19 grapevine varieties had a certain degree of change, including mesophyll tissue disorder, cell gap expansion, mesophyll tissue water loss and atrophy and plasmolysis, which was consistent with the observation results of previous researchers in response to high temperature stress in *Citrus* [37], *Vitis* [26] and *Azalea* [38]. The research shows that the greater the proportion of palisade tissue and the P/S in the leaf structure, the more beneficial it is to enhance the utilization rate of water and light energy in the leaves. In our current study, some varieties with a higher heat tolerance showed a greater proportion of palisade tissue and P/S in the leaf, when compared with the sensitive varieties. It prevents the excessive evaporation of water in the leaves and alleviates the leaf damage caused by high temperature dehydration, which can reflect the heat tolerance of plants [20,39]. In this study, we found that the weak heat-tolerant varieties have relatively lower P/S and higher SR values in 19 grapevine varieties, such as 'Bixiang Wuhe', 'Flame Seedless', 'Shine Muscat' and Thompson Seedless', while the P/S and TPT were also higher and SR was relatively lower in strong heat-tolerant varieties. At the same time, the thicker leaves are more beneficial in preventing water evaporation, and the epidermal cells have the function of regulating the morphological changes of the leaves [40]. This study found that the thickness of the leaves and epidermal cells of the varieties 'Zaoxia Wuhe', 'Centennial Seedless' and 'Kyoho' had strong and relatively higher heat tolerance, which is consistent with the field observation results.

Photosynthesis is extremely sensitive to high temperatures and often ceases before other cell activities are compromised. As an important channel for plant photosynthesis and gas exchange, the morphological and structural changes of leaves are the most obvious characteristics of grapevines affected by high temperature [41,42]. It not only determines the phenotypic structure and function of leaves, but also their adaptive processes in response to high temperature stress, which reflects the degree of high temperature damage and heat tolerance [12,43]. The methods of simulating high temperatures under controlled conditions were widely used in the past to evaluate the heat tolerance of grapevine, but in the field, the grapevines often encountered environmental effects which made the study more complicated, resulting in both the intensity and the timing of daily high temperatures in the summer to vary. However, the temperature drop at night will provide grapevines with opportunities to resume normal growth [18,44]. Studies related to field natural high temperature and indoor-simulated high temperature were also compared. Previous research showed that the PSII activity and heat shock protein changing trends in grapevine leaves were relatively similar, but the degree was different [23,45]. Therefore, in the current study, we chose to analyze the anatomical structure of grapevine leaves grown in the field under the natural high temperature stress in Turpan, combined with the comprehensive observation and comparison of the field phenotypes of the leaves, and the results have more reference significance for revealing the process of grapevine response to high temperature. At the same time, the leaf cell structure reflects the high temperature tolerance of 19 grapevine varieties, which is also essentially consistent with the situation under natural high temperature stress. This can be used as a reference for research on how to choose grapevine varieties and make them resistant to heat in the Turpan grape production industry. In addition, the heat tolerance of plants is controlled by a variety of heat tolerance genes, which are specifically reflected in the morphological and anatomical structures, tissue cells, photosynthetic organs and the physiological and biochemical aspects of plants. However, this study only observed and studied the relationship between the anatomical structure of leaves and heat tolerance, and further research is needed to provide a more scientific theoretical basis for revealing the process of grapevine response to high temperature.

5. Conclusions

Due to the challenge of accurately representing the heat tolerance of grapevines in their natural environment through indoor high-temperature climate simulations, this study employed a comprehensive comparative analysis based on the leaf tissue structure of different grapevine varieties. Additionally, field observations of leaf phenotypes were conducted under natural high-temperature conditions in Turpan. As a result, three grapevine varieties with high temperature tolerance and five grapevine varieties with high temperature sensitivity were identified. This research not only provides valuable insights for grape production in hot and arid regions, but also contributes to understanding the mechanisms of heat tolerance and aiding in grapevine variety selection.

Author Contributions: Conceptualization, J.W. and X.W.; methodology, H.Z. and Q.Z.; software, J.W., C.Z., V.Y. and F.Z.; validation, J.W., Q.Z., H.Z. and X.W.; investigation, R.A., Y.M. and X.L.; writing—review and editing: J.W., H.Z., V.Y., Q.Z. and X.W.; visualization, J.W. and V.Y.; supervision, Q.Z. and X.W.; funding acquisition, J.W. and X.W. All authors have read and agreed to the published version of the manuscript.

Funding: This research was funded by the Youth Science and Technology Backbone Innovation Ability Training Project of Xinjiang Academy of Agricultural Sciences (xjnkq-2021010), by the Xinjiang Uygur Autonomous Region Tianshan Talents Training Program—Young top-notch scientific and technological talents (2022TSYCJC0036), by the Xinjiang Uygur Autonomous Region Innovation Environment Construction Special Project (PT2314) and by the Xinjiang Uygur Autonomous Region Tianchi Talent—Special Expert Project (Xiping Wang, 2022).

Data Availability Statement: The data that support the findings of this study are available upon request from the corresponding author.

Acknowledgments: We would like to thank supportive and scientific staff in research lab and grape research farm of Xinjiang Academy of Agriculture Sciences.

Conflicts of Interest: The authors declare that they have no conflict of interest.

References

1. Kim, J.S.; Jeon, B.W.; Kim, J. Signaling Peptides Regulating Abiotic Stress Responses in Plants. *Front. Plant Sci.* **2021**, *12*, 704490. [CrossRef]
2. Upadhyay, A.; Upadhyay, A.K. Global transcriptome analysis of heat stress response of grape variety 'Fantasy Seedless' under different irrigation regimens. *Vitis* **2021**, 143–151. [CrossRef]
3. Zha, Q.; Xi, X.J.; He, Y.N.; Jiang, A.L. Comprehensive evaluation of heat resistance in 68 Vitis germplasm resources. *Vitis* **2018**, *57*, 75–81.
4. Liu, M.; Fang, Y.L. Effects of heat stress on physiological indexes and ultrastructure of grapevines. *Sci. Agric. Sin.* **2020**, *53*, 1444–1458.
5. Coneva, V.; Frank, M.H.; Balaguer, M.A.D.L.; Li, M.; Sozzani, R.; Chitwood, D.H. Genetic Architecture and Molecular Networks Underlying Leaf Thickness in Desert-Adapted Tomato *Solanum pennellii*. *Plant Physiol.* **2017**, *175*, 376–391. [CrossRef]
6. Ogburn, R.M.; Edwards, E.J. Chapter 4—The Ecological Water-Use Strategies of Succulent Plants. In *Advances in Botanical Research*; Academic Press: Cambridge, MA, USA, 2010; Volume 55, pp. 179–225. [CrossRef]
7. Poorter, H.; Niinemets, Ü.; Poorter, L.; Wright, I.J.; Villar, R. Causes and consequences of variation in leaf mass per area (LMA): A meta-analysis. *New Phytol.* **2009**, *182*, 565–588. [CrossRef]
8. Wright, I.J.; Reich, P.B.; Westoby, M.; Ackerly, D.D.; Baruch, Z.; Bongers, F.; Cavender-Bares, J.; Chapin, T.; Cornelissen, J.H.C.; Diemer, M.; et al. The worldwide leaf economics spectrum. *Nature* **2004**, *428*, 821–827. [CrossRef]

9. Smith, S.D.; Monson, R.K.; Anderson, J.E. *Physiological Ecology of North American Desert Plants*; Springer: Berlin/Heidelberg, Germany, 1997. [CrossRef]
10. Szenteleki, K.; Ladanyi, M.; Gaal, M.; Zanathy, G.; Bisztray, G. Climatic risk factors of central hungarian grape growing regions. *Appl. Ecol. Environ. Res.* **2012**, *10*, 87–105. [CrossRef]
11. Asproudi, A.; Petrozziello, M.; Cavalletto, S.; Guidoni, S. Grape aroma precursors in cv. Nebbiolo as affected by vine microclimate. *Food Chem.* **2016**, *211*, 947–956. [CrossRef]
12. Zha, Q.; Xi, X.J.; He, Y.N.; Jiang, A.L. Research on high temperature stress response of different table grape cultivars. *Acta Agric. Shanghai* **2018**, *34*, 77–83. [CrossRef]
13. Xu, H.; Song, B.; Gu, Z.F.; Bi, Y.F.; Wei, B. Advances in heat tolerance mechanisms of plants. *Jiangsu J. Agr. Sci.* **2020**, *36*, 43–50. [CrossRef]
14. IPCC. *Climate Change 2021: The Physical Science Basis*; Contribution of Working Group I to the Sixth Assessment Report of the Intergovernmental Panel on Climate Change; Cambridge University Press: Cambridge, UK, 2021.
15. Venios, X.; Korkas, E.; Nisiotou, A.; Banilas, G. Grapevine Responses to Heat Stress and Global Warming. *Plants* **2020**, *9*, 1754. [CrossRef]
16. Zhang, H.; Zhu, J.; Gong, Z.; Zhu, J.-K. Abiotic stress responses in plants. *Nat. Rev. Genet.* **2021**, *23*, 104–119. [CrossRef]
17. Champa, W.H.; Gill, M.; Mahajan, B.; Aror, N.; Bedi, S. Brassinosteroids improve quality of table grapes (*Vitis vinifera* L.) cv. flame seedless. *Trop. Agric. Res.* **2015**, *26*, 368. [CrossRef]
18. Divi, U.K.; Rahman, T.; Krishna, P. Gene expression and functional analyses in brassinosteroid-mediated stress tolerance. *Plant Biotechnol. J.* **2015**, *14*, 419–432. [CrossRef]
19. Zha, Q.; Xi, X.J.; He, Y.N.; Jiang, A.L. High temperature in field: Effect on the leaf tissue structure of grape varieties. *Chin. Agric. Sci. Bull.* **2019**, *35*, 74–77. [CrossRef]
20. Qiu, Y.M.; Wang, W.; Hu, X.; Sun, M.M.; Ni, J.Z. Realationship between leaf anatomical structure and heat resistance of Rhododendron simsii. *For. Enviromental Sci.* **2021**, *37*, 69–81.
21. Bueno, A.; Alfarhan, A.; Arand, K.; Burghardt, M.; Deininger, A.-C.; Hedrich, R.; Leide, J.; Seufert, P.; Staiger, S.; Riederer, M. Effects of temperature on the cuticular transpiration barrier of two desert plants with water-spender and water-saver strategies. *J. Exp. Bot.* **2019**, *70*, 1613–1625. [CrossRef]
22. Ding, X.; Zhong, H.X.; Wang, X.P.; Song, J.Y.; Wu, J.Y.; Liu, G.H.; Zhang, F.C.; Hu, X.; Pan, M.Q.; Wu, X.Y. Observation on leaf anatomical structure and evaluation of drought resistance of grape rootstocks in Xinjiang. *Mol. Plant Breed.* **2022**, 1–18. Available online: https://kns.cnki.net/kcms/detail/46.1068.S.20220429.1131.010.htm (accessed on 16 September 2022).
23. Zha, Q.; Xi, X.; He, Y.; Yin, X.; Jiang, A. Effect of Short-Time High-Temperature Treatment on the Photosynthetic Performance of Different Heat-Tolerant Grapevine Cultivars. *Photochem. Photobiol.* **2021**, *97*, 763–769. [CrossRef]
24. Cramer, G. Abiotic stress and plant responses from the whole vine to the genes. *Aust. J. Grape Wine Res.* **2010**, *16*, 86–93. [CrossRef]
25. Jiao, S.-Z.; Guo, C.; Yao, W.-K.; Zhang, N.-B.; Zhang, J.-Y.; Xu, W.-R. An Amur grape VaHsfC1 is involved in multiple abiotic stresses. *Sci. Hortic.* **2021**, *295*, 110785. [CrossRef]
26. Yan, T.T. Evaluation of heat tolerance in different grape cultivars and proteomic analysis of heat tolerance difference between cultivars. Master's Thesis, South China Agricultural University, Guangzhou, China, 16 July 2020.
27. Wu, J.Y.; Lian, W.J.; Zeng, X.Y.; Liu, Z.G.; Mao, L.; Liu, Y.X.; Jiang, J.F. Physiological response to high temperature and heat tolerance evaluation of different grape cultivars. *Acta Bot. Boreali-Occident. Sin.* **2019**, *39*, 1075–1084. Available online: https://www.proquest.com/docview/2653586083?pq-origsite=gscholar (accessed on 15 June 2023).
28. Wu, J.Y.; Lian, W.J.; Liu, Z.G.; Zeng, X.Y.; Jiang, J.F.; Wei, Y.N. High temperature response of chlorophyll fluorescence parameter and heat tolerance evaluation of different grape cultivars. *J. Northwest AF Univ. (Nat. Sci. Ed.)* **2019**, *47*, 80–88.
29. Li, M.H.; Cherubini, P.; Dobbertin, M.; Arend, M.; Xiao, W.F.; Rigling, A. Responses of leaf nitrogen and mobile carbohydrates in different Quercus species/provenances to moderate climate changes. *Plant Biol.* **2013**, *15*, 177–184. [CrossRef]
30. Yang, J.; Chong, P.; Chen, G.; Xian, J.; Liu, Y.; Yue, Y. Shifting plant leaf anatomical strategic spectra of 286 plants in the eastern Qinghai-Tibet Plateau: Changing gears along 1050–3070 m. *Ecol. Indic.* **2023**, *146*, 109741. [CrossRef]
31. Jin, J.; Yang, L.; Fan, D.; Liu, X.; Hao, Q. Comparative transcriptome analysis uncovers different heat stress responses in heat-resistant and heat-sensitive jujube cultivars. *PLoS ONE* **2020**, *15*, e0235763. [CrossRef]
32. Wahid, A.; Gelani, S.; Ashraf, M.; Foolad, M.R. Heat tolerance in plants: An overview. *Environ. Exp. Bot.* **2007**, *61*, 199–223. [CrossRef]
33. Jiang, J.F.; Zhang, Y.; Tian, Z.S.; Fan, X.C.; Liu, C.H. Observation on microstructure and ultrastucture of leaves in *Vitis* L. *Acta Bot. Boreali-Occident. Sin.* **2012**, *32*, 1365–1371.
34. Fu, L.J.; Li, C.X.; Su, S.Y.; Li, Y.H.; Zhou, Y. Screening of cucumber germplasms in seedling stage and the construction of evaluation system for heat tolerance. *Plant Physiol. J.* **2020**, *56*, 1593–1604. [CrossRef]
35. Li, M.; Su, H.; Li, Y.Y.; Li, J.P.; Li, J.C.; Zhu, Y.L.; Song, Y.H. Analysis of heat tolerance of wheat with different genotypes and screening of identification indexes in Huang-huai-hai region. *Sci. Agric. Sin.* **2021**, *54*, 3381–3393. [CrossRef]
36. Guo, Y.; Zhang, S.H.; Li, Y.; Zhang, X.F.; Wang, G.P. Leaf morphology, anatomical structure and drought resistance evaluation of 238 Chinese chestnut cultivars (lines). *Acta Hortic. Sin.* **2020**, *47*, 1033–1046. [CrossRef]
37. Huang, L.; Sun, Y.Q.; Hao, L.H.; Dang, C.H.; Zhu, Y.; Wang, H.X.; Cheng, D.J.; Zhang, Y.X.; Zheng, Y.P. Effects of high temperatures on leaf structures and physiological metabolism of north highbush blueberry. *Acta Hortic. Sin.* **2016**, *43*, 1044–1056. [CrossRef]

38. Ma, W.; Liang, W.; Zhao, B. Effect of Relative Air Humidity and High Temperature on the Physiological and Anatomical Responses of Two Rhododendron Cultivars. *Hortscience* **2019**, *54*, 1115–1123. [CrossRef]
39. Tian, M.; Yu, G.; He, N.; Hou, J. Leaf morphological and anatomical traits from tropical to temperate coniferous forests: Mechanisms and influencing factors. *Sci. Rep.* **2016**, *6*, 19703. [CrossRef]
40. Kim, S.; Hwang, G.; Kim, S.; Thi, T.N.; Kim, H.; Jeong, J.; Kim, J.; Kim, J.; Choi, G.; Oh, E. The epidermis coordinates thermoresponsive growth through the phyB-PIF4-auxin pathway. *Nat. Commun.* **2020**, *11*, 1053. [CrossRef]
41. Xie, Z.S.; Guo, X.J.; Cao, H.M. Effect of root restriction on vegetative growth and leaf anatomy of Kyoho grapevines cultivar. *Afr. J. Agric. Res.* **2013**, *8*, 1304–1309. [CrossRef]
42. Wu, J.; Xu, G.; Lian, W.; Chen, Y.; Li, H.; Liu, Y.; Jiang, J.; Wen, J. Physiological response to high temperature and evaluation of heat tolerance of different grape cultivars. *Agric. Biotechnol.* **2022**, *11*, 47–54+85. [CrossRef]
43. Liu, Y.; Jiang, J.; Fan, X.; Zhang, Y.; Wu, J.; Wang, L.; Liu, C. Identification of Heat Tolerance in Chinese Wildgrape Germplasm Resources. *Horticulturae* **2020**, *6*, 68. [CrossRef]
44. Ahammed, G.J.; Li, X.; Liu, A.; Chen, S. Brassinosteroids in Plant Tolerance to Abiotic Stress. *J. Plant Growth Regul.* **2020**, *39*, 1451–1464. [CrossRef]
45. Zha, Q.; Xi, X.; Jiang, A.; Tian, Y. High Temperature Affects Photosynthetic and Molecular Processes in Field-Cultivated *Vitis vinifera* L. × *Vitis labrusca* L. *Photochem. Photobiol.* **2016**, *92*, 446–454. [CrossRef]

horticulturae

MDPI

Article

Morpho-Physiological Responses of Three Italian Olive Tree (*Olea europaea* L.) Cultivars to Drought Stress

Sara Parri [1,*], Marco Romi [1], Yasutomo Hoshika [2], Alessio Giovannelli [2], Maria Celeste Dias [3], Francesca Cristiana Piritore [1,†], Giampiero Cai [1] and Claudio Cantini [4]

[1] Department of Life Sciences, University of Siena, Via Mattioli 4, 53100 Siena, Italy; marco.romi@unisi.it (M.R.); francescacristiana.piritore@univr.it (F.C.P.); cai@unisi.it (G.C.)
[2] Research Institute on Terrestrial Ecosystem (IRET), National Research Council (CNR), Via Madonna del Piano 10, 50019 Sesto Fiorentino, Italy; yasutomo.hoshika@cnr.it (Y.H.); alessio.giovannelli@cnr.it (A.G.)
[3] Centre for Functional Ecology, Department of Life Sciences, University of Coimbra, Calçada Martim de Freitas, 3000-456 Coimbra, Portugal; celeste.dias@uc.pt
[4] Institute for BioEconomy (IBE), National Research Council (CNR), Strada Provinciale Aurelia Vecchia 49, 58022 Follonica, Italy; claudio.cantini@ibe.cnr.it
* Correspondence: sara.parri@student.unisi.it
† Current Address: Department of Neurosciences, Biomedicine, and Movement Sciences, University of Verona, Piazzale Ludovico Antonio Scuro 10, 37124 Verona, Italy.

Abstract: Water scarcity in agriculture can limit crop production and trigger the need for more effective water resource management. As a result, it is critical to identify new crop genotypes that are more drought tolerant and perform better under low irrigation or even rain-fed conditions. The olive tree is a high-value crop that is well adapted to dry Mediterranean conditions. However, different genotypes may have developed different mechanisms of tolerance to water stress. To investigate such mechanisms, we examined three Italian olive cultivars ('Giarraffa', 'Leccino', and 'Maurino') grown in a greenhouse under drought stress. We found that single genotypes responded differently to the drought, though not all parameters revealed significant differences. The first major difference among the cultivars was in transpiration: the lower stomatal density and stomatal conductance of 'Giarraffa' allow this cultivar to use water more conservatively. In parallel with the reduction in stomatal and mesophyll conductance, the drought-stressed group of 'Giarraffa' maintained the electron transport rate and effective efficiency levels of photosystem II similar to those of the control until the fourth week of stress. The fluorescence parameters revealed the earlier closure of reaction photosynthetic centres in 'Leccino'. Finally, the higher rate of electrolyte leakage in 'Maurino' indicated a significant ions loss in this cultivar when it was subjected to the drought. Both water management under stress conditions and the effect of drought on photosynthesis make 'Giarraffa' interesting to researchers studying its use in breeding or water-saving programmes.

Keywords: gas exchange; mesophyll conductance; locally adapted cultivars; pigments; stomatal density; chlorophyll fluorescence; water content

1. Introduction

Droughts have a negative impact on plant growth and productivity, as evapotranspiration exceeds the amount of water absorbed by the roots, leading to drought stress. A low soil moisture level and a high air vapor pressure deficit determine the intensity of the water deficit in leaf tissues [1]. Plants are constantly threatened by water stress due to natural climate change, but anthropogenic climate change will exacerbate water deficits for plants and crops [2]; the IPCC 2021 report indicates that climate warming is leading to an increased atmospheric evaporation demand and decreased soil moisture availability, which could lead to more frequent and severe droughts in semi-arid regions [3]. In light of this, the cultivation of crop plants will require more irrigation water to maintain the yield and productivity [4].

Droughts have a significant impact on plant traits, including biomass production, fruit yield, growth, and development. Photosynthesis is particularly sensitive to drought stress, as it reduces the rates of carbohydrate synthesis and accumulation [5]. Plants use a variety of defence mechanisms to cope with droughts, including the production of osmoprotectants, the synthesis of proteins, and changes in metabolic processes, hormone levels, and gene expression. To maintain cell turgor and reduce water loss, plants close their stomata [4], which results in a decrease in stomatal conductance (g_s), in association with decreases in the leaf water potential (Ψ) and relative water content (RWC) [6,7]. Stomata closure immediately causes a reduction in the amount of CO_2 in the substomatal cavity, which then slows the rate of photosynthesis: in most cases, photosynthesis is completely stopped due to stomatal closure before the metabolism is affected [8]. However, as RWC and g_s severely decline during a drought, additional metabolic restrictions can take place [9]. For instance, Flexas and Medrano [10] found that the Ribulose-1,5-bisphosphate (RuBP) regeneration capacity decreases as the ATP production decreases. Furthermore, droughts can cause an impairment of the carboxylation capacity due to Ribulose-1,5-bisphosphate carboxylase/oxygenase (Rubisco) content reduction only at very low g_s values (<0.01 mol H_2O m^{-2} s^{-1}) [8].

The increase in mesophyll diffusion resistance to CO_2 has also gained recognition as a limitation to the photosynthetic process under stress conditions [8,11,12]; for example, under salt stress conditions, the changes in mesophyll conductance (g_m) have been shown to be as rapid as those in g_s [11]. When the rate of light absorption exceeds the capacity to photosynthesise, this results in irreversible photoinhibition [13]. In addition, a low transpiration rate caused by stomata closure under drought conditions leads to an increase in the leaf temperature, which may exacerbate photoinhibition [14]. As a result, the photosynthetic pigments may degrade, and the PSII efficiency may decline [6].

Olive trees (*Olea europaea* L.) are one of the oldest domesticated plants in the Mediterranean basin and are still extremely important in agriculture, economy, and culture [4]. According to the International Olive Council (IOC), the global olive oil production total for the 2021/22 crop year was 3.1 million tons. IOC member countries produced 2.9 million tons (93.9% of the global total). EU countries have suffered a significant reduction in production as a result of droughts and adverse weather [15]. Italy is one of Europe's top producers of olive oil, with over a million hectares dedicated to olive cultivation [16]. Italy has over 500 cultivars of olive trees registered in the National Olive Oil register, making it a country rich in olive biodiversity. However, the number is likely to rise as analyses (mostly genetic) continue. These cultivars differ morphologically, as well as in terms of the yield and quality of the oil they produce. 'Frantoio', 'Leccino', 'Taggiasca', and 'Coratina' are among the most common and well-known cultivars, but many others characterise various Italian regions. This biodiversity contributes to Italian oil being a unique product that is valued around the world [17–19]. Olive trees require specific environmental conditions in order to grow and produce. These include hot, dry summers, and mild, but cool winters, with ideal annual temperatures of 15–25 °C. High temperatures can harm flowering and fruiting. Olive trees require sunlight to grow a healthy canopy and produce oil. Although olive trees are drought-tolerant, they require adequate irrigation during critical periods, such as flower differentiation, fruit setting, and fruit swelling. Olive plants can grow in a variety of soil types so long as the soil is well drained and there is no standing water. The coastal and hilly areas of central and southern Italy (Puglia, Calabria, Sicily, Tuscany, Umbria, and Lazio) are the best for olive cultivation. The olive tree is the model woody plant used for the study of drought responses and tolerance [20], and it has been studied under a range of environmental stress conditions, such as UV-B [21] and salt stress [22].

Specifically, several traits have been described as critical for drought tolerance, such as a small stomata, waxy leaf surface, narrow xylem vessels, non-photochemical quenching activation, and rapid osmotic adjustment [4,20,23]. Nevertheless, irrigation practices have spread throughout Europe due to the correlation between crop productivity and increased water availability [24]. Although olive trees are generally considered to be drought-tolerant, only a few studies have investigated the differences in the physiological responses of olive cultivars to limited water availability. However, the large number of olive cultivars currently listed in world repositories strongly supports the presence of significant differences in physiological responses to droughts [7,25,26]. These are related to the specific environment in which cultivars are grown and adapted, emphasizing the importance of studying the diversity of olive plants and promoting the more thorough characterization of various stress tolerance mechanisms [27–29]. The comparison of various cultivars will allow the identification of both the basic, ubiquitous plant tolerance mechanisms and the potential interactions between diverse mechanisms for drought tolerance, as well as identify potential strategies for improving olive tree growth and productivity in water-stressed environments.

We investigated three different olive cultivars grown in Italy, among the many catalogued in the Italian National Registry, to perform thorough analysis. The cultivar Giarraffa, native to the arid region of Sicily, is tolerant to drought as well as to UV light [21]. The cultivar, Leccino, is more diffused worldwide and shows good tolerance to drought, cold, and bacteria (*Xylella fastidiosa*) [30]. The cultivar, Maurino, which is autochthonous in Tuscany and cold-tolerant, showed a peculiar morphological trait (wilting of leaves) when it was exposed to dry conditions (Claudio Cantini, personal communication). Based on a GBS-derived SNP catalogue of 94 Italian cultivars, it was proposed that 'Maurino' and 'Leccino' are members of a cluster population descended from local oleasters. 'Giarraffa', on the other hand, is clearly distinct from other Italian cultivars and was most likely introduced from Spain and Morocco [31]. Our hypothesis was that the three cultivars would respond differently to drought due to their long-term adaptation to different environments. To test this hypothesis, fully drought-stressed plants were evaluated for physiological, morphological, and biochemical parameters, as well as their soil water content. This study aimed to assess and distinguish the drought tolerance of three Italian olive cultivars by understanding how changes in water distribution and management could affect their physiological responses. The comparison of various cultivars will allow the identification of both the basic, ubiquitous plant tolerance mechanisms and the potential interactions between diverse mechanisms of drought tolerance. Furthermore, comparing different olive cultivars can provide useful data on the most effective drought response. As a result, we sought to investigate the mechanisms by which different olive cultivars tolerate or adapt to drought stress, as well as identify major parameters to be used to classify the olive tree drought response within germplasm collections. The findings will be useful in understanding the resilience of olive trees to droughts and developing smart management strategies for olive production in the face of climate change and water scarcity.

2. Materials and Methods

2.1. Plant Growth Conditions and Drought Stress Treatment

Certified, 18 month-old olive trees (*Olea europaea* L., cultivars Leccino, Maurino, and Giarraffa) were provided by "Spoolivi" (Società Pesciatina di Orticoltura, Pescia, PT, Italy). The plants were grown in 4 L pots with a substrate of 50% peat and 50% pumice [32]. Upon arrival at the university laboratories, the plants were transferred to a growth chamber in which illumination was provided by LEDs for flowering and growth (TLED secret Jardin—SRL AGOMOON, Manage, Belgium). The photoperiod was 12 h of light and 12 h of dark. After one week of adaptation to the general environmental conditions with steady watering, 20 plants of each cultivar were split in two 10-plant groups to be used as a control (CTRL) or to be subjected to drought stress (DS). The CTRL groups were fully irrigated (500 mL of water per week), while the DS groups were totally deprived of water for 4 weeks. The experimental period consisted of an increasing water deficit divided into 5 time points: t0, t1, t2, t3, and t4, corresponding to the onset of the withholding irrigation and the first, second, third, and fourth weeks of irrigation deprivation, respectively. The pots inside the chamber were rotated every week to avoid any positional effects [29]. Temperature and humidity were recorded hourly with the EBI 20-th1 datalogger (Ebro): the temperature was 27.5 °C, and the humidity was 51.1% (in both cases, the data are averaged throughout the day and night). The minimum and the maximum values reached by temperature were 22.3 °C and 31.8 °C, respectively; 29.8% and 70.1% were the minimum and the maximum values of recorded humidity, respectively.

2.2. Soil Water Content

The soil water content (SWC) was evaluated according to Bilskie [33]. Four soil samples were collected for each group and immediately weighed to obtain the wet mass (m_{wet}). Then, samples were put in the oven for 24 h at 105 °C, and then weighed again (m_{dry}). Finally, soil water content was calculated as:

$$SWC = (m_{wet} - m_{dry})/m_{dry}$$

2.3. Relative Water Content of Leaves and Stems

The relative water contents of leaves (leaf RWC) and stems (stem RWC) were calculated as described by El Yamani et al. [6]. Fully expanded and mature leaves at each time point were cut below the petiole and immediately placed in pre-weighed plastic tubes. The leaves were weighed along with the tubes to obtain the fresh weight (FW). Stems were harvested only at t0, t2, and t4. For both the leaves and stems, tubes were filled with distilled water, and samples were incubated for 24 h at 4 °C in the dark. Afterwards, the leaves and stems were removed from the tubes and dried with paper towels to absorb excess water. The samples were weighed to determine the turgid weight (TW). Finally, the samples were placed in paper bags and heated in an oven at 80 °C for 48 h. The samples were weighed to determine the dry weight (DW). The RWC of leaves and stems was calculated as:

$$RWC\ (\%) = (FW - DW)/(TW - DW) \times 100$$

2.4. Stomatal Density

Stomatal density was measured according to Xu et al. [34]. Briefly, 5 mature leaves per group were selected at t0, t2, and t4. The abaxial epidermis of the leaf was coated with clear nail polish. Once dried, the film was peeled off the leaf and placed on a slide over a drop of water. The samples were examined with the Zeiss Axiophot light microscope (Oberkochen, Germany). Six images were taken for each leaf sample (thus, there was a total of 30 images per group). The images were analysed with ImageJ. The stomatal density was then calculated as the number of stomata per leaf area.

2.5. Leaf Gas Exchange and Chlorophyll Fluorescence

The LiCor-6800 instrument (LICOR, Lincoln, NB, USA) equipped with a leaf chamber fluorometer was used to assess gas exchange and chlorophyll fluorescence. The light-saturated net photosynthetic rate (A) and g_s were recorded throughout the experiment (t0 to t4). During the gas exchange measurements, the following conditions were set: photosynthetic active radiation (PAR) at 1600 μmol m^{-2} s^{-1} [35,36], CO_2 at 400 ppm, block temperature at 28 °C, and relative humidity at 60% inside a leaf cuvette. In addition, CO_2 assimilation rate curves against the intercellular CO_2 concentration (i.e., A/C_i curves) were obtained at t2 by using the following 12 CO_2 concentration steps: 400, 200, 50, 100, 300, 400, 600, 800, 1000, 1200, 1400, an d1600 ppm. A-C_i curves data were studied using the approach described by Ethier and Livingston [37] to obtain the maximum carboxylation efficiency (V_{cmax}) and maximum rate of electron transport (J_{max}). A/C_i and A/C_c curves are provided in the Supplementary Material (Figure S1). This assumes that g_m was constant throughout the CO_2 range. V_{cmax} and J_{max} values were standardised at 25 °C using a temperature dependency of those parameters [38]. Michaelis–Menten constants for CO_2 (Kc) and O_2 (Ko) were derived according to the approach by Bernacchi et al. [38]. To calculate the g_m values, the variable J method was applied for calculating the A/C_c curves, as the use of an independent methodology should be preferable for preventing the propagation of errors or assumptions [39]. The variable J method was applied to calculate g_m [40] based on the point measurement of A at 400 ppm of Ca with fluorescence measurements:

$$g_m = A/(C_i - \Gamma^* [ETR + 8 (A + Rd)]/[ETR - 4 (A + Rd)])$$

where: Rd is daytime respiration, which was obtained from the previous study on olive leaves (1.39 μmol m^{-2} s^{-1} [41]), Γ^* is the CO_2 compensation point to photorespiration, which was calculated using the Rubisco specificity factor estimated for evergreen woody species [42]. The fluorescence of chlorophyll was evaluated throughout the experiment (t0 to t4) using the same apparatus with the activated fluorometer (rectangular flash with a red target of 8000 μmol m^{-2} s^{-1}, a duration of 1000 ms, and an output rate of 100 Hz [43]). Light-adapted leaves were used to obtain the effective efficiency of PSII (ΦPSII) and the electron transport rate (ETR) according to Gilbert et al. [44]. The concentration of CO_2 at the chloroplast envelope (C_c) was estimated using the g_m value.

$$C_c = C_i - A/g_m$$

Relative photosynthetic limitations were calculated according to Grassi and Magnani [45] as follows:

$$L_s = (g_{tot}/[g_s/1.6]\ \delta A/\delta C_c)/(g_{tot} + \delta A/\delta C_c)$$
$$L_m = (g_{tot}/g_m\ \delta A/\delta C_c)/(g_{tot} + \delta A/\delta C_c)$$
$$L_b = g_{tot}/(g_{tot} + \delta A/\delta C_c)$$

where L_s, L_m, and L_b are the relative limitations of stomatal diffusion, mesophyll diffusion, and biochemical limitation, respectively. g_{tot} is the total CO_2 conductance ($g_{tot} = [(g_s/1.6)\ g_m]/[(g_s/1.6) + g_m]$), 1.6 is the ratio of the diffusion coefficients for water vapor to CO_2, and $\delta A/\delta C_c$ indicates an initial slope of A/C_c curves that was estimated using a range of 0–150 μmol m^{-2} s^{-1} of C_c. In addition, the maximum quantum yield of PSII (F_v/F_m) was evaluated for dark-adapted leaves covered with aluminium foil for at least 20 min. One fully expanded leaf per five to six plants per treatment were used for each gas exchange and fluorescence parameter.

2.6. Electrolyte Leakage

The cell membrane permeability to solutes was assessed by measuring the electrolyte leakage (EL) according to ben Abdallah et al. [46]. Two leaf discs with a 0.5 cm diameter were cut from a fresh leaf and placed inside capped tubes filled with 10 mL deionised water. The samples were incubated for 3 h at 37 °C. Immediately after incubation, the conductivity of the solution was measured to obtain the electrical conductivity E_1 value. Then, the samples were heated at 95 °C for 30 min before measuring the conductivity again (E_2). The EL was calculated as:

$$EL = E_1/E_2 \times 100 \quad (1)$$

2.7. Malondialdehyde Content

Malondialdehyde (MDA) content was used to quantify lipid peroxidation in the leaves. Frozen leaves (0.1 g) were ground with 1.5 mL of 0.1% (*w*/*v*) trichloroacetic acid (TCA). The samples were centrifuged at 10,000× *g* for 10 min at 4 °C. Then, 0.25 mL of supernatant was mixed with 1 mL of 20% (*w*/*v*) TCA containing 0.5% (*w*/*v*) thiobarbituric acid for the positive control; the same supernatant was mixed with 1 mL of 20% (*w*/*v*) TCA alone for the negative control. The samples were incubated for 30 min at 95 °C; then, the extracts were immediately cooled on ice and centrifuged (10,000× *g* for 10 min at 4 °C). Supernatants from positive and negative controls were read at 600, 532, and 400 nm using a microplate reader, EnSpire (PerkinElmer, Waltham, MA, USA). MDA equivalents were calculated according to Hodges et al. [47], and then normalised to dry weight.

2.8. Pigments Quantification

Chlorophyll a, chlorophyll b, and carotenoids were extracted from pure acetone and quantified spectrophotometrically according to Lichtenthaler [48]. Briefly, 1.5 mL of cold 100% acetone was added to ground frozen leaves (500 mg). After 2 to 3 min of agitation, the samples were centrifuged (15,000× *g* for 15 min at 4 °C) and the supernatants were collected in a new tube. The pellets were extracted again twice. The pool of extracts was then read at 662, 645, and 470 nm with a Shimadzu UV-1280 spectrophotometer. The pigment content was then normalised to dry weight.

2.9. Statistical Analysis

Each analysis included at least five biological replicates (n = 5). For all the parameters recorded at different time points, the effects of the drought treatment (S), cultivar (C), and their interaction (C × S) were evaluated using 2-way Repeated ANOVA. For the MDA content, g_m, L_s, L_m, L_b, V_{cmax}, and J_{max}, which were measured only at t2, the effects of the treatment (S), cultivar (S), and their interaction (C × S) were analysed by using 2-way ANOVA. At each time point, post hoc analysis was performed using Tukey HSD test. ANOVA and post hoc tests were performed using the Systat 11 statistical package (Systat Software Inc., Richmond, CA, USA). At each time point, the bar graphs show the mean and the standard error of the recorded parameters and significant differences according to the post hoc test ($p < 0.01$). In the pie chart, the percentages shown are the averages of five values calculated independently for each limitation (L_s, L_m, and L_b). Eight parameters were taken into account to create the correlogram. First, each parameter was expressed as the ratio of the DS group value to the CTRL group value at each time point. Then, Rstudio (ver. 4.2.2, R core team, Vienna, Austria, 2022) was used for correlation according to the time course.

3. Results

3.1. Drought Effects on Plant Water Status and Biochemical Responses

The results are organised as follows in this first section. Table 1 shows the variability of nine physiological parameters as well as the significance of two factors (cultivar and treatment) and their mutual interaction as determined in the ANOVA tests. In addition, only the data where the interaction between cultivar and treatment (C × S) was significant are fully described.

Table 1. The effects of factors "cultivar" (C), "treatment" (S), and their interaction (C × S), as well as their statistical significance, on the following parameters: soil water content (SWC), relative water content of leaves (leaf RWC) and stems (stem RWC), stomatal density (SD), stomatal conductance (g_s), electrolyte leakage (EL), lipid peroxidation (as measured by malondialdehyde content, MDA), photosynthetic pigments content of chlorophyll a and b (Chl a + b), and carotenoids (Car). Each value represents the mean ± standard deviation. Different superscripts indicate statistical differences ($p < 0.05$).

	SWC (%)	Leaf RWC (%)	Stem RWC (%)	SD (n./mm^2)	g_s (mol m^{-2} s^{-1})	EL (%)	MDA (mmol/kg DW)	Chl a + b (µg/mg DW)	Car (µg/mg DW)
Cultivar (C)									
Giarraffa	89 ± 52 ^{a}	64.0 ± 16.1	69.2 ± 17.9	34.4 ± 7.6 ^{c}	0.086 ± 0.039 ^{b}	26.3 ± 12.5 ^{b}	5.28 ± 0.98 ^{b}	3.93 ± 0.82	0.70 ± 0.15
Leccino	76 ± 50 ^{b}	60.5 ± 15.1	66.7 ± 17.0	38.4 ± 5.0 ^{b}	0.070 ± 0.040 ^{c}	27.4 ± 13.3 ^{b}	3.78 ± 0.50 ^{c}	3.66 ± 0.71	0.65 ± 0.13
Maurino	76 ± 49 ^{b}	61.3 ± 17.6	69.4 ± 20.0	48.4 ± 5.4 ^{a}	0.112 ± 0.051 ^{a}	37.7 ± 21.1 ^{a}	7.26 ± 1.33 ^{a}	3.49 ± 1.93	0.63 ± 0.32
p-value	0.014	0.080	0.051	<0.001	<0.001	<0.001	<0.001	0.837	0.649
Treatment (S)									
CTRL	117 ± 26 ^{a}	82.0 ± 8.6 ^{a}	88.0 ± 2.1 ^{a}	40.1 ± 9.3	0.119 ± 0.037 ^{a}	21.9 ± 4.8 ^{b}	4.93 ± 1.67 ^{b}	3.43 ± 1.17	0.61 ± 0.19
DS	26 ± 21 ^{b}	50.9 ± 10.8 ^{b}	64.4 ± 9.8 ^{b}	40.7 ± 7.6	0.052 ± 0.029 ^{b}	41.2 ± 20.1 ^{a}	5.94 ± 1.72 ^{a}	4.10 ± 1.31	0.73 ± 0.23
p-value	<0.001	<0.001	<0.001	0.203	<0.001	<0.001	0.004	0.116	0.143
C × S									
p-value	0.715	0.041	0.001	<0.001	<0.001	<0.001	0.422	0.636	0.824

The first parameter we monitored was the soil water content (SWC), which was found to be strongly affected by the DS treatments. 'Giarraffa' exhibited a higher value of SWC (89% compared to 76% of the other cultivars). However, the significance of the interaction (C × S) was not relevant, meaning that the SWC was not affected by the cultivar growing in the soil, but only by treatment. A graph containing SWC data is available in the Supplementary Material (Figure S2).

Drought stress also had a significant effect (p-value < 0.001) on the leaf RWC data, but in this case, the cultivar and the interaction C × S were also significant (Table 1). As shown in Figure 1, the drought-stressed groups of 'Giarraffa' and 'Leccino' retained less water than 'Maurino' DS did after the first week of stress. While the difference between the control and stressed groups was significant for all cultivars at t1, 'Giarraffa' DS did not differ from its respective control one week later.

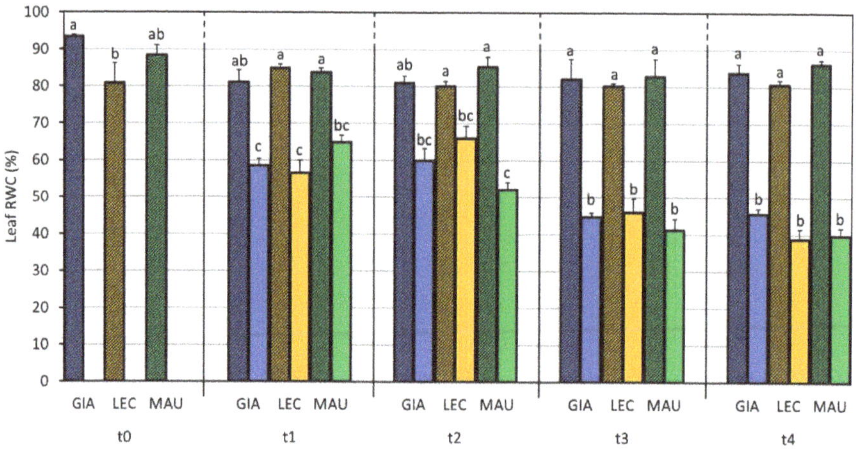

Figure 1. Leaf relative water content (leaf RWC) measured in control and drought olive cultivars from t0 to t4. At t0, the stressed and control samples were still one group. The bars represent mean ± standard error. Values for 'Giarraffa' (GIA) are blue, 'Leccino' (LEC) values are orange, and 'Maurino' (MAU) values are green. Striped bars indicate control samples. For each time point, different letters denote statistical significance (p-value < 0.01) according to Tukey's multiple post hoc tests considering both cultivar and treatment.

Table 1 shows that, considering both the CTRL and DS groups, the 'Giarraffa' stem contains more water than the cultivar Leccino does (69.2% and 66.7%, respectively). The higher value recorded in the cultivar Maurino is probably due to the lack of samples at t4. Drought stress reduced stem RWC, resulting in a reduction to 64.4% in the stressed samples compared to 88.0% in the controls. Figure 2 shows that all cultivars lost stem water as the study progressed, with 'Maurino' DS having a lower and significantly different stem RWC value than that of 'Giarraffa' DS at t2. The stems stored more water than the leaves did as the stress worsened, with from 10% more water at t0 up to 20% at t4.

Stomatal density (SD) affects the rate of exchange between plants and the external environment; since the sampled, mature, fully expanded leaves had developed before the experiment, the drought treatment had no effects on this parameter (Table 1). However, the cultivar (C) had a significant impact on it, with the average numbers of stomata per mm^2 being 34.4 for 'Giarraffa', 38.4 for 'Leccino', and 48.4 for 'Maurino'.

The stomatal conductance, i.e., the water vapour flux through a leaf sample, showed a significant difference for each factor considered (C, S, and C × S) (Table 1). The cultivar Maurino showed the highest values of g_s (0.112 mol m^{-2} s^{-1}), and it had a higher g_s even when comparing only the control groups of the cultivars, as shown in Figure 3. The drought stress had a strong effect (p-value < 0.001), and all the stressed groups had lower and significantly different g_s value compared to that of the controls from t2, but 'Giarraffa' DS already differed from the respective control at t1. The lowest value of g_s was reached by all stressed groups of the cultivars at t4.

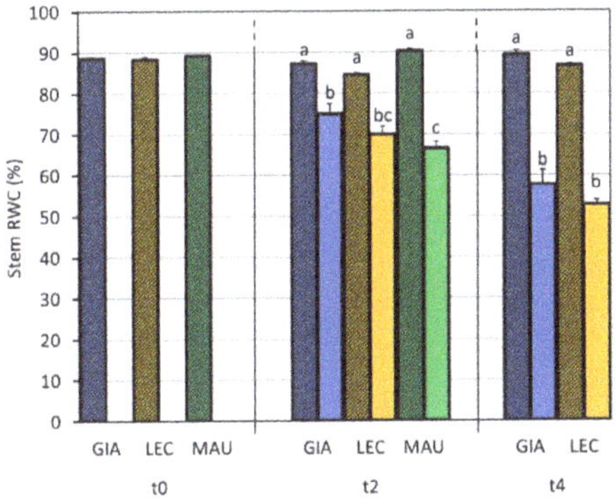

Figure 2. Stem relative water content (stem RWC) measured in control and drought olive cultivars analysed at t0, t2, and t4. At t0, the stressed and control samples were still one group. The bars represent mean ± standard error. Values for 'Giarraffa' (GIA) are blue, 'Leccino' (LEC) values are orange, and 'Maurino' (MAU) values are green. Striped bars indicate control samples. For each time point, different letters denote statistical significance (p-value < 0.01) according to Tukey's multiple post hoc tests considering both cultivar and treatment. MAU values at t4 are not shown due to technical problems.

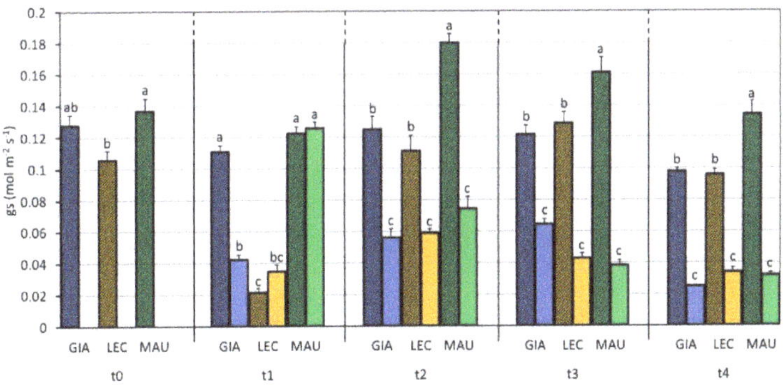

Figure 3. Stomatal conductance (g_s) in control and drought-stressed olive cultivars from t0 to t4. At t0, the stressed and control samples were still one group. The bars represent mean ± standard error. The values for 'Giarraffa' (GIA) are blue, those for 'Leccino' (LEC) are orange, and 'Maurino' (MAU) values are green. Strip bars refer to control samples. Within each time point, different letters denote statistical significance (p-value < 0.01) according to Tukey's multiple post hoc tests considering both cultivar and treatment.

Electrolyte leakage (EL) is the loss of electrolytes from cells or tissues and can be caused by a variety of factors, including physical injuries and diseases. This parameter was found to be strongly affected by the factors, C and S, and their interaction (Table 1). As shown in Figure 4, the rate of EL increased progressively and significantly from t1 to t4, and the stress condition resulted in a two-fold increase in the value in the stressed samples (41.2%) compared to that of the control samples (21.9%). 'Maurino' was the cultivar that was most strongly affected, already showing a consistent increase in the EL value at t2. After an additional week of stress (t3), the 'Giarraffa' DS (40.2%) and 'Leccino' DS (42.6%) groups differed significantly compared to their respective controls. At t4, the stressed group of each cultivar scored the highest EL value.

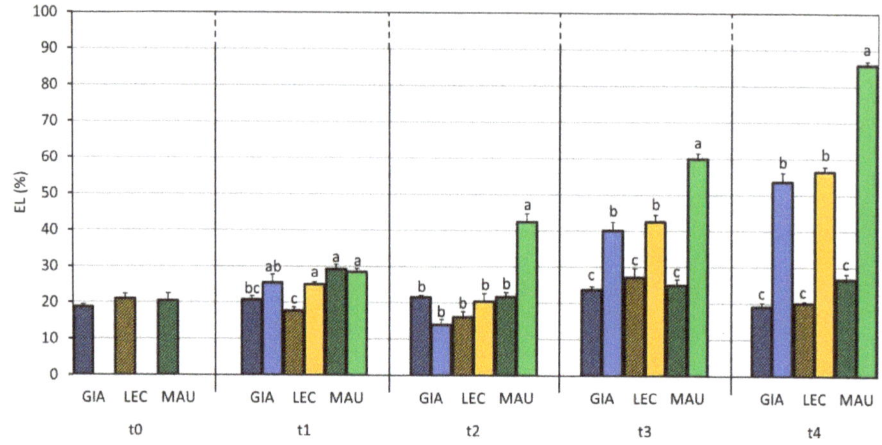

Figure 4. Electrolyte leakage (EL) in control and drought-stressed olive cultivars from t0 to t4. At t0, the stressed and control samples were still one group. The bars represent mean ± standard error. The values for 'Giarraffa' (GIA) are in blue, those for 'Leccino' (LEC) are orange, and 'Maurino' (MAU) values are green. Strip bars refer to control samples. Within each time point, different letters denote statistical significance (p-value < 0.01) according to Tukey's multiple post hoc tests considering both cultivar and treatment.

The malondialdehyde (MDA) content is proportional to the intensity of lipid peroxidation and was analysed only at t2 for the sole purpose of saving the leaf samples. The cultivar, Leccino, showed the lowest MDA content, followed by 'Giarraffa' (5.28 mmol kg^{-1} dw) and 'Maurino' (7.26 mmol kg^{-1} dw). The drought-stressed samples exhibited a slightly higher value than the control samples did, but no statistical difference was observed regarding the C × S interaction. MDA data are not shown, but are available in the Supplementary Material (Figure S3). Table 1 also shows the content of major photosynthetic pigments. Carotenoids (Car) and chlorophylls a + b (Chl a + b) were unaffected by the variables considered and remained constant throughout the experiment (Supplementary Material, Figures S4 and S5).

3.2. Impact of Water Deficit on the Photosynthetic Process

Table 2 shows the variability of ten photosynthetic parameters, as well as the significance of two factors (cultivar and treatment) and their mutual interaction, as determined via 2-way Repeated ANOVA. Like the previous section, only the data where the interaction between the cultivar and treatment (C × S) was significant are fully described.

As shown in Table 2, net CO_2 assimilation (A) is significantly influenced by the cultivar, the drought, and their interaction. The cultivar Maurino showed a higher value of A than 'Giarraffa' and 'Leccino' did, even when comparing the control groups (Figure 5). Drought stress generally reduced A, but the earliest reduction occurred in 'Giarraffa' DS at t1. From t2, the stressed groups of all cultivars significantly differed from their respective controls.

Table 2. Effects of the factors "cultivar" (C), "treatment" (S), and their interaction (C × S), and their statistical significance on the following parameters: net photosynthetic rate (A), effective efficiency of photosystem II (Φ PSII), maximum efficiency of photosystem II (F_v/F_m), electron transport rate (ETR), mesophyll conductance (g_m), stomatal limitation (L_s), mesophyll conductance limitation (L_m), biochemical limitation (L_b), maximum rate of carboxylation (V_{cmax}), and maximum electron transport rate (J_{max}). Each value is the mean ± standard deviation. Different superscripts indicate statistical differences ($p < 0.05$).

	A (μmol m^{-2} s^{-1})	Φ PSII	F_v/F_m	ETR (μmol m^{-2} s^{-1})	g_m (mol m^{-2} s^{-1})	L_s (%)	L_m (%)	L_b (%)	V_{cmax} (μmol m^{-2} s^{-1})	J_{max} (μmol m^{-2} s^{-1})
Cultivar (C)										
Giarraffa	7.79 ± 3.87 ^{b}	0.149 ± 0.032 ^{b}	0.797 ± 0.048	100.6 ± 21.5 ^{b}	0.110 ± 0.043	44.5 ± 10.7 ^{a}	27.5 ± 10.4	28.0 ± 17.9	57.1 ± 13.4	74.9 ± 16.2 ^{ab}
Leccino	7.14 ± 3.78 ^{b}	0.154 ± 0.034 ^{ab}	0.797 ± 0.055	100.1 ± 19.3 ^{b}	0.139 ± 0.077	43.9 ± 5.2 ^{ab}	21.6 ± 8.0	34.5 ± 12.7	55.8 ± 12.1	87.4 ± 13.4 ^{a}
Maurino	10.66 ± 5.48 ^{a}	0.166 ± 0.039 ^{a}	0.800 ± 0.047	111.7 ± 26.1 ^{a}	0.112 ± 0.053	38.9 ± 11.6 ^{b}	27.5 ± 11.1	33.6 ± 14.7	49.1 ± 23.0	68.8 ± 19.4 ^{b}
p-value	<0.001	<0.001	0.860	<0.001	0.061	0.026	0.295	0.196	0.520	0.054
Treatment (S)										
CTRL	11.38 ± 3.40 ^{a}	0.172 ± 0.031 ^{a}	0.823 ± 0.010 ^{a}	113.9 ± 19.6 ^{a}	0.156 ± 0.058 ^{a}	34.8 ± 6.1 ^{b}	21.9 ± 6.9 ^{b}	43.3 ± 6.1 ^{a}	52.8 ± 12.4	76.2 ± 17.9
DS	4.96 ± 3.46 ^{b}	0.137 ± 0.031 ^{b}	0.768 ± 0.061 ^{b}	91.9 ± 21.1 ^{b}	0.087 ± 0.037 ^{b}	50.7 ± 6.2 ^{a}	30.7 ± 11.4 ^{a}	18.6 ± 11.5 ^{b}	54.9 ± 21.9	75.2 ± 18.6
p-value	<0.001	<0.001	<0.001	<0.001	<0.001	<0.001	0.015	<0.001	0.706	0.945
C × S *p-value*	<0.001	0.001	0.091	0.034	0.732	0.067	0.140	0.271	0.419	0.058

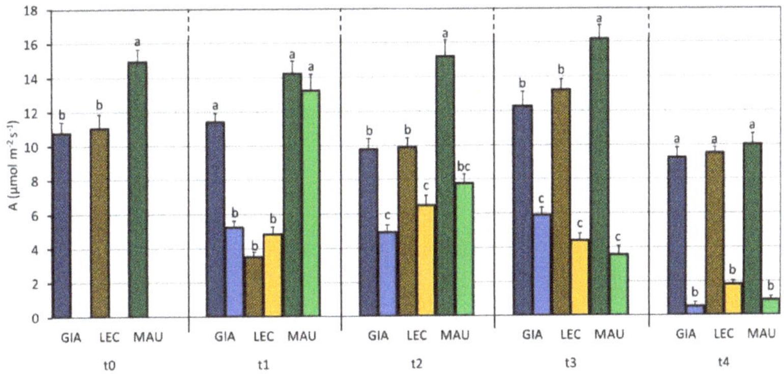

Figure 5. Net CO_2 assimilation (A) in control and drought-stressed olive cultivars from t0 to t4. At t0, the stressed and control samples were still one group. The bars represent mean ± standard error. The values for 'Giarraffa' (GIA) are in blue, those for 'Leccino' (LEC) are orange, and 'Maurino' (MAU) values are green. Strip bars refer to control samples. Within each time point, different letters denote statistical significance (p-value < 0.01) according to Tukey's multiple post hoc tests considering both cultivar and treatment.

Figure 6 shows the actual PSII efficiency. It was found that ΦPSII was affected by the cultivar, the drought stress, and their interaction (Table 2). The cultivar, Leccino, showed the first difference between the control and stressed groups at t1. From t2, the significant difference between the control and stressed groups appeared in the cultivar, Maurino. The 'Giarraffa' DS group value was significantly lower compared to its respective control one only at t4.

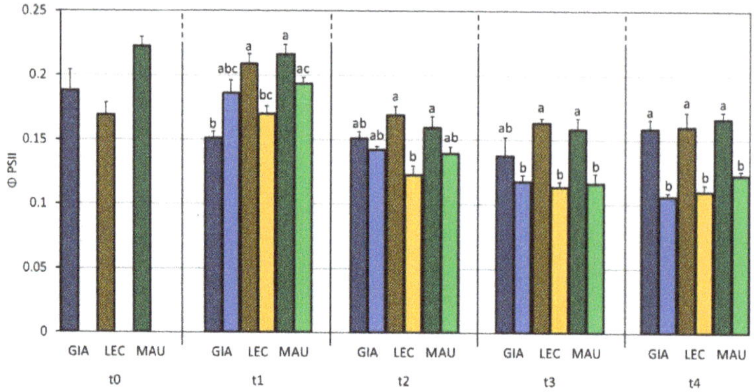

Figure 6. Effective efficiency of PSII (ΦPSII) in control and drought-stressed olive cultivars from t0 to t4. At t0, the stressed and control samples were still one group. The bars represent mean ± standard error. The values for 'Giarraffa' (GIA) are blue, those for 'Leccino' (LEC) are orange, and 'Maurino' (MAU) values are green. Strip bars refer to control samples. Within each time point, different letters denote statistical significance (*p*-value < 0.01) according to Tukey's multiple post hoc tests considering both cultivar and treatment.

On the contrary, no differences among the cultivars were found for F_v/F_m. However, drought stress affected the maximum efficiency of PSII, but the cultivars behaved similarly in relation to the stress (C × S, *p*-value = 0.091). All the cultivars showed a decrease in F_v/F_m starting from t3, with the lowest values in the DS groups (0.68 for 'Giarraffa' DS, 0.69 for 'Leccino' DS, and 0.70 for 'Maurino' DS) at t4. A graph is available in the Supplementary Material (Figure S6).

Conversely, the cultivar (C), the drought stress (S) and their interaction (C × S) significantly affected the ETR (Table 2). As shown in Figure 7, drought stress slowed down the ETR of all the stressed groups, but 'Leccino' showed the first difference at t2 ('Leccino' CTRL 110.7 μmol m^{-2} s^{-1}; 'Leccino' DS 82.6 μmol m^{-2} s^{-1}, $p < 0.01$), when the stressed groups of the other two cultivars were still comparable to their respective controls. Significant differences between the CTRL and DS groups appeared in 'Maurino' at t3 and in 'Giarraffa' only at t4.

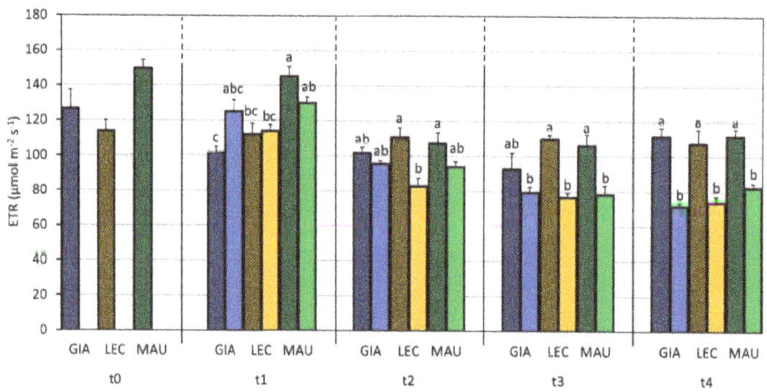

Figure 7. Electron transport rate (ETR) in control and drought-stressed olive cultivars from t0 to t4. At t0, the stressed and control samples were still one group. The bars represent mean ± standard error. The values for 'Giarraffa' (GIA) are blue, those for 'Leccino' (LEC) are orange, and 'Maurino' (MAU) values are green. Strip bars refer to control samples. Within each time point, different letters denote statistical significance (*p*-value < 0.01) according to Tukey's multiple post hoc tests considering both cultivar and treatment.

Unlike stomatal conductance, Table 2 shows that the mesophyll conductance exhibited no differences among the cultivars, even if a significant decrease in g_m is determined by stress, resulting in 0.156 ± 0.058 mol m^{-2} s^{-1} in the controls and 0.087 ± 0.037 mol m^{-2} s^{-1} in the stressed groups. A graph containing g_m values calculated at t0, t2, and t4 is available in the Supplementary Material (Figure S7).

As a result, drought stress significantly affected the stomatal (L_s), mesophyll conductance (L_m), and biochemical (L_b) limitations of photosynthesis at t2 (Table 2). Specifically, after two weeks of stress, the contribution of L_s and L_m was 16% and 9% higher, respectively, in the stressed groups compared to that in the control group, which occurred at the expense of biochemical limitation, which decreased. 'Giarraffa' and 'Maurino' showed a significant higher stomatal limitation in the stressed group compared to that of their respective controls, while the L_s values in the 'Leccino' DS and CTRL groups are similar, as shown in Figure 8. As suggested by the low biochemical limitation of the stressed groups, V_{cmax} and J_{max}, the maximum carboxylation rate and the maximum electron transport rate, respectively, did not seem to be significantly affected by the stress treatment (Table 2). The data are plotted in the Supplementary Material (Figures S8 and S9).

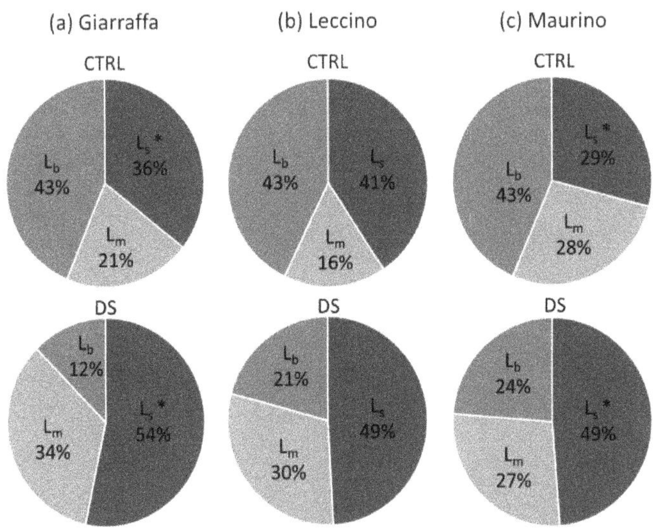

Figure 8. Percentages of stomatal limitation (L_s), mesophyll diffusion resistance (L_m), and biochemical limitation (L_b) on photosynthetic process in control and drought-stressed olive cultivars, calculated for 'Giarraffa' (**a**), 'Leccino' (**b**), and 'Maurino' (**c**) at t2. The percentages shown for each group are the means of five values calculated independently for each limitation (L_s, L_m, and L_b). The asterisks (*) denote statistical significance (p-value < 0.01) according to Tukey's multiple post hoc test. Only the significant difference between control and stressed samples of the same cultivar is highlighted.

3.3. *Correlation*

Figure 9 shows correlograms generated for each cultivar. When the cultivars were examined together, they gave very similar results. Firstly, there is a positive correlation between all the parameters, except EL. Therefore, the main difference is the magnitude of the correlation. Looking at the correlations between the water-related parameters (the SWC, the RWC, and g_s), the cultivars behaved differently: the cultivar Maurino had the strongest positive correlation between all these parameters. 'Leccino' showed the same strong positive correlation between the SWC and g_s, but the value was lower when the leaf RWC was correlated with the SWC and g_s. In the cultivar, Giarraffa, the correlations of both leaf RWC and SWC with g_s were weak, while that between the SWC and leaf RWC was stronger than it was in 'Leccino'. Concerning the correlation between the water-related parameters and photosynthesis, 'Maurino' showed the strongest correlation between the SWC, leaf RWC, g_s, and A. The same strong positive correlation was found between A and the SWC in 'Leccino'. The correlations between the leaf RWC and A (in 'Leccino') and the leaf RWC, SWC, and A (in 'Giarraffa') are weaker. Moreover, ETR and ΦPSII were strongly positively correlated with g_s in 'Maurino' and 'Leccino', while they were more positively correlated with the leaf RWC in 'Giarraffa'.

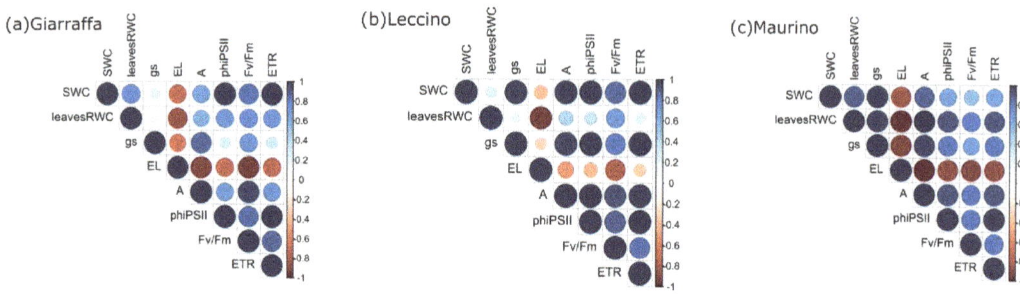

Figure 9. Correlogram of the 8 parameters evaluated at each time point in 'Giarraffa' (**a**), 'Leccino' (**b**), and 'Maurino' (**c**). Each parameter is expressed as the ratio of the DS group to the CTRL group value at each time point and correlates to the time course (t1, t2, t3, and t4). The size of the dots corresponds to the value of the correlation coefficient according to the right-positioned scale, while the colour indicates the direction of change (blue for positive correlation; red for negative correlation).

4. Discussion

Olive trees are one of the most cultivated crops in the Mediterranean region and will face frequent droughts due to climate change. Understanding how water deficit affects olive cultivars differently will help us to select those that are best suited to future climate change [49]. This study examined the inter-varietal differences in drought stress responses among three olive cultivars in Italy (cultivars Maurino, Leccino, and Giarraffa). The main variables investigated were the plant–water relationship and photosynthetic traits. The analysis of relative water content revealed the distribution of water within plants, and physiological analysis aimed to show how drought affects photosynthesis light reactions.

The first basic variable we investigated was the amount of water in the soil, which could also indicate how well different plants absorb water through their roots. Indeed, after only one week of stress, the soil water content of all the stressed plants was half of that of the controls, indicating that the cultivars remove comparable amounts of water from the soil. This parameter could not distinguish between the cultivars, which is consistent with the findings of other studies. In the study by Oddo et al. [50], for example, the three Italian cultivars tested (cultivars Giarraffa, Biancolilla, and Nocellara del Belice) showed no significant differences, and the report indicates a 55% reduction from the value of the controls during the first week of stress. There were no discernible differences in the decrease in the SWC in plants stressed for two and three weeks compared to that of the plants stressed for one week. Melaouhi et al. [51] studied the response of two-year-old 'Arbequina' and 'Empeltre' cultivars to mild (30% field capacity) and severe (50% field capacity) stress and found that both treatments resulted in a significant reduction in the SWC, but no differences or interactions between cultivars or treatments and cultivars were observed. Using the SWC as a benchmark, the current study confirmed that drought stress had an equal impact on all the cultivar groups, with the severity directly related to the duration of the water shortage. Some differences emerge if the available water rather than the soil water content is considered. In other-than-soil media (such as peat and pumice used in this case), the available water in the pots is the difference between the volumetric soil water content determined at psi = −1 kPa and −10 kPa [52]. Data reported by Pardossi et al. [53] defined the range (57/39%) of the volumetric soil water content as the available water for peat: pumice 1:1. Considering the gravimetric soil water content of this study, we assume that the control groups of the three cultivars have the maximum amount of available water during the whole period. The stressed groups of 'Leccino' and 'Maurino' nearly wilted at t1, while 'Giarraffa' reached this state one week later. From t2 onwards, the water available to the stressed groups of all cultivars was very limited.

Although the cultivars retained a similar amount of water in the soil, they were consistently different in terms of transpiration. Firstly, each cultivar had a unique default stomatal density. This morphological difference is perfectly reflected in a different rates of stomatal conductance. The cultivar, Maurino, for example, has the highest stomatal density and the highest stomatal conductance, even when it is grown in well-watered conditions, which is followed by 'Leccino', and then 'Giarraffa', which has the lowest stomatal density and the earliest decrease in stomatal conductance. Oddo et al. [50], on the other hand, found that 'Giarraffa' retained high g_s values after seven days of drought stress (0.4 mol m^{-2} s^{-1}). In this case, 'Maurino' appeared to have stomatal characteristics that were less effective in preventing water loss than those of the controls of the other two cultivars, but this did not result in an overall lower leaf relative water content, except for at t2. As suggested by Bosabalidis and Kofidis [25], the higher stomatal density of 'Maurino' may allow the more precise regulation of transpiration. The drought tolerance of plants is generally associated with their ability to maintain a high leaf RWC under drought stress conditions [54]. Given that stomatal conductance regulates the plant water status [55], the simultaneous decrease in the RWC and g_s in 'Giarraffa' at t1 results in a smaller difference between the stressed and control groups at t2 compared to that of the other two cultivars. On the contrary, in 'Leccino' and 'Maurino', stomatal conductance decreased one week later than the leaf RWC did, which is unlike what has been described by Boussadia et al. [56], Guerfel et al. [7], and Lawlor [9]. During periods of water scarcity, the stem could serve as a water reservoir [57]. At t2, however, the slightly higher water content in the stems of 'Maurino' under the control condition did not provide any support to counteract the RWC loss in the leaves at the same time point. The differences between cultivars in water-related parameters became apparent after one or two weeks of stress, but were completely lost as the stress level increased (at t3 and t4). Together with the tendency to maintain higher water contents, the rapid reduction in the stomatal conductance and the low stomatal density could allow 'Giarraffa' to use water resources more conservatively and make this cultivar interesting to researchers conducting future water management studies.

Water stress generally reduces stomatal conductance, which decreases the quantity of CO_2 taken up by the leaves, limiting net carbon assimilation (A) [58], and thus, affecting plant growth. All the olive cultivars studied showed a decrease in parameter A, supporting the conclusion reached by Sofo et al. [59]. The authors of that case study looked at the cultivars, Biancolilla and Coratina, and found that ten days of total water depletion (roughly between t1 and t2 in the current study) resulted in a decrease in A of 11.8 to 3 mol m^{-2} s^{-1} and 14.1 to 1.1 mol m^{-2} s^{-1}, respectively. In the current study, drought-stressed 'Giarraffa' showed an early decrease in A at t1, coinciding with stomatal closure (as indicated by the decrease in g_s) and with a decrease in the leaf RWC. Instead, both 'Leccino' and 'Maurino' cultivars showed a delayed decrease in A compared to the timing of the decrease in the leaf RWC. Guerfel et al. [7] used the leaf water potential to assess the plant water status in two olive cultivars ('Chemlai' and 'Chetoui') subjected to 21 days of drought. They found a relationship between the leaf water potential and A (R^2 0.96 for 'Chemlai' and 0.79 for 'Chetoui'), which can be compared to the behaviour of the stressed group of 'Giarraffa'. Mesophyll conductance is another important diffusive parameter that regulates the transfer of CO_2 from sub-stomatal cavities to the chloroplasts within a leaf [58]. The reduction in g_m was found to be associated with a decrease in A in drought-stressed plants, which played a crucial role in reducing CO_2 availability in the mesophyll, as confirmed by a relative increase in mesophyll conductance limitation under water stress conditions. The underlying mechanisms involved in the regulation of g_m under water stress conditions are not fully understood and deserve further investigation. However, previous studies have reported that g_m may be affected by increasing abscisic acid (ABA) concentrations as the soil dries [12,60] or by changes in the leaf structure (such as cell wall thickness) as an acclimation response to the drought [61].

According to Grassi and Magnani [45], the contribution of stomatal, mesophyll conductance, and biochemical limitations can explain the decrease in net CO_2 assimilation under drought stress. In this study, the limitations were calculated after two weeks of the drought study. The main factors limiting photosynthesis are stomatal and mesophyll conductance, confirming that diffusive limitations rather than biochemical impairments play a significant role under intermediate stress conditions [8,45]. In particular, water stress increased the overall contribution of the stomatal limitation to photosynthesis in stressed 'Giarraffa' and 'Maurino' plants at t2. The lack of a biochemical impairment was further demonstrated by the absence of any effects on the maximum electron transport rate (J_{max}) and the maximum carboxylation capacity (V_{cmax}) at least up to t2, as found in drought-stressed *Q. ilex* by Hoshika et al. [12,62] and in salt-stressed olive trees by Centritto et al. [11]. Chlorophyll fluorescence parameters provide more information about the functionality of photosynthetic machinery during stress. The decrease in the ETR and ΦPSII observed in 'Leccino' at t2 anticipates the PSII impairment shown by the strong reduction in F_v/F_m at t3. The cultivar, Giarraffa, which shows decreases in the ETR and ΦPSII only at t4, seems to avoid a photosystem impairments thanks to the earlier decreases in g_s and A [63]. The cultivar Maurino showed an intermediate pattern, revealing the first PSII impairments at t3.

Measuring the pigment levels can help assess the health of a plant and determine whether it is under abiotic stress [64]. Indeed, Dias et al. [27] found that the cultivars, Cobrancosa, Cordovil de Serpa, and Cordovil de Castelo Branco, showed a decrease in the chlorophyll and carotenoid contents, which are most likely due to oxidative stress damage under drought stress. However, no stress-related differences in the pigment content were observed in this study, indicating that pigment impairments were not present throughout the experiment. Marino et al. [65] found similar results: after a dry summer without irrigation, 10-year-old olive trees ('Leccino') showed no significant changes in the chlorophyll and carotenoid contents. In our case study, the lack of differences in lipid peroxidation between the control and stressed plants suggest that damage to the lipid bilayer of cell membranes does not occur in the early stages of drought stress. Since no lipid peroxidation was observed (at least until t2), the increase in EL may be due to a preliminary response of the plants to drought stress rather than cell damage. According to Demidchik et al. [66], EL is primarily caused by the efflux of K^+ through specific channels activated by ROS. K^+ release can cause programmed cell death (PCD) or decrease anabolism in favour of catabolic processes, resulting in energy release. All the cultivars studied increased their EL in response to stress, but 'Maurino' showed a stronger and faster response. The cultivar, Maurino, could theoretically achieve a faster EL response due to the increased water loss and transpiration rate, at least until t2. The rapid reduction of gas exchange as the SWC and RWC decreased has been shown to give an advantage to the defence of photosynthetic process under drought stress in 'Giarraffa', that even with a lower A compared to that of the control, maintained a higher electron transport rate and effective photosystem II efficiency until the last week of the stress study.

5. Conclusions

The conservation of biodiversity is becoming increasingly important in both natural and agricultural environments as climate change has a stronger impact on plants. The widely cultivated olive trees are represented by a broad range of cultivars, each with their own genetic background that may play a role in olive cultivar adaptation to various abiotic stresses such as drought. The three Italian olive cultivars tested in this study differed significantly in their responses to drought stress in growth chamber experiments; though, not all the parameters revealed significant differences. Some defence mechanisms are not used by any of the olive cultivars, whereas others are implemented first by a specific cultivar, resulting in an anticipatory and protective response. The data collected thus far suggest that

Giarraffa is likely the most tolerant cultivar, but the molecular processes underlying this evidence are not fully understood. 'Giarraffa' showed the earliest stomatal response, but this is not the determining factor for a significantly better water saving capacity than that of the other cultivars. To summarise, the physiological results described here suggest that the genetic basis of water deficit responses may differ between olive cultivars and that specific genotypes (such as 'Giarraffa') may be used in breeding programmes to develop preferred cultivars for drought-prone regions. Taking into account both water-related parameters and photosynthesis, we suggest using fluorescent chlorophyll parameters (such as ΦPSII and ETR) as markers to discriminate the drought resistance of cultivars within large germplasm collections, and also, considering the ease with which these data can be collected for a high number of plants. However, the more precise cataloguing of drought-tolerant olive cultivars necessitates additional testing. We have already planned biochemical research to track proteins (such as dehydrins and osmotin) and molecules (such as proline) related to the water balance in the leaves and stems. These may provide a valid alternative to stomatal closure to conserve water during drought and will give a more accurate picture of the specific response of each cultivar.

Supplementary Materials: The following supporting information can be downloaded at: https://www.mdpi.com/article/10.3390/horticulturae9070830/s1. Figure S1: A/C_i and A/C_c curve; Figure S2: SWC; Figure S3: MDA; Figure S4: Chl a + b; Figure S5: Car; Figure S6: F_v/F_m; Figure S7: g_m; Figure S8: V_{cmax}; Figure S9: J_{max}.

Author Contributions: Conceptualization, S.P., M.R., G.C. and C.C.; methodology, S.P., M.R., M.C.D. and F.C.P.; data elaboration, S.P. and Y.H.; writing—original draft preparation, S.P.; writing—review and editing, A.G., Y.H., M.C.D., G.C. and C.C.; supervision, G.C. and C.C. All authors have read and agreed to the published version of the manuscript.

Funding: This research was funded by the European Union Next-GenerationEU (PIANO NAZIONALE DI RIPRESA E RESILIENZA (PNRR)—MISSIONE 4 COMPONENTE 2, INVESTIMENTO 1.4—MUR Notice No. 3138 of 16/12/2021, Directorial Decree of Granting of Funding No. 1032 of 17/06/2022, CN000022). This manuscript reflects only the authors' views and opinions, neither the European Union nor the European Commission can be considered responsible for them.

Data Availability Statement: Data available on request due to restriction.

Acknowledgments: We thank the Tuscany Region for supporting the Ph.D. grant of S.P. in the framework of 'Pegaso'—POR FSE TOSCANA 2014–2020 Program GiovaniSì. We also thank Caterina Margheriti, Laura Palma, Fabio Tullio, and Sonia Piattelli (University of Siena) for the highly appreciated technical assistance.

Conflicts of Interest: The authors declare no conflict of interest.

References

1. Chaves, M.M.; Flexas, J.; Pinheiro, C. Photosynthesis under Drought and Salt Stress: Regulation Mechanisms from Whole Plant to Cell. *Ann. Bot.* **2009**, *103*, 551–560. [CrossRef] [PubMed]
2. Vergni, L.; Todisco, F. Spatio-Temporal Variability of Precipitation, Temperature and Agricultural Drought Indices in Central Italy. *Agric. For. Meteorol.* **2011**, *151*, 301–313. [CrossRef]
3. Douville, H.; Raghavan, K.; Renwick, J. Water Cycle Changes. In *Climate Change 2021: The Physical Science Basis*; Contribution of Working Group I to the Sixth Assessment Report of the IPCC; IPCC: Geneva, Switzerland, 2021; ISBN 978-1-00-915789-6.
4. Brito, C.; Dinis, L.T.; Moutinho-Pereira, J.; Correia, C.M. Drought Stress Effects and Olive Tree Acclimation under a Changing Climate. *Plants* **2019**, *8*, 232. [CrossRef] [PubMed]
5. Galmés, J.; Conesa, M.A.; Ochogavía, J.M.; Perdomo, J.A.; Francis, D.M.; Ribas-Carbó, M.; Savé, R.; Flexas, J.; Medrano, H.; Cifre, J. Physiological and Morphological Adaptations in Relation to Water Use Efficiency in Mediterranean Accessions of Solanum Lycopersicum. *Plant Cell Environ.* **2011**, *34*, 245–260. [CrossRef] [PubMed]
6. El Yamani, M.; Sakar, E.H.; Boussakouran, A.; Rharrabti, Y. Physiological and Biochemical Responses of Young Olive Trees (*Olea europaea* L.) to Water Stress during Flowering. *Arch. Biol. Sci.* **2019**, *71*, 123–132. [CrossRef]
7. Guerfel, M.; Baccouri, O.; Boujnah, D.; Chaïbi, W.; Zarrouk, M. Impacts of Water Stress on Gas Exchange, Water Relations, Chlorophyll Content and Leaf Structure in the Two Main Tunisian Olive (*Olea europaea* L.) Cultivars. *Sci. Hortic.* **2009**, *119*, 257–263. [CrossRef]

8. Flexas, J.; Bota, J.; Loreto, F.; Cornic, G.; Sharkey, T.D. Diffusive and Metabolic Limitations to Photosynthesis under Drought and Salinity in C3 Plants. *Plant Biol.* **2004**, *6*, 269–279. [CrossRef]
9. Lawlor, D.W. Limitation to Photosynthesis in Water-Stressed Leaves: Stomata vs. Metabolism and the Role of ATP. *Ann. Bot.* **2002**, *89*, 871–885. [CrossRef]
10. Flexas, J.; Medrano, H. Drought-Inhibition of Photosynthesis in C_3 Plants: Stomatal and Non-Stomatal Limitations Revisited. *Ann. Bot.* **2002**, *89*, 183–189. [CrossRef]
11. Centritto, M.; Loreto, F.; Chartzoulakis, K. The Use of Low [CO_2] to Estimate Diffusional and Non-Diffusional Limitations of Photosynthetic Capacity of Salt-Stressed Olive Saplings. *Plant Cell Environ.* **2003**, *26*, 585–594. [CrossRef]
12. Hoshika, Y.; Paoletti, E.; Centritto, M.; Gomes, M.T.G.; Puértolas, J.; Haworth, M. Species-Specific Variation of Photosynthesis and Mesophyll Conductance to Ozone and Drought in Three Mediterranean Oaks. *Physiol. Plant.* **2022**, *174*, e13639. [CrossRef] [PubMed]
13. Petridis, A.; Therios, I.; Samouris, G.; Koundouras, S.; Giannakoula, A. Effect of Water Deficit on Leaf Phenolic Composition, Gas Exchange, Oxidative Damage and Antioxidant Activity of Four Greek Olive (*Olea europaea* L.) Cultivars. *Plant Physiol. Biochem.* **2012**, *60*, 1–11. [CrossRef] [PubMed]
14. Kitao, M.; Lei, T.T.; Koike, T.; Tobita, H.; Maruyama, Y. Susceptibility to Photoinhibition of Three Deciduous Broadleaf Tree Species with Different Successional Traits Raised under Various Light Regimes. *Plant Cell Environ.* **2000**, *23*, 81–89. [CrossRef]
15. The World of Olive Oil. Available online: https://www.internationaloliveoil.org/the-world-of-olive-oil/ (accessed on 3 March 2023).
16. FAOSTAT. Available online: https://www.fao.org/faostat/en/#data/QCL (accessed on 3 March 2023).
17. Alba, V.; Montemurro, C.; Sabetta, W.; Pasqualone, A.; Blanco, A. SSR-Based Identification Key of Cultivars of *Olea europaea* L. Diffused in Southern-Italy. *Sci. Hortic.* **2009**, *123*, 11–16. [CrossRef]
18. Albertini, E.; Torricelli, R.; Bitocchi, E.; Raggi, L.; Marconi, G.; Pollastri, L.; Di Minco, G.; Battistini, A.; Papa, R.; Veronesi, F. Structure of Genetic Diversity in *Olea europaea* L. Cultivars from Central Italy. *Mol. Breed.* **2011**, *27*, 533–547. [CrossRef]
19. Muzzalupo, I.; Vendramin, G.G.; Chiappetta, A. Genetic Biodiversity of Italian Olives (*Olea europaea*) Germplasm Analyzed by SSR Markers. *Sci. World J.* **2014**, *2014*, e296590. [CrossRef] [PubMed]
20. Diaz-Espejo, A.; Fernández, J.E.; Torres-Ruiz, J.M.; Rodriguez-Dominguez, C.M.; Perez-Martin, A.; Hernandez-Santana, V. The Olive Tree under Water Stress. In *Water Scarcity and Sustainable Agriculture in Semiarid Environment: Tools, Strategies, and Challenges for Woody Crops*; Academic Press: Cambridge, MA, USA, 2018; ISBN 978-0-12-813164-0.
21. Piccini, C.; Cai, G.; Dias, M.C.; Araújo, M.; Parri, S.; Romi, M.; Faleri, C.; Cantini, C. Olive Varieties under Uv-b Stress Show Distinct Responses in Terms of Antioxidant Machinery and Isoform/Activity of Rubisco. *Int. J. Mol. Sci.* **2021**, *22*, 11214. [CrossRef]
22. Rossi, L.; Francini, A.; Minnocci, A.; Sebastiani, L. Salt Stress Modifies Apoplastic Barriers in Olive (*Olea europaea* L.): A Comparison between a Salt-Tolerant and a Salt-Sensitive Cultivar. *Sci. Hortic.* **2015**, *192*, 38–46. [CrossRef]
23. Fernández, J.E. Understanding Olive Adaptation to Abiotic Stresses as a Tool to Increase Crop Performance. *Environ. Exp. Bot.* **2014**, *103*, 158–179. [CrossRef]
24. Carr, M.K.V. The Water Relations and Irrigation Requirements of Olive (*Olea europaea* L.): A Review. *Exp. Agric.* **2013**, *49*, 597–639. [CrossRef]
25. Bosabalidis, A.M.; Kofidis, G. Comparative Effects of Drought Stress on Leaf Anatomy of Two Olive Cultivars. *Plant Sci.* **2002**, *163*, 375–379. [CrossRef]
26. Sofo, A.; Dichio, B.; Xiloyannis, C.; Masia, A. Lipoxygenase Activity and Proline Accumulation in Leaves and Roots of Olive Trees in Response to Drought Stress. *Physiol. Plant.* **2004**, *121*, 58–65. [CrossRef] [PubMed]
27. Dias, M.C.; Correia, S.; Serôdio, J.; Silva, A.M.S.; Freitas, H.; Santos, C. Chlorophyll Fluorescence and Oxidative Stress Endpoints to Discriminate Olive Cultivars Tolerance to Drought and Heat Episodes. *Sci. Hortic.* **2018**, *231*, 31–35. [CrossRef]
28. Faraloni, C.; Cutino, I.; Petruccelli, R.; Leva, A.R.; Lazzeri, S.; Torzillo, G. Chlorophyll Fluorescence Technique as a Rapid Tool for in Vitro Screening of Olive Cultivars (*Olea europaea* L.) Tolerant to Drought Stress. *Environ. Exp. Bot.* **2011**, *73*, 49–56. [CrossRef]
29. Pierantozzi, P.; Torres, M.; Bodoira, R.; Maestri, D. Water Relations, Biochemical—Physiological and Yield Responses of Olive Trees (*Olea europaea* L. Cvs. Arbequina and Manzanilla) under Drought Stress during the Pre-Flowering and Flowering Period. *Agric. Water Manag.* **2013**, *125*, 13–25. [CrossRef]
30. de Pascali, M.; Vergine, M.; Sabella, E.; Aprile, A.; Nutricati, E.; Nicolì, F.; Buja, I.; Negro, C.; Miceli, A.; Rampino, P.; et al. Molecular Effects of Xylella Fastidiosa and Drought Combined Stress in Olive Trees. *Plants* **2019**, *8*, 437. [CrossRef]
31. D'Agostino, N.; Taranto, F.; Camposeo, S.; Mangini, G.; Fanelli, V.; Gadaleta, S.; Miazzi, M.M.; Pavan, S.; di Rienzo, V.; Sabetta, W.; et al. GBS-Derived SNP Catalogue Unveiled Wide Genetic Variability and Geographical Relationships of Italian Olive Cultivars. *Sci. Rep.* **2018**, *8*, 15877. [CrossRef]
32. Tosca, A.; Valagussa, M.; Martinetti, L.; Frangi, P. Biochar and Green Compost as Peat Alternatives in the Cultivation of Photinia and Olive Tree. *Acta Hortic.* **2021**, *1305*, 257–262. [CrossRef]
33. Bilskie, J. *Soil Water Status: Content and Potential*; Campbell Scientific, Inc.: Logan, UT, USA, 2001. Available online: https://www.google.com/url?sa=t&rct=j&q=&esrc=s&source=web&cd=&ved=2ahUKEwjj_97LrZiAAxV5yAIHHVo3CJIQFnoECA4QAQ&url=https%3A%2F%2Fs.campbellsci.com%2Fdocuments%2Ffr%2Ftechnical-papers%2Fsoilh20c.pdf&usg=AOvVaw3jbEEwVoaHKEhJVP65Jm9a&opi=89978449 (accessed on 19 July 2023).

34. Xu, Z.; Zhou, G. Responses of Leaf Stomatal Density to Water Status and Its Relationship with Photosynthesis in a Grass. *J. Exp. Bot.* **2008**, *59*, 3317–3325. [CrossRef]
35. Diaz-Espejo, A.; Nicolás, E.; Fernández, J.E. Seasonal Evolution of Diffusional Limitations and Photosynthetic Capacity in Olive under Drought. *Plant Cell Environ.* **2007**, *30*, 922–933. [CrossRef]
36. Rodriguez-Dominguez, C.M.; Buckley, T.N.; Egea, G.; de Cires, A.; Hernandez-Santana, V.; Martorell, S.; Diaz-Espejo, A. Most Stomatal Closure in Woody Species under Moderate Drought Can Be Explained by Stomatal Responses to Leaf Turgor. *Plant Cell Environ.* **2016**, *39*, 2014–2026. [CrossRef] [PubMed]
37. Ethier, G.J.; Livingston, N.J. On the Need to Incorporate Sensitivity to CO_2 Transfer Conductance into the Farquhar-von Caemmerer-Berry Leaf Photosynthesis Model. *Plant Cell Environ.* **2004**, *27*, 137–153. [CrossRef]
38. Bernacchi, C.J.; Singsaas, E.L.; Pimentel, C.; Portis, A.R.; Long, S.P. Improved Temperature Response Functions for Models of Rubisco-Limited Photosynthesis. *Glob. Change Biol.* **2001**, *21*, 253–259. [CrossRef]
39. Pons, T.L.; Flexas, J.; Von Caemmerer, S.; Evans, J.R.; Genty, B.; Ribas-Carbo, M.; Brugnoli, E. Estimating Mesophyll Conductance to CO_2: Methodology, Potential Errors, and Recommendations. *J. Exp. Bot.* **2009**, *60*, 2217–2234. [CrossRef]
40. Loreto, F.; Harley, P.C.; Marco, G.D.; Sharkey, T.D. Estimation of Mesophyll Conductance to CO_2 Flux by Three Different Methods. *Plant Physiol.* **1992**, *98*, 1437–1443. [CrossRef] [PubMed]
41. Marchi, S.; Guidotti, D.; Sebastiani, L.; Tognetti, R. Changes in Assimilation Capacity during Leaf Development in Broad- Leaved *Prunus Persica* and Sclerophyllous *Olea europaea*. *J. Hortic. Sci. Biotechnol.* **2007**, *82*, 69–78. [CrossRef]
42. Galmés, J.; Flexas, J.; Keys, A.J.; Cifre, J.; Mitchell, R.A.C.; Madgwick, P.J.; Haslam, R.P.; Medrano, H.; Parry, M.A.J. Rubisco Specificity Factor Tends to Be Larger in Plant Species from Drier Habitats and in Species with Persistent Leaves. *Plant Cell Environ.* **2005**, *28*, 571–579. [CrossRef]
43. Loriaux, S.D.; Avenson, T.J.; Welles, J.M.; Mcdermitt, D.K.; Eckles, R.D.; Riensche, B.; Genty, B. Closing in on Maximum Yield of Chlorophyll Fluorescence Using a Single Multiphase Flash of Sub-Saturating Intensity. *Plant Cell Environ.* **2013**, *36*, 1755–1770. [CrossRef]
44. Gilbert, M.E.; Pou, A.; Zwieniecki, M.A.; Holbrook, N.M. On Measuring the Response of Mesophyll Conductance to Carbon Dioxide with the Variable J Method. *J. Exp. Bot.* **2012**, *63*, 413–425. [CrossRef]
45. Grassi, G.; Magnani, F. Stomatal, Mesophyll Conductance and Biochemical Limitations to Photosynthesis as Affected by Drought and Leaf Ontogeny in Ash and Oak Trees. *Plant Cell Environ.* **2005**, *28*, 834–849. [CrossRef]
46. Ben Abdallah, M.; Trupiano, D.; Polzella, A.; De Zio, E.; Sassi, M.; Scaloni, A.; Zarrouk, M.; Ben Youssef, N.; Scippa, G.S. Unraveling Physiological, Biochemical and Molecular Mechanisms Involved in Olive (*Olea europaea* L. Cv. Chétoui) Tolerance to Drought and Salt Stresses. *J. Plant Physiol.* **2018**, *220*, 83–95. [CrossRef] [PubMed]
47. Hodges, D.M.; DeLong, J.M.; Forney, C.F.; Prange, R.K. Improving the Thiobarbituric Acid-Reactive-Substances Assay for Estimating Lipid Peroxidation in Plant Tissues Containing Anthocyanin and Other Interfering Compounds. *Planta* **1999**, *207*, 604–611. [CrossRef]
48. Lichtenthaler, H.K. Chlorophylls and Carotenoids: Pigments of Photosynthetic Biomembranes. *Methods Enzymol.* **1987**, *148*, 350–382.
49. Ennajeh, M.; Vadel, A.M.; Cochard, H.; Khemira, H. Comparative Impacts of Water Stress on the Leaf Anatomy of a Drought-Resistant and a Drought-Sensitive Olive Cultivar. *J. Hortic. Sci. Biotechnol.* **2010**, *85*, 289–294. [CrossRef]
50. Oddo, E.; Virgilio, F.; Grisafi, F. Effects of Water Deficit on the Leaf Water Relations of Pot-Grown Olive Cultivars. *Plant Stress* **2008**, *2*, 56–63.
51. Melaouhi, A.; Baraza, E.; Escalona, J.M.; El-AouOuad, H.; Mahjoub, I.; Bchir, A.; Braham, M.; Bota, J. Physiological and Biochemical Responses to Water Deficit and Recovery of Two Olive Cultivars (*Olea europaea* L., Arbequina and Empeltre Cvs.) under Mediterranean Conditions. *Theor. Exp. Plant Physiol.* **2021**, *33*, 369–383. [CrossRef]
52. Pardossi, A.; Incrocci, L.; Incrocci, G.; Malorgio, F.; Battista, P.; Bacci, L.; Rapi, B.; Marzialetti, P.; Hemming, J.; Balendonck, J. Root Zone Sensors for Irrigation Management in Intensive Agriculture. *Sensors* **2009**, *9*, 2809–2835. [CrossRef]
53. Pardossi, A.; Marzialetti, P.; Spe, C.V.; Bibbiani, C. I substrati e la coltivazione delle piante in contenitore. *Fertilitas Agrorum* **2009**, *3*, 22–31.
54. Flower, D.J.; Ludlow, M.M. Contribution of Osmotic Adjustment to the Dehydration Tolerance of Water-Stressed Pigeon Pea (*Cajanus cajan* (L.) Millsp.) Leaves. *Plant Cell Environ.* **1986**, *9*, 33–40. [CrossRef]
55. Jones, H.G.; Tardieu, F. Modelling Water Relations of Horticultural Crops: A Review. *Sci. Hortic.* **1998**, *74*, 21–46. [CrossRef]
56. Boussadia, O.; Mariem, F.B.; Mechri, B.; Boussetta, W.; Braham, M.; Hadj, S.B.E. Response to Drought of Two Olive Tree Cultivars (Cv Koroneki and Meski). *Sci. Hortic.* **2008**, *116*, 388–393. [CrossRef]
57. Traversari, S.; Francini, A.; Traversi, M.L.; Emiliani, G.; Sorce, C.; Sebastiani, L.; Giovannelli, A. Can Sugar Metabolism in the Cambial Region Explain the Water Deficit Tolerance in Poplar? *J. Exp. Bot.* **2018**, *69*, 4083–4097. [CrossRef] [PubMed]
58. Larcher, W. *Physiological Plant Ecology: Ecophysiology and Stress Physiology of Functional Groups*; Springer Science & Business Media: Berlin/Heidelberg, Germany, 2003.
59. Sofo, A.; Dichio, B.; Montanaro, G.; Xiloyannis, C. Photosynthetic Performance and Light Response of Two Olive Cultivars under Different Water and Light Regimes. *Photosynthetica* **2009**, *47*, 602–608. [CrossRef]

60. Brunetti, C.; Gori, A.; Marino, G.; Latini, P.; Sobolev, A.P.; Nardini, A.; Haworth, M.; Giovannelli, A.; Capitani, D.; Loreto, F.; et al. Dynamic Changes in ABA Content in Water-Stressed Populus Nigra: Effects on Carbon Fixation and Soluble Carbohydrates. *Ann. Bot.* **2019**, *124*, 627–644. [CrossRef]
61. Clemente-Moreno, M.J.; Gago, J.; Díaz-Vivancos, P.; Bernal, A.; Miedes, E.; Bresta, P.; Liakopoulos, G.; Fernie, A.R.; Hernández, J.A.; Flexas, J. The Apoplastic Antioxidant System and Altered Cell Wall Dynamics Influence Mesophyll Conductance and the Rate of Photosynthesis. *Plant J. Cell Mol. Biol.* **2019**, *99*, 1031–1046. [CrossRef] [PubMed]
62. Hoshika, Y.; Fares, S.; Pellegrini, E.; Conte, A.; Paoletti, E. Water Use Strategy Affects Avoidance of Ozone Stress by Stomatal Closure in Mediterranean Trees—A Modelling Analysis. *Plant Cell Environ.* **2020**, *43*, 611–623. [CrossRef]
63. Maxwell, K.; Johnson, G.N. Chlorophyll Fluorescence-a Practical Guide. *J. Exp. Bot.* **2000**, *51*, 659–668. [CrossRef]
64. Jaleel, C.A.; Llorente, B.E. Drought Stress in Plants: A Review on Water Relations. *Biosci. Res.* **2009**, *6*, 20–27.
65. Marino, G.; Pallozzi, E.; Cocozza, C.; Tognetti, R.; Giovannelli, A.; Cantini, C.; Centritto, M. Assessing Gas Exchange, Sap Flow and Water Relations Using Tree Canopy Spectral Reflectance Indices in Irrigated and Rainfed *Olea europaea* L. *Environ. Exp. Bot.* **2014**, *99*, 43–52. [CrossRef]
66. Demidchik, V.; Straltsova, D.; Medvedev, S.S.; Pozhvanov, G.A.; Sokolik, A.; Yurin, V. Stress-Induced Electrolyte Leakage: The Role of K+-Permeable Channels and Involvement in Programmed Cell Death and Metabolic Adjustment. *J. Exp. Bot.* **2014**, *65*, 1259–1270. [CrossRef]

horticulturae

MDPI

Article

Effects of 10 Dwarfing Interstocks on Cold Resistance of 'Tianhong 2' Apple

Junli Jing, Mingxiao Liu, Baoying Yin, Bowen Liang, Zhongyong Li, Xueying Zhang, Jizhong Xu * and Shasha Zhou *

College of Horticulture, Hebei Agricultural University, Baoding 071001, China; 18730277283@163.com (J.J.); liumingxiao0518@163.com (M.L.); by40901123@126.com (B.Y.); lbw@hebau.edu.cn (B.L.); yylzy@hebau.edu.cn (Z.L.); zhangxueying1996@163.com (X.Z.)

* Correspondence: xjzhxw@126.com (J.X.); zhoushasha198609@126.com (S.Z.)

Abstract: The lack of dwarf stock with good cold resistance has affected the production of apples in northern China. Annual dormant branches of 'Tianhong 2' apple were grafted on 10 different dwarf interstocks (the rootstocks were the seedlings of *Malus hupehensis* var. *Pingyiensis*) as test materials. Among these 10 interstocks, Huang 6, 244, NO.1, 53, 24-5, ZC9-3, Jizhen 1 were newly developed by us (Apple Research Group of Hebei Agricultural University), and three interstocks with different degrees of cold resistance (GM256–with strongest cold resistance, SH40–with stronger cold resistance, M9–with poor cold resistance) were used as controls. The semi-lethal temperature (LT50) and related physiological indexes of the branches in the overwintering process were studied. Based on the comprehensive physiological indexes, the effects of 10 interstocks on cold resistance of the 'Tianhong 2' apple were analyzed. The results showed that the effects of 10 kinds of interstocks on the cold resistance of 'Tianhong 2' apple were quite different. The order of effects on cold resistance from strong to weak was as follow: GM256 > Huang 6 > 244 > NO.1 > 53 > 24-5 > ZC9-3 > Jizhen 1 > SH40 > M9. The purpose of this study was to screen out the interstocks with strong cold resistance, in order to provide some basis for the selection and utilization of interstocks with strong cold resistance in apple cultivation to further promote the development of the apple industry in China.

Keywords: apple; dwarfing interstocks; cold resistance

1. Introduction

The apple (*Malus domestica* Borkh) is a nutritious fruit and is consumed worldwide, providing an important food source [1]. China is in a dominant position in apple production globally with both the largest apple growing area and the largest export of fresh apples [2]. However, in many apple-producing areas of China (such as high-latitude and high-altitude areas), there are different degrees of low-temperature freezing injury. Freezing injury affects the growth, development, and yield of apple trees, and can even lead to tree death, thus posing a major threat to the development of the apple industry [3]. Low-temperature stress is a key limiting factor affecting plant growth, physiology, and metabolism, affecting overall yields [4,5]. Therefore, it is particularly important to improve the cold resistance of apple trees. Rootstock is an important factor in conferring apple cold resistance [6]. Grafting desirable apple varieties on rootstocks with inherent strong cold resistance can improve the cold resistance of the mature fruit trees. In the northern apple-producing areas of China, the improper application of dwarf rootstock has led to poor cold-resistant varieties with frequent occurrence of freezing injury [7]. Therefore, the cultivation of apple cold-resistant dwarf rootstocks has attracted much attention. Screening dwarf rootstocks with strong cold resistance is necessary to develop superior varieties to alleviate and prevent overwintering freezing injury of apple trees. Evaluating and identifying the mechanism of cold resistance in apple rootstocks and breeding dwarf rootstocks with strong cold resistance, suitable for northern China, are necessary to promote the development of the apple industry and expand apple cultivation. To do this, we have bred multiple rootstocks with dwarfing potential, and evaluated their various characteristics, including cold resistance.

'Tianhong 2' (*Malus domestica* Borkh. cv. 'Tianhong 2') apple is a short-branch Fuji variant (*Malus domestica* Borkh. cv. Fuji). This variety was bred by the Apple Research Group of Hebei Agricultural University in 1994 and certified as a new variety by the Forest Tree Variety Approval Committee of Hebei Province, China in 2005 [8]. The fruit of 'Tianhong 2' is round and the fruit shape index can reach more than 0.9. The average single fruit weight is approximately 260 g. The fruit surface is smooth, and the surface color is bright red. Its pulp is yellow-white, crispy, and juicy. The taste is sweet and sour. The fruit contains 14.5%~16.8% soluble solids and has good aroma. On the whole, the quality is superior. The fruit matures between late October and early November in Baoding, Hebei Province, China and can be stored for 13 months. This variety is widely cultivated in Hebei Province, especially in Baoding (located in northern China). In this area, fruit growers usually choose vigorous rootstocks adapted to local climate as the base stock to improve the comprehensive resistance of 'Tianhong 2' apple trees, and graft dwarf interstocks on the base stock to dwarf the trees. Selecting suitable, highly cold-resistant dwarfing interstock will improve the yield and quality of the 'Tianhong 2' apple.

In this study, annual dormant branches of 'Tianhong 2' apple were grafted onto 10 different interstocks and tested for cold resistance. The tested interstocks include seven newly developed rootstocks with dwarf potential that were cultivated by the Apple Research Group of Hebei Agricultural University: Jizhen 1, Huang 6, 244, NO. 1, 53, 24-5, and ZC9-3. Three other interstocks that are widely cultivated in China were used as controls: SH40 and GM256, which exhibit strong cold resistance, and M9 with poor cold resistance. All interstocks were grafted on apomictic *Malus hupehensis* var. *Pingyiensis* as the base rootstock.

Semi-lethal temperature (LT50) is an important index widely used to evaluate plant cold resistance [9–11]. The cold resistance of apples may be closely related to many indicators. For example, reactive oxygen species such as superoxide anion are harmful substances and are produced at high levels in plants under low-temperature stress [12]. SOD and POD are important protective enzymes that can remove reactive oxygen species as part of plant cold resistance [13]. Proline, soluble sugar, and soluble protein are osmotic regulators that are essential players in the process of plant cold resistance [14]. Starch [15] and malondialdehyde (MDA) [16] are also important indicators of cold resistance of plants. Because many factors affect the cold resistance of fruit trees, it may be insufficient to measure cold resistance using a single index. Principal component analysis can be used to comprehensively evaluate cold resistance using multiple indicators. This type of analysis has been successfully applied to assess cold resistance of apples [7], grapes [9], and other crops. In recent years, there have been many reports on the cold resistance of apples. Some studies utilized apple rootstocks grown in an artificial low-temperature environment to measure different indicators and evaluate cold resistance [7,17]. Other studies measured cold resistance-related indicators to assess cold resistance in apple rootstocks subjected to natural overwintering and low temperature [7].

In this study, the semi-lethal temperature (LT50) and physiological indexes related to cold resistance were measured by conductance, and the different responses of various indexes to low-temperature stress were studied, which provided theoretical reference for studying the physiological characteristics of plant cold resistance during the process of overwintering. Principal component and cluster analysis were used to comprehensively evaluate the effects of cold resistance of these 10 different interstocks to 'Tianhong 2' apple. Selecting some new interstocks with strong cold resistance can provide some certainty in the choice and use of interstocks with strong cold hardiness for apple cultivation, further promoting the development of the apple industry. In many apple production areas, freezing injury caused great economic losses for the industry, and the hardiness of rootstock and interstock has a great influence on the cold resistance of apple tree. Therefore, this research has great significance for the development of apple industry.

2. Materials and Methods

2.1. Plant Materials

The test materials were grown in the West Campus of Hebei Agricultural University, Baoding City, Hebei Province, China. At the end of March 2019, the annual seedlings of *Malus hupehensis* var. *Pingyiensis* (apomixis) with the same growth vigor were planted in pots (22 cm high, upper diameter 30 cm, and lower diameter 20 cm) as the base rootstocks, and the no budding buds of 10 dwarfing interstocks were grafted on the base rootstock in early April 2019 by the plate budding method. The 10 dwarfing interstocks included: SH40 (with strong cold resistance), GM256 (with strongest cold resistance), and M9 (with poor cold resistance), which are widely used in production and were used as controls in this experiment; Jizhen 1, 53, 244, 24-5, Huang 6, ZC9-3, and NO.1, which were new dwarfing rootstocks cultivated by the Apple Research Group of Hebei Agricultural University. The background information of these seven new rootstocks was shown in Table 1. The no budding buds of 'Tianhong 2' apple were grafted on the interstock in early April 2020. Under unified management, 30 potted trees with the same growth vigor were selected from each rootstock-interstock-scion combination for the experiment. The annual dormant branches of 'Tianhong 2' (the scion) grafted on the 10 interstocks were used as test materials in the winter of 2020–2021.

Table 1. Background information of the seven new rootstocks.

Rootstock	Parents	Breeding Institution
Jizhen 1	seedling progeny of SH40	
53	seedling progeny of SH40	
244	seedling progeny of SH40	
NO. 1	seedling progeny of SH40	Hebei Agricultural University
24-5	*Malus micromalus* × Inner Mongolia11	
Huang 6	seedling progeny of *Malus robusta* Rehd	
ZC9-3	seedling progeny of P22	

2.2. Test Methods

2.2.1. Materials, Treatment and Sampling

The branch samples were taken on 22 October 2020, 1 December 2020, 5 January 2021, and 5 March 2021. The maximum temperature, minimum temperature, and average temperature were recorded on each sampling day (shown in Table 2). Annual dormant branches (about 0.5 cm in diameter) of 'Tianhong 2' that were disease-free, strong, and exhibited consistent growth were selected for sampling. The harvested branches were divided into two parts for every rootstock-interstock-scion combination. One part was cleaned immediately after harvest and the branch bark was removed, quickly frozen in liquid nitrogen, and stored in a −80 °C freezer before the determination of physiological and biochemical indexes. The other part was cleaned with distilled water and deionized water and then divided into six groups and transferred to six plastic sealing bags (sprayed with deionized water to prevent the branches from drying out) for separate artificial low-temperature treatments (Table 3).

Table 2. Temperature on each sampling day during overwintering.

Sampling Date	Natural Wintering Temperature/ °C		
	Maximum Temperature	Minimum Temperature	Average Temperature
22 October 2020	18	2	9.46
1 December 2020	1	−5	−1.08
5 January 2021	−2	−14	−7.33
5 March 2021	10	1	4.29

Table 3. Cold treatments for cold resistance determination.

Sampling Date	Temperature Settings/ °C					
22 October 2020	4	−8	−20	−26	−36	−45
1 December 2020	4	−8	−22	−36	−50	−60
5 January 2021	4	−10	−22	−45	−55	−65
5 March 2021	4	−12	−20	−28	−36	−45

The low-temperature treatment was performed as described by Wilner [18]. A variable temperature, ultra-low-temperature refrigerator was used for the artificial freezing treatments, with six temperature gradients included the temperature at which all samples are alive and the temperature at which all samples are killed (Table 3). The cooling rate of the refrigerator was 6 °C/h, and samples were placed for 12 h after the temperature reached the set temperature. After that, the samples were thawed at 0 °C for 8 h, and then thawed at 4 °C for 24 h. After low-temperature treatment, the relative conductivity of the branches was measured.

Each treatment was performed with three repetitions. After low-temperature treatment, the relative conductivity was determined and the LT50 value was calculated.

2.2.2. Relative Electrolyte Leakage (REL) Measurement and LT50 Calculation

The relative electrolyte leakage (REL) and LT50 were measured and calculated according to the methods of Wilner [18] and Zhang [19], with slight modifications. The low-temperature treated branches were cut into 15 mm branch sections, split in the cross-sectional direction, and put into a test tube containing 10 mL ultrapure water. The test tubes were shaken on a shaking table for 24 h. After shaking, a DDS-307A conductivity meter (Shanghai Yidian Scientific Instrument Limited Liability Company, Shanghai, China) was used to measure the initial conductivity R1 and the blank conductivity R0. Tubes were then incubated in a boiling water bath for 30 min, shaken for 24 h, and then the final conductivity R2 was measured. The relative conductivity E of each treatment was calculated according to the formula E (%) = (R1 − R0)/(R2 − R0) × 100. The treatment temperatures and relative conductivity were fitted with a logistic equation using SPSS 21.0 software to calculate the inflection point temperature, which is the LT50 (C in the following Formula (1)).

$$y = \frac{A}{1 + e^{B(C-X)}} + D \quad (1)$$

In this formula (1). y = relative electrolyte leakage rate (%); X = low-temperature treatment temperature (°C); A is the difference between the highest E and the lowest E, B = slope (%•°C), the slope of the curve at the inflection point temperature; C = inflection point temperature (°C); and D is the lowest value of E.

2.2.3. Determination of Relevant Physiological Indexes

The content of MDA was determined by the thiobarbituric acid (TBA) method [20]; the content of superoxide anion was determined by the hydroxylamine oxidation method [21]; SOD activity was determined by the nitroblue tetrazolium (NBT) photoreduction method [20]; POD activity was determined by the guaiacol colorimetry method [20]; the content of soluble sugar and starch were determined by anthrone colorimetry [22]; the content of soluble protein was determined by Coomassie brilliant blue G-250 staining [23]; and the content of free proline was determined by acid ninhydrin staining [21]. Refer to the corresponding literature for specific steps.

2.2.4. Statistical Analysis

The data were fitted by SPSS 21.0 software using a logistic equation to calculate the LT50 values. Microsoft Excel 2010 software was used to analyze the test data and draw the figures. SPSS 21.0 was used for one-way ANOVA, and the Duncan test ($p < 0.05$) was used for significance analysis. SPSS 21.0 was used for descriptive analysis, principal component analysis, and cluster analysis.

3. Results

3.1. Changes of LT50 of 'Tianhong 2' Apple Branches Grafted on Different Interstocks under Natural Overwintering Process

After the different low-temperature treatments, the relative conductivity of the apple branches grafted on interstocks was measured. The values were fit to a logistic equation to determine the LT50 values for each interstock, as shown in Figure 1. During overwintering stress, the LT50 changes corresponded to the change trend of winter temperature, which decreased firstly and then increased. From 22 October 2020 to 5 January 2021, with the gradual decrease in winter temperature, after adapting to the changes of external cold stress environmental conditions, the cold resistance of the branches grafted on each interstock increased and the corresponding LT50 values decreased, so the lowest LT50 values were obtained in January when the outside air temperature was the lowest and the stress was the most serious. From 5 January 2021 to 5 March 2021, as the temperature increased, the LT50 vales also increased. During overwintering, the branches of 'Tianhong 2' apple grafted on GM256 and Huang 6 showed the strongest cold resistance.

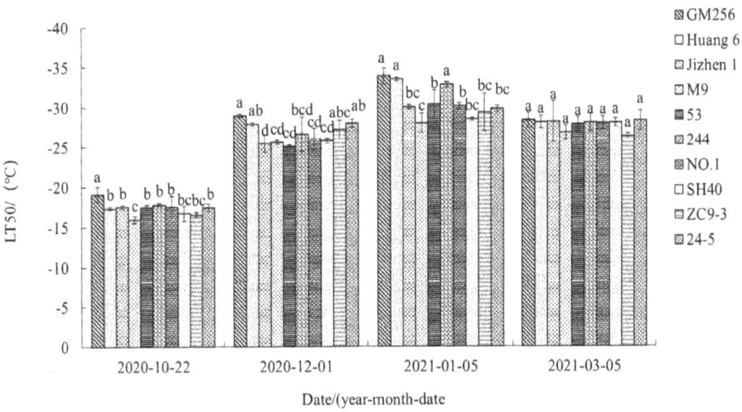

Figure 1. LT50 values of 'Tianhong 2' apple branches grafted on different interstocks. Note: In the figure, the lowercase letters indicate that significant differences exist between branches grafted on different interstocks ($p < 0.05$).

3.2. Changes of Physiological Indexes of 'Tianhong 2' Apple Branches Grafted on Different Interstocks under Natural Overwintering Process

3.2.1. Change of MDA Content

As shown in Figure 2, the change of MDA content in the branches was opposite to the change trend of winter temperature, first increasing and then decreasing. From 22 October 2020 to 5 January 2021, with the decrease in winter temperature and the intensification of low-temperature stress, the MDA content of apple branches increased, and MDA content was highest when the temperature stress was the most serious in January. From 5 January 2021 to 5 March 2021, as temperatures increased, low-temperature stress was alleviated, and the MDA content was reduced. Throughout overwintering, MDA levels remained lower in 'Tianhong 2' branches grafted on GM256 and Huang 6 and relatively higher for branches of 'Tianhong 2' grafted on ZC9-3.

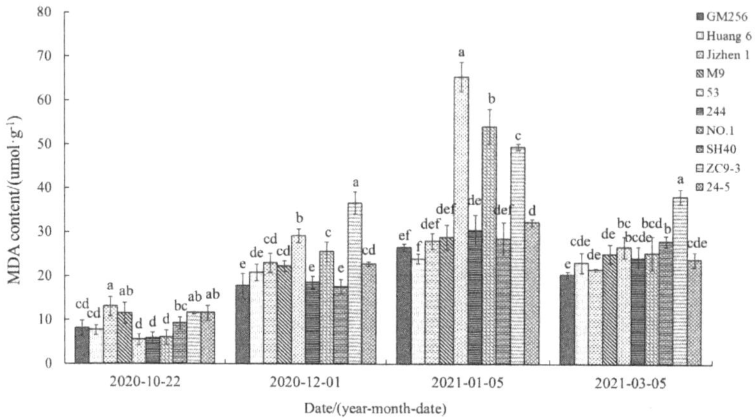

Figure 2. MDA content in 'Tianhong 2' apple branches grafted on different interstocks. Note: In the figure, the lowercase letters indicate that significant differences exist between branches grafted on different interstocks ($p < 0.05$).

3.2.2. Change of Superoxide Dismutase (SOD) Activity

As shown in Figure 3, the change of SOD activity in 'Tianhong 2' apple branches grafted on different interstocks was opposite to the change trend of winter temperature, first increasing and then decreasing. During overwintering, the SOD activity remained higher in the branches of 'Tianhong 2' grafted on GM256, Huang 6, and NO.1 and was relatively lower in the branches of 'Tianhong 2' grafted on M9 and ZC9-3.

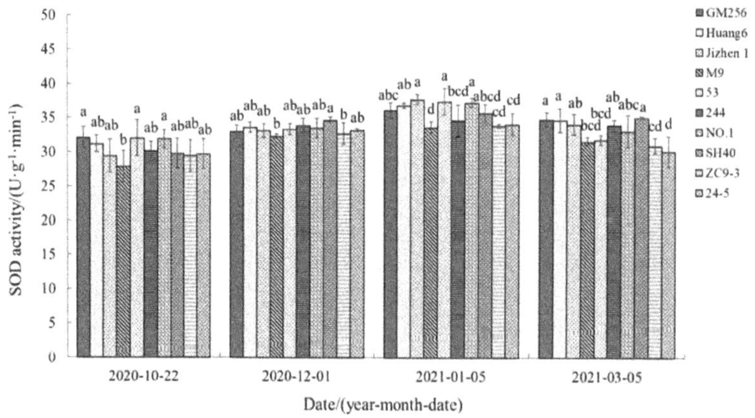

Figure 3. SOD activity in 'Tianhong 2' apple branches grafted on different interstocks. Note: In the figure, the lowercase letters indicate that significant differences exist between branches grafted on different interstocks ($p < 0.05$).

3.2.3. Change of Peroxide Dismutase (POD) Activity

Like SOD activity, POD activity in 'Tianhong 2' apple branches grafted on different interstocks was opposite to the change trend of winter temperature, first increasing and then decreasing, as shown in Figure 4. Throughout overwintering stress, POD activity remained higher for branches of 'Tianhong 2' grafted on GM256 and Huang 6 and relatively lower in branches of 'Tianhong 2' grafted on M9 and Jizhen 1.

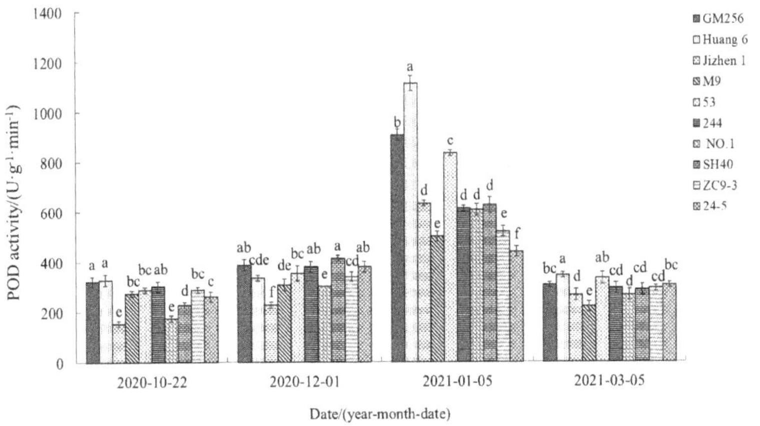

Figure 4. POD activity in 'Tianhong 2' apple branches grafted on different interstocks. Note: In the figure, the lowercase letters indicate that significant differences exist between branches grafted on different interstocks ($p < 0.05$).

3.2.4. Change of Superoxide Anion (O_2^-) Production Rate

As shown in Figure 5, the change of O_2^- production rate of 'Tianhong 2' apple branches grafted on most interstocks was opposite to the change trend of winter temperature, first increasing and then decreasing. Of the tested interstocks, the variation trend of O_2^- production rate in 'Tianhong 2' branches grafted on M9 and 24-5 was slightly different from the others. The O_2^- production rate of branches grafted on M9 continued to increase during the overwintering period but for branches grafted on 24-5 decreased and then increased. These two interstocks exhibited the highest O_2^- production rate in March. The O_2^- production rate remained relatively lower for 'Tianhong 2' branches grafted on GM256 and Huang 6 and relatively higher for branches grafted on M9, Jizhen 1 and ZC9-3.

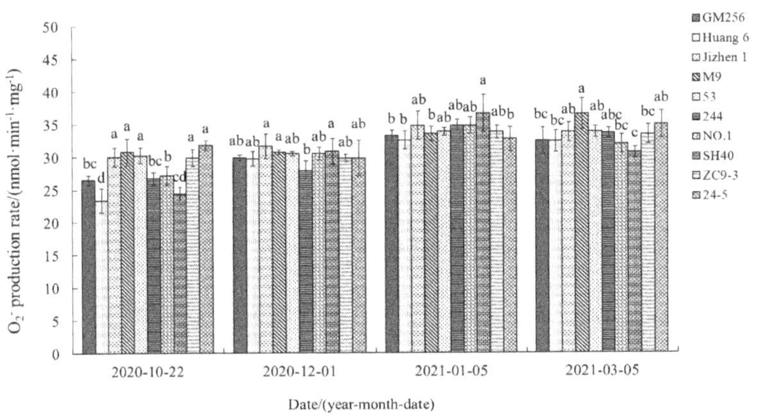

Figure 5. Superoxide anion production rate of 'Tianhong 2' apple branches grafted on different interstocks. Note: In the figure, the lowercase letters indicate that significant differences exist between branches grafted on different interstocks ($p < 0.05$).

3.2.5. Change of Starch Content

As shown in Figure 6, the content of starch increased and then decreased in 'Tianhong 2' branches grafted on GM256, Huang 6, Jizhen 1, M9, 53, 244, 1, SH40, and ZC9-3 interstocks.

The change trend of starch content in branches grafted on 24-5 was slightly different, with a continuous decreasing trend through the overwintering period. At various times throughout overwintering, there were significant differences in starch content in 'Tianhong 2' apple branches grafted on different rootstocks.

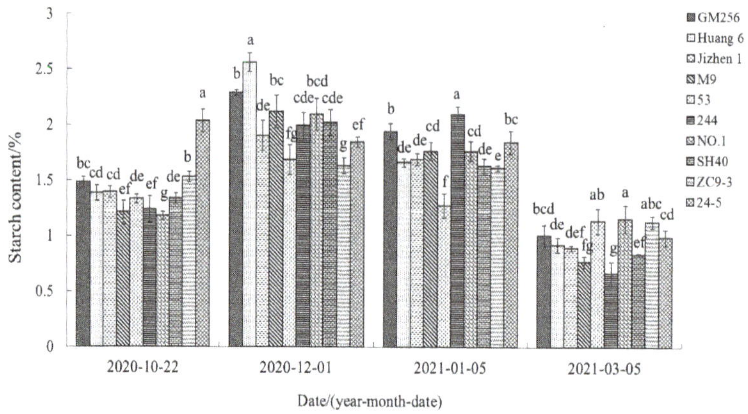

Figure 6. Starch content in 'Tianhong 2' apple branches grafted on different interstocks. Note: In the figure, the lowercase letters indicate that significant differences exist between branches grafted on different interstocks ($p < 0.05$).

3.2.6. Change of Soluble Sugar Content

As shown in Figure 7, the change of soluble sugar content in 'Tianhong 2' apple branches grafted on different interstocks was opposite to the change trend of winter temperature, first increasing and then decreasing. At all time points during overwintering, the soluble sugar content was at a relatively higher level in 'Tianhong 2' branches grafted on 244 and 24-5 and at a relatively lower level in branches of 'Tianhong 2' grafted on M9 and Jizhen 1.

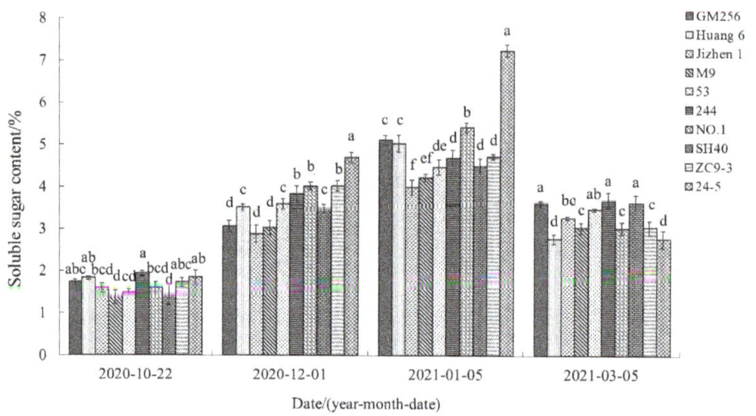

Figure 7. Soluble sugar content in 'Tianhong 2' apple branches grafted on different interstocks. Note: In the figure, the lowercase letters indicate that significant differences exist between branches grafted on different interstocks ($p < 0.05$).

3.2.7. Change of Soluble Protein Content

As shown in Figure 8, the change of soluble protein content was opposite to the change trend of winter temperature, dramatically increasing and then decreasing slightly. There was significant variation in soluble protein content of the different grafted materials, with relatively lower soluble protein content of 'Tianhong 2' branches grafted on NO.1 throughout the measurement process.

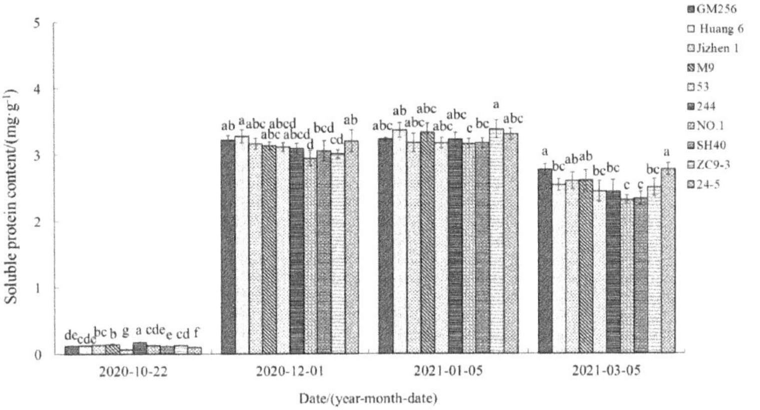

Figure 8. Soluble protein content in 'Tianhong 2' apple branches grafted on different interstocks. Note: In the figure, the lowercase letters indicate that significant differences exist between branches grafted on different interstocks ($p < 0.05$).

3.2.8. Change of Proline Content

As shown in Figure 9, the proline content changed inversely to winter temperature, first increasing and then decreasing. There were significant differences in 'Tianhong 2' apple branches grafted on the different interstocks for all measurements, with consistently lower proline content of 'Tianhong 2' branches grafted on M9.

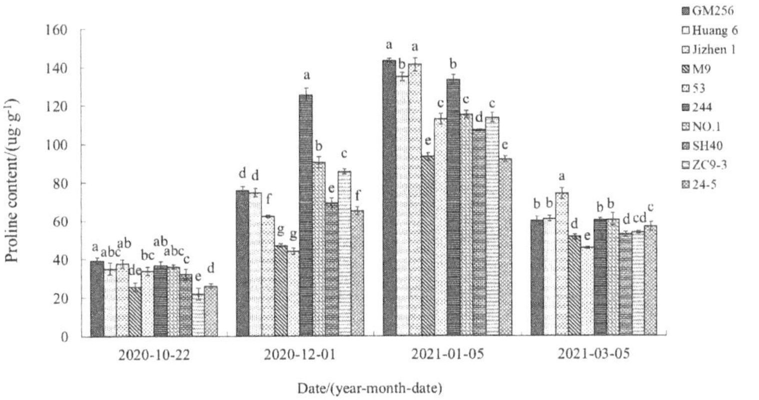

Figure 9. Proline content in 'Tianhong 2' apple branches grafted on different interstocks. Note: In the figure, the lowercase letters indicate that significant differences exist between branches grafted on different interstocks ($p < 0.05$).

3.3. *Comprehensive Evaluation of the Effects of Different Interstocks on the Cold Resistance of 'Tianhong 2' Apple Tree*

3.3.1. Correlation Analysis of Physiological Indexes of 'Tianhong 2' Apple Branches Grafted on Different Interstocks

Correlation analysis of physiological indexes of 'Tianhong 2' apple branches grafted on different interstocks was performed, and the results are shown in Table 4. LT50 values were negatively correlated with seven physiological indicators (SOD and POD enzyme activities, the production rate of superoxide anion, the contents of soluble protein, MDA, proline, and soluble sugar). SOD activity was positively correlated with six physiological indicators (POD activity, the production rate of superoxide anion, the contents of soluble protein, MDA, proline, and soluble sugar). POD activity was positively correlated with five physiological indicators (the production rate of superoxide anion, the contents of soluble protein, MDA, proline, and soluble sugar). The production rate of superoxide anion was positively correlated with four physiological indicators (the contents of soluble protein, MDA, proline, and soluble sugar). The soluble protein content was positively correlated with four physiological indicators (the contents of starch, MDA, proline, and soluble sugar). The MDA content was positively correlated with the contents of proline and soluble sugar. There were also positive correlations between starch and proline content, and between proline and soluble sugar content. The correlations between the indexes and the overlap of some information suggest that the evaluation of different interstocks on the cold resistance of 'Tianhong 2' requires multiple indexes. Principal component analysis is a comprehensive evaluation tool that can assess cold resistance by synthesizing the contributions of multiple indexes.

Table 4. Correlation Analysis between physiological indexes of 'Tianhong 2' apple branches grafted on different interstocks.

Correlation Coefficient	LT50	SOD	POD	O_2^-	Protein	MDA	Starch	Proline	Sugar
LT50	1								
SOD	−0.797 **	1							
POD	−0.630 **	0.713 **	1						
O_2^-	−0.673 **	0.455 **	0.387 *	1					
Protein	−0.931**	0.722 **	0.492 **	0.608 **	1				
MDA	−0.692 **	0.611 **	0.507 **	0.614 **	0.688 **	1			
Starch	−0.164	0.184	0.283	−0.125	0.314 *	0.04	1		
Proline	−0.799 **	0.810 **	0.807 **	0.481 **	0.726 **	0.605 **	0.382 *	1	
Sugar	−0.847 **	0.748 **	0.642 **	0.541 **	0.824 **	0.706 **	0.308	0.790 **	1

Note: * represents significant difference ($p < 0.05$), ** represents extremely significant different ($p < 0.01$).

3.3.2. Principal Component Factor Analysis of Various Indexes of 'Tianhong 2' Apple Branches Grafted on Different Interstocks

Principal component analysis was performed using SPSS; the principal component eigenvalues (Table 5) and the principal component initial factor load matrix (Table 6) were determined; and the feature vectors were calculated (Table 6). As shown in Table 5, the first two principal components with eigenvalues greater than 1 were selected by SPSS, and the cumulative contribution rate of these two components reached 78.562%, indicating that the first two principal components expressed 78.562% of the cold resistance of 'Tianhong 2' apple branches on 10 interstocks. Therefore, these two principal components can be used as comprehensive indexes to evaluate the influence of 10 interstocks on cold resistance of 'Tianhong 2'. As shown in Table 6, the absolute value of factor load of the first principal component is larger, and the order of absolute value of factor load is as follows: LT50 > soluble sugar > proline > soluble protein > SOD > MDA > POD > O_2^-. The results showed that principal component 1 mainly reflected LT50, soluble sugar content, proline content, soluble protein content, SOD enzyme activity, MDA content, POD enzyme activity, and superoxide anion production rate. The starch content of principal component 2 accounted for the largest proportion and was significantly higher than the factor load of principal component 1, indicating that principal component 2 mainly reflected the information of starch content.

Table 5. Component eigenvalue.

Principal Component	Initial Eigenvalue		
	Characteristic Root	Contribution Rate (%)	Cumulative Contribution Rate (%)
1	5.83	64.773	64.773
2	1.241	13.789	78.562

Table 6. Component Matrix.

	Load		Eigenvector	
Biochemical Indexes	Component 1	Component 2	Component 1	Component 2
LT50	−0.942	0.121	−0.390	0.109
Sugar	0.911	0.056	0.377	0.050
Proline	0.902	0.226	0.374	0.203
Protein	0.896	−0.025	0.371	−0.022
SOD	0.869	0.053	0.360	0.048
MDA	0.787	−0.291	0.326	−0.261
POD	0.768	0.232	0.318	0.208
O_2^-	0.672	−0.538	0.278	−0.483
Starch	0.287	0.861	0.119	0.773

3.3.3. Comprehensive Evaluation and Classification of Cold Resistance of 'Tianhong 2' Apple Branches Grafted on Different Interstocks

The characteristic vector value of each principal component was calculated from the eigenvalue and initial load factor, and then the principal component value was calculated from the normalized original data. The comprehensive score was then obtained according to the contribution rate of each principal component. Finally, the cold resistance was assessed based on the comprehensive scores. Interstocks with high comprehensive score have strong cold resistance, and those with low scores have poor cold resistance. The results are shown in Table 7. From strongest to weakest, the influence of each interstock on the cold resistance of 'Tianhong 2' was in the order of: GM256 > Huang 6 > 244 > NO.1 > 53 > 24-5 > ZC9-3 > Jizhen1 > SH40 > M9.

Table 7. Principal component scores, comprehensive scores, ranking, and classification.

Interstock	Principal Component Value		Comprehensive Score	Cold Resistance Order (1, Highest to 10, Lowest)	Classification
	Component 1	Component 2			
GM256	1.420503	2.516214	1.267063	1	Strong tolerance
Huang 6	0.940823	2.989505	1.021622	2	Strong tolerance
Jizhen1	−0.28465	−1.03902	−0.32764	7	Mean tolerance
M9	−2.81883	−1.96704	−2.09708	10	Weak tolerance
53	0.784528	−2.20221	0.204499	5	Relatively resistant
244	0.554611	0.944982	0.489542	3	Relatively resistant
NO. 1	0.660832	0.020357	0.430848	4	Relatively resistant
SH40	−0.95421	0.244786	−0.58431	9	Mean tolerance
ZC9-3	−0.2112	−1.60406	−0.35798	8	Mean tolerance
24-5	−0.09241	0.096477	−0.04656	6	Relatively resistant

3.4. Cluster Analysis of Cold Resistance of 'Tianhong 2' Apple Branches Grafted on Different Interstocks

Cluster analysis was carried out according to the comprehensive scores of principal component evaluation using Ward's method. The results are shown in Figure 10 and

Table 6. The 10 interstocks can be divided into four categories. The first category includes GM256 and Huang 6, with the strongest cold resistance; the second category includes 244, NO.1, 53, and 24-5, with strong cold resistance; the third category includes ZC9-3, Jizhen1, and SH40, with medium cold resistance; and the fourth category includes M9, with weak cold resistance.

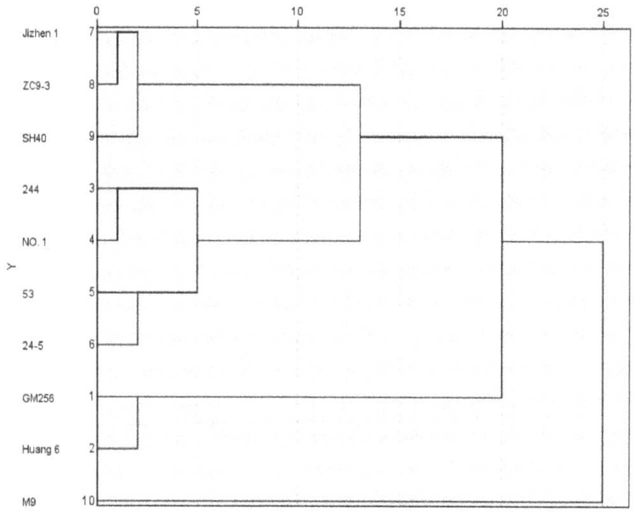

Figure 10. Cluster analysis of cold resistance of branches grafted on 10 interstocks.

4. Discussion

4.1. Relationship between LT50 and Cold Resistance

Under low-temperature stress, the permeability of plant cell membranes increases and intracellular electrolytes leak out, changing the conductivity. Thus, the determination of relative conductivity is an effective method to measure the cold resistance of plants [7], with electrical conductivity negatively correlated with plant cold resistance [24]. Measurements of conductivity can be analyzed by logistic equations to calculate the inflection point temperature and estimate the semi-lethal temperature [25]. LT50 values change with the change of external temperature, and LT50 values should be measured in the coldest month during overwintering to reliably assess apple cold resistance. In this study, the LT50 values of branches of 'Tianhong 2' grafted on GM256, Huang 6, and 244 measured in the coldest month of January were significantly lower than those of other interstocks, indicating strongest cold resistance. The LT50 values of 'Tianhong 2' branches grafted on M9 were the highest, indicating the weakest cold resistance.

4.2. Relationship between MDA Content and Cold Resistance

Under low-temperature stress, the degree of cell membrane damage and plant cold resistance can be evaluated by measuring MDA content [26]. The content of MDA in mango increased with the decrease in temperature [27]. In this study, with the decrease in temperature, the MDA content of 'Tianhong 2' apple branches grafted on different interstocks showed a gradual upward trend. MDA is the product of membrane lipid peroxidation, and the increase in MDA content indicates that low-temperature conditions increase reactive oxygen species, which intensifies the peroxidation and plant membrane damage [28]. The study on cold resistance showed that varieties with stronger cold resistance had higher MDA content, while those with weaker cold resistance had lower MDA content [29]. In this study, throughout the overwintering process, the MDA content was lowest in branches of 'Tianhong 2' grafted on GM256 and Huang 6 and highest in branches of 'Tianhong 2' grafted on ZC9-3.

4.3. Relationship between Reactive Oxygen Species, Antioxidant System, and Cold Resistance

Under low-temperature stress, the higher amounts of reactive oxygen species are produced in plants, and the excessive accumulation of reactive oxygen species will lead to membrane phospholipid peroxidation and damage to the plant [30]. SOD and POD are important components of the enzymatic defense system and important protective enzymes in plants. These enzymes can remove reactive oxygen species produced during cold injury, reduce stress damage, and protect normal plant growth [31,32]. A previous study showed that SOD and POD enzyme activities were highest and superoxide anion content was lowest in varieties with strong cold resistance after low-temperature stress treatment; the content of superoxide anion in varieties with poor cold resistance was always at a high level [33]. In this study, SOD and POD enzyme activities were higher and superoxide anion production rate was lower in rootstock-interstock-scion combinations with stronger cold resistance. SOD and POD activities were lower and superoxide anion production rate was higher in rootstock-interstock-scion combinations with poorer cold resistance. This may indicate that plants with higher protective enzyme activity can remove more reactive oxygen species in vivo, thus reduce damage to the membrane system caused by low temperature and enhancing cold tolerance of plants.

4.4. Relationship between the Content of Osmotic Adjustment Substancse and Cold Resistance

Soluble sugar, soluble protein, and proline are common osmoregulatory substances in plants [14]. The accumulation of soluble sugar can protect plants from low-temperature freezing injury and reduce the damage to plants. When subjected to low-temperature stress, plants can increase the soluble sugar content to resist the effects of the cold [34]. In this study, the soluble sugar content in all apple branches increased with the decrease in temperature and decreased with the increase in temperature. Albina found that low-temperature stress can cause a change in the soluble protein content in cells, and the change of soluble protein content is closely related to cold resistance [35]. In this study, with the decrease in temperature, the soluble protein content in all apple branches increased, and there were significant variations between the contents of soluble protein in 'Tianhong 2' apple branches grafted on different interstocks, affecting their cold resistance. Proline enhances plant cold resistance by increasing plant water retention capacity [14]. In this study, with the decrease in temperature, the proline content increased in 'Tianhong 2' apple branches grafted on all different interstocks. The proline content was higher in branches of rootstock-interstock-scion combinations with stronger cold resistance. This result is consistent with the findings of Zhang et al. [19].

4.5. Relationship between Starch Content and Cold Resistance

In this study, the starch content first increased and then decreased with the decrease in temperature. In the early stage, starch accumulates in plants, so the starch content showed an increasing trend. With low-temperature stress, plants transformed starch into soluble sugar to enhance cold resistance, so the starch content decreased. This is consistent with the findings of Bertrand et al. [36], that the starch content of alfalfa decreases with the progress of low-temperature acclimation. However, other studies did not observe a decrease in starch content in response to low-temperature changes [37], suggesting that this process may depend on the characteristics of different plant species.

5. Conclusions

The tested interstocks exhibited different effects on the cold resistance of 'Tianhong 2' apple. The apple interstocks exhibited cold resistance in the order of: GM256 > Huang 6 > 244 > NO.1 > 53 > 24-5 > ZC9-3 > Jizhen1 > SH40 > M9. The cold resistance of these 'Tianhong 2' interstocks can be divided into four categories. GM256 and Huang 6 exhibit the strongest cold resistance; 244, NO.1, 53, and 24-5 exhibit strong cold resistance; ZC9-3, Jizhen1, and SH40, exhibit medium cold resistance; and M9 showed weak cold resistance.

Author Contributions: J.J. wrote the paper, M.L. carried out the experiment, B.Y. conducted the data analysis, B.L. and Z.L. managed the material, J.X. provided the material, X.Z. and S.Z. designed the experiment. All authors have read and agreed to the published version of the manuscript.

Funding: This research was funded by the National Natural Science Foundation of China (32002008), the Natural Science Foundation of Hebei Province (C2020204015), the Key Research and Development Program of Hebei Province (19226317D), Basic scientific research funds for colleges and universities in Hebei Province (KY2021058), and the Key Research and Development Plan Project of Hebei Province (21326308D-02-03).

Data Availability Statement: Data sharing not applicable: No new data were created or analyzed in this study. Data sharing is not applicable to this article.

Acknowledgments: We thank Zhixue Hu for the grafting of the apple trees.

Conflicts of Interest: The authors have no conflicting interests, and all authors have approved the manuscript and agree with its submission to Horticulturae.

Abbreviations

MDA	Malondialdehyde
REL	Relative electrolyte leakage
LT50	Semi-lethal temperature
SOD	Superoxide dismutase
POD	Peroxidase
$\cdot O_2^-$	Superoxide anion free radical
TBA	Thiobarbituric acid
NBT	Nitroblue tetrazolium

References

1. Peng, Z.; Fu, D. Effects of 1-methylcyclopropene treatment on the quality of red 'Fuji' apples fruit during short-term storage. *Food Qual. Saf.* **2022**, *7*, fyac074. [CrossRef]
2. Wang, N.; Wolf, J.; Zhang, F. Towards sustainable intensification of apple production in China—Yield gaps and nutrient use efficiency in apple farming systems. *J. Integr. Agric.* **2016**, *15*, 716–725. [CrossRef]
3. Shen, X.; Ping, Y.; Bao, C.; Liu, C.; Tahir, M.; Li, X.; Song, Y.; Xu, W.; Ma, F.; Guan, Q. Mdm-miR160-MdARF17-MdWRKY33 module mediates freezing tolerance in apple. *Plant J.* **2023**, *114*, 262–278. [CrossRef] [PubMed]
4. Aydin, S.S.; Büyük, I.; Aras, S. Relationships among lipid peroxidation, sod enzyme activity, and sod gene expression profile in lycopersicum esculentum l. exposed to cold stress. *Genet. Mol. Res.* **2013**, *12*, 3220–3229. [CrossRef] [PubMed]
5. Viswanathan, C.; Zhu, J.K. Molecular genetic analysis of cold–regulated gene transcription. *Philos. Trans. R. Soc. Ser. B* **2002**, *357*, 877–886. [CrossRef]
6. Marini, R.P.; Anderson, J.L.; Autio, W.R.; Barritt, B.H.; Unrath, R. Performance of 'Gala' apple trees on 18 dwarfing rootstocks: Ten-year summary of the 1994 nc-140 rootstock trial. *J. Am. Pomol. Soc.* **2006**, *60*, 69–83.
7. Wang, Y.X.; Ya, H.U.; Chen, B.H.; Zhu, Y.F.; Dawuda, M.M.; Svetla, S. Physiological mechanisms of resistance to cold stress associated with 10 elite apple rootstocks. *J. Integr. Agric.* **2018**, *17*, 857–866. [CrossRef]
8. Zhou, S.; Shen, Z.; Yin, B.; Liang, B.; Li, Z.; Zhang, X.; Xu, J. Effects of Dwarfifing Interstock Length on the Growth and Fruit of Apple Tree. *Horticulturae* **2023**, *9*, 40. [CrossRef]
9. Wang, Z.; Wang, Y.; Wu, D.; Hui, M.; Han, X.; Xue, T.; Yao, F.; Gao, F.; Cao, X.; Li, H.; et al. Identification and Regionalization of Cold Resistance of Wine Grape Germplasms (*V. vinifera*). *Agriculture* **2021**, *11*, 1117. [CrossRef]

10. Melo, P.M.D.; Calvin, F.P.; Sage, R.F. Winter cold-tolerance thresholds in field-grown miscanthus hybrid rhizomes. *J. Exp. Bot.* **2015**, *14*, 4415–4425.
11. Lee, J.; Yu, D.; Lee, J.; Kim, S.; Lee, H. Comparison of mid-Winter cold-hardiness and soluble sugars contents in the shoots of 21 highbush blueberry (*Vaccinium corymbosum*) cultivars. *J. Hortic. Sci. Biotechnol.* **2013**, *88*, 727–734. [CrossRef]
12. Cansev, A.; Gulen, H.; Eris, A. The activities of catalase and ascorbate peroxidase in olive (*Olea europaea* L. cv. Gemlik) under low temperature stress. *Hortic. Environ. Biotechnol.* **2011**, *52*, 113–120. [CrossRef]
13. Sheng, L.; Sun, X.; Mo, C.; Hao, M.; Wei, X.; Ma, A. Relationship between antioxidant enzymes and sclerotial formation of *Pleurotus tuber-regium* under abiotic stress. *Appl. Microbiol. Biotechnol.* **2023**, *107*, 1391–1404. [CrossRef] [PubMed]
14. Saruhan, N.; Turgut-Terzi, R.; Kadioglu, A. The effects of exogenous polyamines on some biochemical changes during drought stress in *Ctenanthe setosa* (Rosc.) Eichler. *Acta Biol. Hung.* **2018**, *57*, 221–229. [CrossRef] [PubMed]
15. Sun, X.; Sun, Z.; Saleh, A.S.M.; Lu, Y.; Zhang, X.; Ge, X.; Shen, H.; Yu, X.; Li, W. Effects of various microwave intensities collaborated with different cold plasma duration time on structural, physicochemical, and digestive properties of lotus root starch. *Food Chem.* **2023**, *405*, 134837. [CrossRef]
16. Wang, J.W.; Ma, L.Y.; Gómez-Del-Campo, M.; Zhang, D.S.; Deng, Y.; Jia, Z.K. Youth tree behavior of olive (*Olea europaea* L.) cultivars in wudu, China: Cold and drought resistance, growth, fruit production, and oil quality. *Sci. Hortic.* **2018**, *236*, 106–122. [CrossRef]
17. Hu, Y.; Wang, Y.; Zhu, Y.; Chen, B. Multivariate statistical analyses and predictive model of cold resistance associated with eleven crabapples and fuji apple. *Cryoletters* **2018**, *39*, 235–244.
18. Wilner, J.J. Note on an electrolytic procedure for differentiating between frost injury of roots and shoots in woody plants. *Can. J. Plant Sci.* **1959**, *39*, 512–513. [CrossRef]
19. Zhang, B.Q.; Yang, L.T.; Li, Y.R. Physiological and Biochemical characteristics related to cold resistance in sugarcane. *Sugar Tech* **2015**, *17*, 49–58. [CrossRef]
20. Du, F.; Shi, H.; Zhang, X.; Xu, X. Responses of Reactive Oxygen Scavenging Enzymes, Proline and Malondialdehyde to Water Deficits among Six Secondary Successional Seral Species in Loess Plateau. *PLoS ONE* **2014**, *9*, e98872. [CrossRef]
21. Wu, Z.; Wang, X.; Wang, X.; Yan, C.; Ma, C.; Liu, L.; Dong, S. Effects of the Soil Moisture Content on the Superoxide Anion and Proline Contents in Soybean Leaves. *Legume Res.* **2022**, *45*, 315–318. [CrossRef]
22. Qian, J.; Zhou, J.; Di, B.; Liu, Y.; Zhang, G.; Yang, X. Using electrical impedance tomography for rapid determination of starch and soluble sugar contents in *Rosa hybrida*. *Sci. Rep.* **2021**, *11*, 2871. [CrossRef]
23. Shahi, S.; Katiyar, R.K.; Bhatnagar, A.K.; Singh, A.B. Soluble and nonsoluble protein assay for antigenic extracts from pollen and seeds of mustard (*Brassica* spp.). *Allergy Asthma Proc.* **2008**, *29*, 78–87. [CrossRef] [PubMed]
24. Ji, L.; Li, P.; Su, Z.; Li, M.; Guo, S. Cold-tolerant introgression line construction and low-temperature stress response analysis for bell pepper. *Plant Signal. Behav.* **2020**, *15*, e1773097. [CrossRef]
25. Lyons, J.M. Chilling injury in plants. *Annu. Rev. Plant Physiol.* **1973**, *24*, 445–466. [CrossRef]
26. Campos, P.S.; nia Quartin, V.; chicho Ramalho, J.; Nunes, M.A. Electrolyte leakage and lipid degradation account for cold sensitivity in leaves of *Coffea* sp. plants. *J. Plant Physiol.* **2003**, *160*, 283–292. [CrossRef]
27. Zhang, Y.; Li, Y.; Yang, J.; Yang, X.; Chen, S.; Xie, Z.; Zhang, M.; Huang, Y.; Zhang, J.; Huang, X. Genome-Wide Analysis and Expression of Cyclic Nucleotide-Gated Ion Channel (CNGC) Family Genes under Cold Stress in Mango (*Mangifera indica*). *Plants* **2023**, *12*, 592. [CrossRef]
28. Sato, Y.; Masuta, Y.; Saito, K.; Murayama, S.; Ozawa, K. Enhanced chilling tolerance at the booting stage in rice by transgenic overexpression of the ascorbate peroxidase gene, OsAPXa. *Plant Cell Rep.* **2011**, *30*, 399–406. [CrossRef]
29. Saadati, S.; Baninasab, B.; Mobli, M.; Gholami, M. Cold Tolerance in Olive Leaves of Three Cultivars Related to Some Physiological Parameters during Cold Acclimation and De-Acclimation Stages. *J. Agric. Sci. Technol.* **2020**, *22*, 1313–1326.
30. Boyer, J.S. Grain yields with limited water. *J. Exp. Bot.* **2004**, *55*, 2385–2394. [CrossRef]
31. Javadian, N.; Karimzadeh, G.; Mahfoozi, S.; Ghanati, F. Cold-induced changes of enzymes, proline, carbohydrates, and chlorophyll in wheat. *Russ. J. Plant Physiol.* **2010**, *57*, 540–547. [CrossRef]
32. Türkan, İ.; Bor, M.; Özdemir, F.; Koca, H. Differential responses of lipid peroxidation and antioxidants in the leaves of drought-tolerant *P. acutifolius* Gray and drought-sensitive *P. vulgaris* L. subjected to polyethylene glycol mediated water stress. *Plant Sci.* **2005**, *168*, 223–231. [CrossRef]
33. Sun, B.; Liu, G.; Phan, T.; Yang, L.; Li, Y.; Xing, Y. Effects of Cold Stress on Root Growth and Physiological Metabolisms in Seedlings of Different Sugarcane Varieties. *Sugar Tech* **2017**, *19*, 165–175. [CrossRef]
34. Bao, J.; Zhang, S. Changes in germination, storage materials and antioxidant enzyme activities in pear (*Pyrus betulaefolia* Bge. and *Pyrus calleryana* Dcne.) stock seeds during cold stratification. *Seed Sci. Technol.* **2011**, *39*, 655–659. [CrossRef]
35. Albina, A.; Wang, Q.Y.; Ludmila, K.; Peter, N. Is microtubule disassembly a trigger for coldacclimation? *Plant Cell Physiol.* **2003**, *44*, 676–686.

36. Bertrand, A.; Bipfubusa, M.; Claessens, A.; Rocher, S.; Castonguay, Y. Effect of photoperiod prior to cold acclimation on freezing tolerance and carbohydrate metabolism in alfalfa (*Medicago sativa* L.). *Plant Sci.* **2017**, *264*, 122–128. [CrossRef]
37. Castonguay, Y.; Bertrand, A.; Michaud, R.; Laberge, S. Cold-Induced Biochemical and Molecular Changes in Alfalfa Populations Selectively Improved for Freezing Tolerance. *Crop Sci.* **2011**, *51*, 213. [CrossRef]

horticulturae

MDPI

Article

Optimization of In Vitro Regeneration Protocol of Tomato cv. MT1 for Genetic Transformation

Shiuli Ahmed [1,2], Wan Aina Sakeenah Wan Azizan [1], Md. Abdullah Yousuf Akhond [3], Abdul Shukor Juraimi [4], Siti Izera Ismail [5], Razu Ahmed [6] and Muhammad Asyraf Md Hatta [1,*]

1 Department of Agriculture Technology, Faculty of Agriculture, Universiti Putra Malaysia, Serdang 43400, Selangor, Malaysia
2 Biotechnology Division, Bangladesh Agricultural Research Institute, Gazipur 1701, Bangladesh
3 Research Wing, Bangladesh Agricultural Research Institute, Gazipur 1701, Bangladesh
4 Department of Crop Science, Faculty of Agriculture, Universiti Putra Malaysia, Serdang 43400, Selangor, Malaysia
5 Department of Plant Protection, Faculty of Agriculture, Universiti Putra Malaysia, Serdang 43400, Selangor, Malaysia
6 Horticulture Research Centre, Bangladesh Agricultural Research Institute, Gazipur 1701, Bangladesh
* Correspondence: m.asyraf@upm.edu.my

Abstract: The tomato (*Solanum lycopersicum* L.) is a major crop of global economic significance. The characterization of genes associated with agriculturally important traits is often performed using genetic transformation. To achieve an efficient transformation protocol, three components are required, namely, a regenerable target tissue, a DNA delivery method, and a robust transformant selection system. The present study was conducted to optimize the in vitro regeneration protocol for the tomato cv. MT1. The regeneration capacity of hypocotyl and cotyledon explants was evaluated using a total of 20 concentration combinations of two plant growth regulators (PGRs) added into the basal MSB5 medium, namely, 6-benzylaminopurine (BAP) (0, 1, 2, 3, and 4 mg/L) and indole-3-acetic acid (IAA) (0, 0.05, 0.1, and 0.5 mg/L). The optimal PGRs combinations for the cotyledons and hypocotyls were MSB5 supplemented with 2 mg/L BAP and 0.5 mg/L IAA and MSB5 supplemented with 2 mg/L BAP and 0.1 mg/L IAA, respectively. To determine the minimum inhibitory concentration (MIC) of kanamycin, eight different concentrations (0, 50, 75, 100, 125, 150, 175, and 200 mg/L) were added to the MSB5 supplemented with 2 mg/L BAP and 0.5 mg/L IAA. The MIC for the cotyledons and hypocotyls were determined to be 50 mg/L and 100 mg/L, respectively.

Keywords: tomato; *Solanum lycopersicum*; in vitro regeneration; hypocotyl; cotyledon; kanamycin

1. Introduction

The tomato (*Solanum lycopersicum* L.) is one of the most economically important crops and is widely cultivated around the world. In terms of total production, it is the most popular vegetable crop in the world. In 2020, the global production of tomato was 186.82 million metric tons, of which 1.25 million metric tons were produced in Cameron Highlands, Malaysia [1,2]. The crop is regarded as a model plant for dicots and is used for gene functional studies that can be applied to other crops [3,4].

Crop genetic transformation is an important tool for studying the role of genes that govern key traits, including biotic factors, abiotic factors, quality, and yield. There are three major components to achieve an efficient transformation system, namely, a regenerable target tissue, DNA delivery method, and a robust selection method for selecting transformants. *Agrobacterium*–mediated transformation has been the most common method for delivering DNA into the tomato genome [5]. In this approach, hypocotyl [6] and cotyledon [7] have been used as the starting materials and the regeneration efficiency varies among genotypes [8–10].

The tomato cv. MT1 developed by the Malaysian Agricultural Research and Development Institute (MARDI) is grown in lowland areas of Malaysia [11]. Most of the previous studies on regeneration of this cultivar used only cotyledons as explants and zeatin as a plant growth regulator (PGR) [12,13]. 6-Benzylaminopurine (BAP) is cheaper than zeatin. A simple protocol that utilizes other potential explants and alternative growth regulators may reduce the regeneration time and cost and increase its efficiency.

A competent selection system for distinguishing transformed cells from non-transformed cells is required for crop transformation. Typically, antibiotic resistance is utilized for the selection of regenerated putative transformants. For this purpose, antibiotic resistance genes are often incorporated into transformation vector to select plant transformants [14,15]. The neomycin phosphotransferase II (*npt II*) gene, which confers resistance to kanamycin, is the most commonly used selectable marker gene for screening transformants [16]. Through expression of the incorporated antibiotic resistance gene, successful transformants can detoxify a specific antibiotic [16,17]. Thus, the selection of positive transformants is achieved using an antibiotic at the minimum required concentration in the regeneration medium, also known as the selection medium. The toxicity of an antibiotic depends on the organism, species, and concentration [18–20]. Plant regeneration and growth are inhibited at higher concentrations of antibiotics [17,21]. Prior to transformation, the minimum inhibitory concentration (MIC) of the antibiotic for non-transformed plants must be determined to ascertain the level of resistance [22].

In the genetic transformation of tomato cv. MT1, the hygromycin resistance gene is often used as a selectable marker gene [23,24]. Although a study described the use of kanamycin as a selection system in this cultivar [13], no information regarding the determination of the MIC of kanamycin has been documented to date.

Therefore, the current experiments were designed to optimize the regeneration protocol from the cotyledon and hypocotyl explants. The optimal concentration of indole-3-acetic acid (IAA) for rooting media and the MIC of kanamycin for tomato cv. MT1 were determined.

2. Materials and Methods

2.1. Preparation of Plant Material

Seeds of tomato cv. MT1 were presoaked for 1 h in sterile distilled water. The seeds were then rinsed well with sterile distilled water and soaked in 70% ethanol for 30 s, then rinsed again with sterile distilled water. The seeds were then soaked in a 2% sodium hypochlorite (bleach) solution, added with 25 µL of Tween-20 per 100 mL of bleach solution. The seeds were stirred continuously in the bleach solution for 15 min. The bleach-treated seeds were then rinsed four times with sterile distilled water and germinated on full strength MS (Murashige and Skoog) medium [25]. The explants (cotyledons and hypocotyls) were prepared aseptically from 13–15-day-old seedlings. Cotyledonary nodes with shoot tips and epicotyls were removed. The cotyledons were cut at both ends, maintaining an explant size of 0.5 cm (approximately). The hypocotyls explants were also cut into 0.5 cm (approximately) pieces. Two cotyledons and 2–3 hypocotyl explants were collected from each seedling and placed on the media. In a shoot induction experiment for optimization of the regeneration protocol, the explants from the hypocotyls and cotyledons were referred as H and C, respectively.

2.2. Culture Media for Shoot Induction

Murashige and Skoog media, including Gamborg B5 vitamins (MSB5), were used as a basal medium [25,26] supplemented with different concentrations of BAP and IAA. Through a literature search on the regeneration of tomato using BAP and IAA [27], concentrations of BAP in the amount of 0, 1, 2, 3, and 4 mg/L and IAA of 0, 0.05, 0.1, and 0.5 mg/L were selected and used in the current study, respectively. All 20 possible combinations of the BAP and IAA concentrations were used for shoot regeneration (Table 1). For root induction, five concentrations of IAA (0, 0.25, 0.5, 0.75, and 1.0 mg/L) were used with the

MS medium to determine the optimal concentration of IAA for robust and effective root induction.

The combination of MSB5 + 2 mg/L BAP + 0.5 mg/L IAA (T18) was found to be the best medium from the experiment on the optimization of the regeneration protocol. Therefore, it was used as a basal/control medium for subsequent experiments. The regeneration medium was supplemented with different concentrations of kanamycin (0, 50, 75, 100, 125, 150, 175, and 200 mg/L).

Table 1. List of treatment combinations of plant growth regulators (PGRs) and explants.

Treatment	BAP + IAA (mg/L)	Hypocotyls (E1) on Corresponding Treatment	Cotyledons (E2) on Corresponding Treatment
T1	0 + 0	HT1	CT1
T2	1 + 0	HT2	CT2
T3	2 + 0	HT3	CT3
T4	3 + 0	HT4	CT4
T5	4 + 0	HT5	CT5
T6	0 + 0.05	HT6	CT6
T7	0 + 0.1	HT7	CT7
T8	0 + 0.5	HT8	CT8
T9	1 + 0.05	HT9	CT9
T10	2 + 0.05	HT10	CT10
T11	3 + 0.05	HT11	CT11
T12	4 + 0.05	HT12	CT12
T13	1 + 0.1	HT13	CT13
T14	2 + 0.1	HT14	CT14
T15	3 + 0.1	HT15	CT15
T16	4 + 0.1	HT16	CT16
T17	1 + 0.5	HT17	CT17
T18	2 + 0.5	HT18	CT18
T19	3 + 0.5	HT19	CT19
T20	4 + 0.5	HT20	CT20

H = Hypocotyl, C = Cotyledon.

2.3. Culture and Subculture

Both the cotyledon and hypocotyl explants were cultured on MSB5 medium supplemented with different concentration combinations of BAP and IAA. The plates with the explants were incubated at 25 ± 2 °C, maintaining 16/8 h light/dark stages. The explants were sub-cultured every two weeks onto fresh regeneration medium containing the same plant growth regulators (PGRs) combinations. The cultures were maintained using fluorescent light of 20,000–25,000 lux intensity at 25 ± 2 °C and a cycle of 16/8 (light/dark) hours.

Both the cotyledon and hypocotyl explants were cultured on the best regeneration medium, MSB5 + 2 mg/L BAP + 0.5 mg/L IAA (T18) supplemented with different concentrations of kanamycin. The plates with the explants were incubated at 25 ± 2 °C, maintaining 16/8 h light/dark stages. The explants were sub-cultured every 2 weeks onto similar fresh regeneration medium. Induced shoots were transferred to MS medium for elongation. After two weeks on the elongation medium, the shoots were transferred to root-inducing media. Ten shoots were used per replication for each treatment. The plantlet were kept in the root-inducing media for three weeks.

2.4. Shoot Elongation, Rooting, and Acclimatization

Individual shoots > 0.5 cm were excised after regeneration and transferred to fresh medium for elongation. Three types of media (MSB5 + 2 mg/L BAP + 0.1 mg/L IAA, MSB5 + 1 mg/L BAP + 0.1 mg/L IAA, and MSB5) were used for shoot elongation.

Shoots derived from the hypocotyl and cotyledon explants in the optimization of the regeneration protocol experiment were referred to as SH and SC, respectively. The shoots were transferred to root-inducing media after two weeks on the elongation medium. Five concentrations of IAA (0.0, 0.25, 0.5, 0.75, and 1.0 mg/L) were used in MS medium for the root initiation experiment. Ten shoots were used per replication for each treatment. The plantlets were kept in the root-inducing media for three weeks.

The shoots derived from three types of media (MSB5 + 2 mg/L BAP + 0.5 mg/L IAA, MSB5 + 2 mg/L BAP + 0.5 mg/L IAA + 50 mg/L kanamycin, and MSB5 + 2 mg/L BAP + 0.5 mg/L IAA + 75 mg/L kanamycin) were separated into three groups (S1, S2, and S3, respectively). Three media (T1 = MS, T2 = MS + 0.75 mg/L IAA, and T3 = MS +1 mg/L IAA) were used for root induction on the three types of shoots.

The rooted plantlets were kept in the greenhouse without removal from the rooting medium for five days to adapt to the environment and then transferred to autoclaved soil medium (Nursery King, Tray substrate, 010FV03, The Netherlands). The pots with plantlets were kept covered with transparent plastic bags for one week. After one week, the covers were removed, and the plantlets were kept in the greenhouse. Well-established plants were transferred to an open field three weeks after transfer to the soil medium.

2.5. *Experimental Design, Data Collection, and Statistical Analysis*

The experiment was set up in a completely randomized design (CRD) with four replications; in each replication, 16 explants were inoculated in a sterile petri dish containing the regeneration medium. Each experiment was repeated three times. The average data from those three experiments was used for analysis.

The selected parameters for collecting data were percent (%) of explants producing direct shoots, type of induced organ, number of shoots per explant, days to root, number of roots per shoot, root length (cm), root diameter (cm), and % survival of plants in soil. The data were collected by closely observing the cultures. The types of induced organ were recorded through visual observation. The induced organs with a growing tip were considered to be shoots (S = shoot), and the induced organs with no growing tip were recorded as leaves (L = leaf only); explants that had induced roots (R) with or without a shoot and/or leaf were also recorded. The priority of induced organs by a specific explant in response to a specific growth regulator combination was recorded and indicated with sequence of S, L, and R (RS/SR/SL/SRL). The % of explants producing shoots and number of shoots per explant was counted at the completion of six weeks of culture, when the individual shoots were clearly identifiable with growing shoot tips [27]. Only directly regenerated shoots formed without a callus were counted. The number of roots and the root length and root diameter were counted and measured with a slide caliper after three weeks in the root-induction media. Three weeks after planting, the survival rate (%) of the rooted shoots in the soil was recorded.

The data were analyzed using SAS software version 9.3. A CRD 2-factor analysis was performed to compare the individual and interaction effects of two factors. The type of explant (cotyledon and hypocotyl) was considered one factor and the growth regulator combinations (BAP and IAA) were considered another factor for the regeneration experiment. For the MIC of the kanamycin experiment, the explant was considered as one factor and the concentration of kanamycin was considered another factor. Tukey's HSD test was used to compare means at $p < 0.05$.

3. Results and Discussion

3.1. *Regeneration of Shoot*

3.1.1. Percentage of Explants Producing Direct Shoots

Significantly different responses were observed among the different types of explants to different concentration combinations of the growth regulators (Figure 1 and Table 2). The percentage of explants forming direct shoots without forming a callus varied from 0 to 98.50%. The highest percentage (98.50%) of hypocotyls and cotyledons producing shoots was observed on T14 (MSB5 + 2 mg/L BAP + 0.1 mg/L IAA) and T18 (MSB5 + 2 mg/L BAP + 0.5 mg/L IAA), respectively (Figure 1). Shoot regeneration was not observed on the hypocotyl explants treated with T8 (MSB5 + 0 mg/L BAP + 0.5 mg/L IAA). A similar observation was recorded for the cotyledon explants treated with T1 (basal MSB5 medium with no growth regulators), T6 (MSB5 + 0 mg/L BAP + 0.05 mg/L IAA), T7 (MSB5 + 0 mg/L BAP + 0.1 mg/L IAA), and T8 (MSB5 + 0 mg/L BAP + 0.5 mg/L IAA). Only 31.00% of the hypocotyls produced shoots on T1 (basal MSB5 medium with no growth regulators).

In general, a higher number of shoots was observed in the hypocotyl explants in comparison to the cotyledons. A statistical analysis revealed that both the single and interaction effects were highly significant for the percentage of explants that regenerated shoots. This suggests a significant difference in the shoot regeneration capacity of the hypocotyl and cotyledon explants. Similarly, Billah et al. [28] reported that a higher percentage of hypocotyls regenerated shoots than cotyledons. Zaman et al. [29] discovered a contrasting observation, wherein a greater percentage of shoots was regenerated from cotyledons rather than hypocotyls. In a previous study conducted by Sarker et al. [30], it was demonstrated that the optimal combination for shoot regeneration in tomato plants was MS medium supplemented with 1 mg/L BAP + 0.1 mg/L IAA from cotyledon explants, which is different from the present study and is likely due to varietal response.

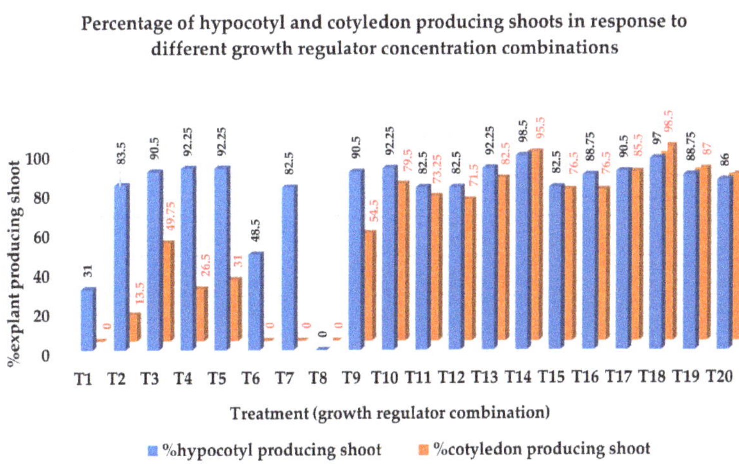

Figure 1. Percentage (%) of hypocotyl and cotyledon explants producing shoots on different concentration combinations of BAP and IAA in MSB5 medium.

Table 2. Response of MT1 explants to different concentration combinations of BAP and IAA.

Treatment	% Explants Producing Direct Shoots	% Explants Producing Root	No. of Shoots per Explant	Type of Induced Organ
H = Hypocotyl	79.63 a ± 2.73	22.19 a ± 3.97	2.73 a ± 0.13	-
C = Cotyledon	54.28 b ± 3.95	17.11 b ± 3.94	2.59 a ± 0.19	-
Level of significance	**	**	ns	-
MSD value	1.06	0.6	0.14	-
CV (%)	5.04	9.74	17.08	-
HT1	31.00 k ± 2.45	62.50 b ± 2.55	1.00 ef ± 0	RS
HT2	83.50 d–g ± 1.50	7.81 cd ± 1.56	2.25 de ± 0.25	SR
HT3	90.50 a–e ± 2.02	0 e ± 0	2.25 de ± 0.25	S
HT4	92.25 a–d ± 1.75	0 e ± 0	3.25 cd ± 0.25	SCL
HT5	92.25 a–d ± 1.75	0 e ± 0	3.25 cd ± 0.25	SCL
HT6	48.50 j ± 1.50	100 a ± 0	2.25 de ± 0.25	RS

Table 2. *Cont.*

Treatment	% Explants Producing Direct Shoots	% Explants Producing Root	No. of Shoots per Explant	Type of Induced Organ
HT7	82.50 e–h ± 1.50	100 a ± 50	3.25 cd ± 0.25	RS
HT8	0 m ± 0	100 a ± 0	0.00 f ± 0	R
HT9	90.50 a–e ± 2.02	6.25 d ± 0	3.25 cd ± 0.25	SR
HT10	92.25 a–d ± 1.75	0 e ± 0	2.25 de ± 0.25	S
HT11	82.50 e–h ± 1.50	0 eh ± 0	2.25 de ± 0.25	LCS
HT12	82.50 e–h ± 1.50	0 e ± 0	2.25 de ± 0.25	LCS
HT13	92.25 a–d ± 1.75	9.38 cd ± 1.80	2.25 de ± 0.25	SRL
HT14	98.50 a ± 1.50	10.94 cd ± 1.56	5.25 ab ± 0.25	SRL
HT15	82.50 e–h ± 1.50	6.25 d ± 0	2.25 de ± 0.25	SRL
HT16	88.75 b–f ± 1.75	6.25 d ± 0	3.25 cd ± 0.25	SRL
HT17	90.50 a–e ± 2.02	12.50 c ± 2.55	3.25 cd ± 0.25	SRL
HT18	97.00 ab ± 1.73	7.81 cd ± 1.56	4.25 bc ± 0.25	SRL
HT19	88.75 b–f ± 1.75	7.81 cd ± 1.56	3.25 cd ± 0.25	SRLC
HT20	86.00 efg ± 1.00	6.25 d ± 0	3.25 cd ± 0.25	SRLC
CT1	0 m ± 0	20.31 c ± 0	00 f ± 0	R
CT2	13.50 l ± 1.50	0 e ± 0	1.00 ef ± 0	S
CT3	49.75 j ± 2.66	0 e ± 0	2.25 de ± 0.25	S
CT4	26.50 k ± 1.50	0 e ± 0	2.25 de ± 0.25	LSC
CT5	31.00 k ± 2.45	0 e ± 0	2.25 de ± 0.25	LSC
CT6	0 m ± 0	100 a ± 0	00 f ± 0	R
CT7	0 m ± 0	100 a ± 0	00 f ± 0	R
CT8	0 m ± 0	100 a ± 0	00 f ± 0	R
CT9	54.50 j ± 2.87	0 e ± 0	2.25 de ± 0.25	SL
CT10	79.50 f–i ± 1.50	0 e ± 0	4.25 bc ± 0.25	SL
CT11	73.25 hi ± 1.75	0 e ± 0	3.25 cd ± 0.25	LS
CT12	71.50 i ± 2.02	0 e ± 0	3.25 cd ± 0.25	LS
CT13	82.50 e–h ± 1.50	6.25 d ± 0	2.25 de ± 0.25	SRL
CT14	95.50 abc ± 1.50	0 e ± 0	4.25 bc ± 0.25	SL
CT15	76.50 ghi ± 1.50	0 e ± 0	3.25 cd ± 0.25	LCS
CT16	76.50 ghi ± 1.50	0 ei ± 0	4.25 bc ± 0.25	LCS
CT17	85.50 d–g ± 1.50	10.94 cd ± 1.56	4.25 bc ± 0.25	SRL
CT18	98.50 a ± 1.50	9.38 cd ± 1.80	6.25 a ± 0.25	SRL
CT19	87.00 c–f ± 2.44	7.81 cd ± 1.56	3.25 cd ± 0.25	LSRC
CT20	84.00 d–g ± 1.73	7.81 cd ± 1.56	3.25 cd ± 0.25	LSRC
MSD value	9.28	5.42	1.28	-
CV (%)	4.91	9.77	17.01	-
Level of (E × T) significance	**	**	**	-

Means followed by the same letters in a column are not significantly different at a 5% level. MSD = minimum significant difference, CV = coefficient of variation, H = hypocotyl, C = cotyledon. Note: ns—not significant at $p > 0.05$ and **—significant at $p \leq 0.01$; ±—standard error (n = 4); (as ANOVA). C = callus, R = root, S = shoot, L = leaf; the sequence of C, R, S, and L indicates the precedence type of organ.

3.1.2. Percentage of Explants Producing Roots

The highest percentage of root formation (100%) in both types of explants was observed in response to three media combinations, i.e., T6, T7, and T8 (T6 = MSB5 + 0 mg/L BAP + 0.05 mg/L IAA, T7 = MSB5 + 0 mg/L BAP + 0.1 mg/L IAA, and T8 = MSB5 + 0 mg/L BAP + 0.5 mg/L IAA), containing IAA only with MSB5 (Table 2). Both the hypocotyl and cotyledon explants produced roots in the basal MSB5 medium without any PGRs, but the percentage was higher for hypocotyl (62.50%) (Table 2). Root induction was inhibited in response to media containing higher concentrations (≥2 mg/L) of BAP (cytokinin). On average, the percentage of root formation was higher in the hypocotyl explants than in the cotyledon explants. The results reveal that both explants, when cultured on media with no or a low dose (1 mg/L) of cytokinin, produced roots simultaneously with shoots. Similarly, a higher percentage of root formation from hypocotyl explants than from cotyledons was observed by Jamous and Abu-Qaoud [31]. In another study, Jawad et al. [32] observed simultaneous root formation from explants on MS in combination with cytokinin and auxin.

3.1.3. Number of Shoots per Explant

Both single and interaction effects were highly significant for the number of shoots per explant (Table 2). The difference between the hypocotyl and cotyledon explants for the number of shoots produced was clearly observed (Figure 2). The maximum number of shoot regeneration (6.25 ± 0.25) was recorded from the cotyledons on T18 (MSB5 + 2 mg/L BAP + 0.5 mg/L IAA), followed by the hypocotyls (5.25 ± 0.25) on T14 (MSB5 + 2 mg/L BAP + 0.1 mg/L IAA). Billah et al. [28] reported that a similar PGRs combination (2 mg/L BAP + 0.5 mg/L IAA) resulted in the maximum number of shoots per cotyledon explant, whereas, under the same treatment, the hypocotyls formed fewer shoots than the cotyledon explants (Figure 2). The hypocotyls on T18 (MSB5 + 2 mg/L BAP + 0.5 mg/L IAA) and the cotyledons on T10 (MSB5 + 2 mg/L BAP + 0.05 mg/L IAA), T14 (MSB5 + 2 mg/L BAP + 0.1 mg/L IAA), and T16 (MSB5 + 4 mg/L BAP + 0.1 mg/L IAA) produced statistically similar results (4.25 ± 0.25). The hypocotyl explants on T1 (MSB5 + 0 mg/L BAP + 0 mg/L IAA) produced a single shoot per explant. Nevertheless, both types of explants on T8 and cotyledons on T1 (MSB5 + 0 mg/L BAP + 0 mg/L IAA), T6 (MSB5 + 0 mg/L BAP + 0.05 mg/L IAA), and T7 (MSB5 + 0 mg/L BAP + 0.1 mg/L IAA) produced roots but no shoots at all. Similar results were reported by Billah et al., Zaman et al., Ashakiran et al., Jehan and Hassanein, Osman et al., and Gubis et al. [28,29,33–36]. They observed that cotyledon explants produced a higher number of shoots than hypocotyls. Baye et al. [37] found that 2.0 mg/L BAP induced a maximum number of shoots for one of the tomato varieties. A similar result was found by Billah et al. [28] for the effect of the BAP + IAA combination from cotyledon explants. They also discovered that MS medium with 2 mg/L BAP + 0.5 mg/L IAA produced the most shoots per cotyledon explant. Mohamed et al. [38] reported that 2 mg/L BAP gave the maximum number of shoots for both cotyledon and hypocotyl explants.

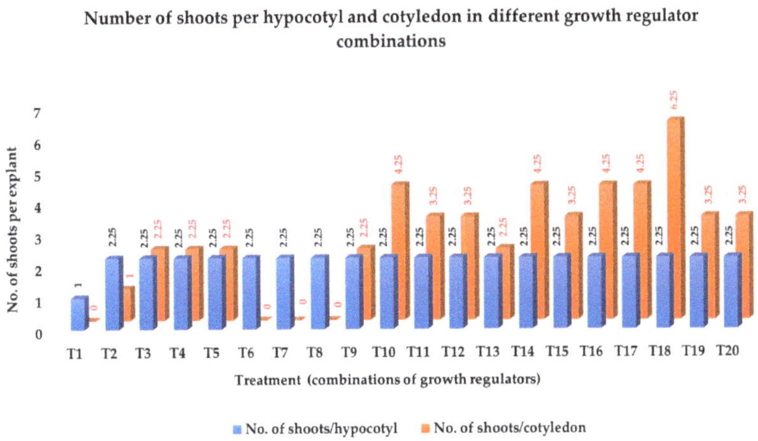

Figure 2. Number of shoots per hypocotyl and cotyledon explants on MSB5 supplemented with different concentration combinations of BAP and IAA.

3.1.4. Type of Induced Organ

Both the hypocotyl and cotyledon explants showed different types of organogenesis. Shoots, roots, leaves (without growing tips), and calli were observed to grow from the explants. The hypocotyl explants (H) produced shoots on all the media except T8 (MSB5 + 0 mg/L BAP + 0.5 mg/L IAA) (Table 2). The hypocotyl explants produced both roots and shoots on T1 (MSB5 + 0 mg/L BAP + 0 mg/L IAA), T6 (MSB5 + 0 mg/L BAP + 0.05 mg/L IAA), and T7 (MSB5 + 0 mg/L BAP + 0.1 mg/L IAA), but the roots were more prominent than the shoots (RS) (Table 2). In T11 (MSB5 + 3 mg/L BAP + 0.05 mg/L IAA) and T12 (MSB5 + 4 mg/L BAP + 0.05 mg/L IAA), the hypocotyl explants produced more leaves than shoots (LS) (Table 2). The cotyledon explants on T1 (MSB5 + 0 mg/L BAP + 0 mg/L IAA), T6 (MSB5 + 0 mg/L BAP + 0.05 mg/L IAA), T7 (MSB5 + 0 mg/L BAP + 0.1 mg/L IAA), and T8 (MSB5 + 0 mg/L BAP + 0.5 mg/L IAA) produced only roots (Table 2 and Figure 3c), whereas on T4 (MSB5 + 3 mg/L BAP + 0 mg/L IAA), T5 (MSB5 + 4 mg/L BAP + 0 mg/L IAA), T11 (MSB5 + 3 mg/L BAP + 0.05 mg/L IAA), T12 (MSB5 + 4 mg/L BAP + 0.05 mg/L IAA), T15 (MSB5 + 3 mg/L BAP + 0.1 mg/L IAA), T16 (MSB5 + 4 mg/L BAP + 0.1 mg/L IAA), T19 (MSB5 + 3 mg/L BAP + 0.5 mg/L IAA), and T20 (MSB5 + 4 mg/L BAP + 0.5 mg/L IAA), both the hypocotyl and cotyledon explants produced more leaves and calli (Table 2 and Figure 3d) over shoots (LCS). The hypocotyl explants on T2 (MSB5 + 1 mg/L BAP + 0 mg/L IAA), T3 (MSB5 + 2 mg/L BAP + 0 mg/L IAA), T9 (MSB5 + 1 mg/L BAP + 0.05 mg/L IAA), T10 (MSB5 + 2 mg/L BAP + 0.05 mg/L IAA), T13 (MSB5 + 1 mg/L BAP + 0.1 mg/L IAA), and T20 (MSB5 + 4 mg/L BAP + 0.5 mg/L IAA) and the cotyledon explants on T9 (MSB5 + 1 mg/L BAP + 0.05 mg/L IAA), T10 (MSB5 + 2 mg/L BAP + 0.05 mg/L IAA), T13 (MSB5 + 1 mg/L BAP + 0.1 mg/L IAA), T14 (MSB5 + 2 mg/L BAP + 0.1 mg/L IAA), T17 (MSB5 + 1 mg/L BAP + 0.5 mg/L IAA), and T18 (MSB5 + 2 mg/L BAP + 0.5 mg/L IAA) produced more shoots and fewer roots and leaves (Table 2 and Figure 3e–h).

These results indicate that the cotyledons on the MSB5 media without cytokinin (BAP) and with or without auxin (IAA) produced only roots but no shoots or leaves, whereas the hypocotyl explants produced more roots than shoots without any PGR and with a low concentration (0.05–0.2 mg/L) of auxin only. Both the cotyledon and hypocotyl explants produced vigorous roots but failed to produce shoots on medium containing a relatively higher concentration of auxin (0.5 mg/L IAA) without cytokinin (BAP). Jawad et al. [39] found simultaneous root and shoot formation in response to a combination of higher cytokinin and low auxin. They also observed explants with no shoots for some PGRs combinations.

Explants with higher concentrations (3–4 mg/L) of BAP (cytokinin) produced more calli and leaves without growing tips. In response to lower concentrations (1–2 mg/L) of BAP, with/without IAA, more shoots were produced over roots and leaves. Cotyledons, as compared to hypocotyls, are more likely to produce leaves in response to higher concentrations (3–4 mg/L) of BAP with/without IAA. Jawad et al. [39] reported callus formation on hypocotyl and nodal explants in response to 2 mg/L BAP + 0.5–1 mg/L NAA and 3 mg/L BAP, while Zaman et al. [29] observed a higher rate of callus formation from both hypocotyl and cotyledon explants in response to higher concentrations of cytokinin with low auxin (3 mg/L BAP + 0.2 mg/L NAA). Another study reported that higher concentrations of BAP did not result in shoot regeneration, while callus formation was observed at a BAP concentration of 4 mg/L [38]. The combination of 2 mg/L BAP and 1.75 mg/L kinetin was found to induce a healthy callus and an entire shoot formation, as reported by Hanur and Krishnareddy [40]. The induction of shoots was observed by Jehan and Hassanein [34] with a concentration of 1–2.5 mg/L BAP, either alone or in combination with 0.5 mg/L NAA. They also discovered that 1 mg/L of IAA, IBA, and NAA did not stimulate root formation.

Figure 3. Regeneration of shoots from cotyledon and hypocotyl explants of tomato cv. MT1 under different concentration combinations of BAP and IAA: (**a**) HT1 induced roots and shoots (scale bar: 1 cm); (**b**) CT1 induced only roots (scale bar: 1 cm); (**c**) CT8 induced only roots (scale bar: 1 cm); (**d**) E1T4 induced more leaves than calli than shoot (scale bar: 1 cm); (**e**) HT14 induced multiple shoots, roots, and leaves (scale bar: 1 cm); (**f**) CT18 induced multiple shoots, roots, and leaves (scale bar: 1 cm); (**g**) multiple shoots regenerated from hypocotyls (scale bar, 0.5 cm); (**h**) multiple shoots regenerated from cotyledons (scale bar, 0.5 cm). H = hypocotyls, C = cotyledons, T1 = MSB5 + 0 mg/L BAP + 0 mg/L IAA, T4 = MSB5 + 3 mg/L BAP + 0.0 mg/L IAA, T8 = MSB5 + 0 mg/L BAP + 0.5 mg/L IAA, T14 = MSB5 + 2 mg/L BAP + 0.1 mg/L IAA, T18 = MSB5 + 2 mg/L BAP + 0.5 mg/L IAA.

3.1.5. Shoot Elongation

The regenerated shoots were initially transferred onto fresh elongation medium containing MS + 2 mg/L BAP + 0.1 mg/L IAA. The PGR combination resulted in the formation of a callus on the leaves and/or near the shoot tips (Figure 4a). This observation persisted even when the concentration of BAP was decreased within the fresh medium (MS + 1 mg/L BAP + 0.1 mg/L IAA). The observed callus formation on the elongation media might be due to the sensitivity of the regenerated shoots from cv. MT1 towards the PGRs combinations, which accelerated the cell division and dedifferentiation of cells, leading to the formation of calli. Subsequently, shoot elongation was carried out using MS with no PGRs, as the shoots cultured on PGR-free medium exhibited normal elongation (Figure 4c). Cruz-Mendivil et al. and Brassard et al. [27,41] achieved successful and improved shoot elongation of regenerated shoots by using MS media without PGRs.

Figure 4. Elongation of regenerated shoots. (**a**) Callus (near the shoot tip) growing on the shoot in MS +2 mg/L BAP +0.1 mg/L IAA; (**b**) callus (on leaves) growing on shoot MS +2 mg/L BAP +0.1 mg/L IAA; (**c**) normal growth of shoots on MS medium without PGR. Scale bar = 2 cm.

3.2. Root Initiation

The root formation varied significantly between the shoots regenerated from the cotyledon and hypocotyl explants in response to different IAA concentrations (Table 3). Significant interaction effects were observed between the shoots derived from the two types of explants, which were treated with different concentrations of IAA with regard to the duration of days required for root initiation, the percentage of shoots that initiated roots, and the number of roots per shoot (Table 3). However, the interaction effects were not significant for root length, root diameter, and the survival rate of the rooted shoots.

Table 3. Effect of different concentrations of IAA on root induction on regenerated shoots from cotyledon and hypocotyl explants.

Treatment	% Shoot Induced Root	No. of Roots per Shoot	Root Length (cm)	Root Diameter (cm)	Survival Rate in Soil (%)
SH = Shoot from hypocotyl	97.50 a ± 0.02	20.45 a ± 2.50	6.70 a ± 0.056	0.38 a ± 0.03	60.00 a ± 6.28
SC = Shoot from cotyledon	61.50 b ± 0.02	8.35 b ± 1.16	5.41 b ± 0.27	0.34 b ± 0.03	55.00 b ± 6.75
Level of significance	**	**	**	*	*
MSD value	2.89	1.58	0.9	0.03	3.99
CV (%)	5.62	16.96	23.11	15.94	10.77
SH I1	87.50 b ± 2.50	3.75 e ± 1.11	5.75 b ± 0.85	0.20 e ± 0.04	12.50 c ± 2.50
SH I2	100 a ± 00	15.00 c ± 1.47	5.88 b ± 0.43	0.35 cd ± 0.03	47.50 b ± 2.50
SH I3	100 a ± 00	20.75 b ± 1.75	10.50 a ± 1.04	0.35 cd ± 0.03	75.00 a ± 2.88
SH I4	100 a ± 00	34.00 a ± 1.68	5.13 b ± 1.01	0.43 bc ± 0.03	82.50 a ± 2.50
SH I5	100 a ± 00	28.75 a ± 1.49	6.25 b ± 0.85	0.58 a ± 0.05	82.50 a ± 2.50
SC I1	27.50 e ± 2.50	1.50 e ± 0.29	5.25 b ± 0.32	0.18 e ± 0.03	5.00 c ± 2.88
SC I2	47.00 d ± 2.50	4.00 e ± 0.91	5.13 b ± 0.43	0.28 de ± 0.03	40.00 b ± 4.08
SC I3	65.00 c ± 2.88	9.50 d ± 0.96	7.06 ab ± 0.41	0.30 cde ± 0	75.00 a ± 2.88
SC I4	80.00 b ± 4.08	14.00 cd ± 0.91	4.88 b ± 0.69	0.40 bcd ± 0	77.50 a ± 2.50
SC I5	87.50 b ± 2.50	12.75 cd ± 0.85	4.75 b ± 0.48	0.53 ab ± 0.03	77.50 a ± 4.78

Table 3. *Cont.*

Treatment	% Shoot Induced Root	No. of Roots per Shoot	Root Length (cm)	Root Diameter (cm)	Survival Rate in Soil (%)
MSD value	11.18	5.47	3.53	0.14	14.35
CV (%)	5.78	15.6	23.95	16.21	10.26
Interaction level (S × I) of significance	**	**	**	**	**

Means in a column that include the same letters are not statistically different at a 5% level using Tukey's HSD test. MSD = minimum significant difference, CV = coefficient of variation, I1= MS + 0 mg/L IAA (control), I2 = MS + 0.25 mg/L IAA, I3 = MS + 0.50 mg/L IAA, I4 = MS + 0.75 mg/L IAA, I5 = MS + 1.00 mg/L IAA. Note: ns—not significant at $p > 0.05$, *—significant at $p \leq 0.05$, and **—significant at $p \leq 0.01$; ±—standard error (n = 4); (as ANOVA).

3.2.1. Percentage of Shoots Producing Roots

All the shoots derived from the hypocotyl explants (100%) initiated roots when treated with IAA, whereas the shoots from the cotyledons differed significantly in response to different concentrations of IAA (Table 3 and Figure 5). The root initiation of cotyledon-derived shoots in response to higher concentrations of IAA (I4 = MS + 0.75 mg/L IAA and I5 = MS + 1 mg/L IAA) was similar to the response of the hypocotyl-derived shoots in media without IAA (I1 = MS + 0.0 mg/L IAA). The percentage of rooted shoots was significantly lower for cotyledon-derived shoots (SC) than for shoots derived from the hypocotyl explants (SH). Jawad et al. [39] reported 100% root formation on regenerated adventitious shoots on MS medium supplemented with IAA, which is similar to the present study. Sarker et al. [30] observed maximum root initiation on MS medium with 0.2 mg/L IAA, while Arulananthu et al. [42] were successful in inducing roots using MS medium without PGR. In the present study, although the roots were initiated on MS medium without PGR, the percentage of shoots producing roots was lower than those on MS media supplemented with IAA.

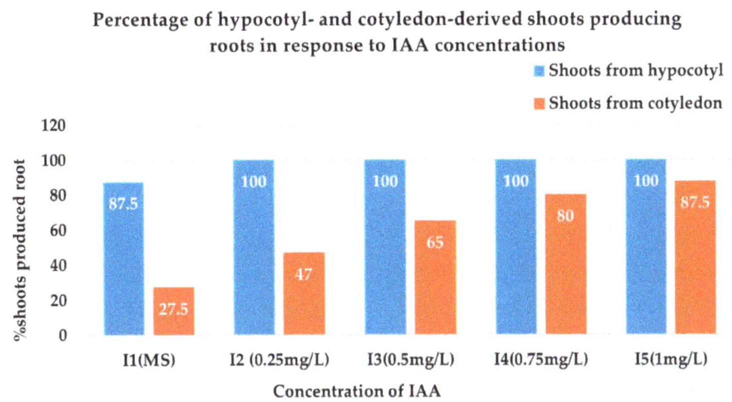

Figure 5. Percentage of hypocotyl- and cotyledon-derived shoots producing roots on MS media supplemented with different IAA concentrations.

3.2.2. Number of Roots per Shoot

The hypocotyl shoots (SH) produced the highest number of roots (34 ± 1.68) for I4 (0.75 mg/L IAA with basal MS medium); thus, the interaction of SHI4 was the best for the number of roots per shoot (Table 3). For both SH and SC, the lowest number of roots per shoot was observed for basal MSB5 medium with no IAA (I1). With an increase in the IAA concentration (I2 < I3 < I4), the number of roots increased, but for I5 (MS + 1.0 mg/L IAA), the number of roots per shoot decreased for both types of shoots (Table 3 and Figure 6). A higher number of roots was observed on the hypocotyl shoots than the cotyledonary shoots.

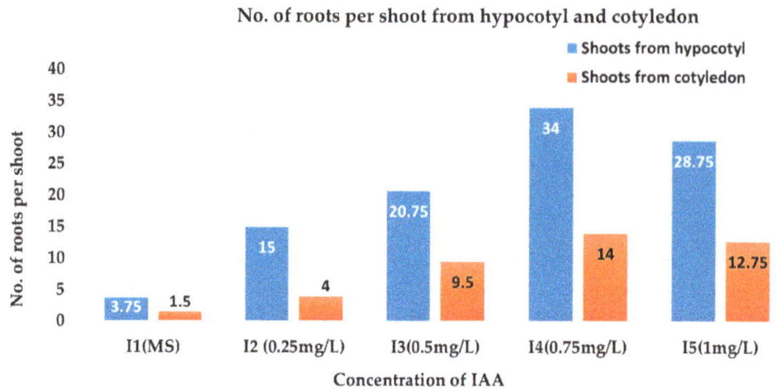

Figure 6. Number of roots per shoot derived from hypocotyl and cotyledon explants on MS media supplemented with different IAA concentrations.

3.2.3. Root Length and Diameter (cm)

The roots with the longest length (10.50 ± 1.04 cm) were observed as a result of exposure to I3 (MS + 0.5 mg/L IAA) (Table 3). The findings indicate that there was a positive correlation between the concentration of IAA and the length of the root, as evidenced by the increase in root length up to I3 (MS + 0.5 mg/L IAA) (Figure 7c). However, a reduction in the root length was observed again in the cases of I4 (MS + 0.75 mg/L IAA) and I5 (MS + 1.0 mg/L IAA) (Table 3). The study found no statistically significant variation in root length across the various shoot types. The experimental results indicate that the treatment with SHI5 (MS + 1.0 mg/L IAA) resulted in the largest diameter of the roots (0.58 ± 0.05 cm), whereas the smallest root diameter (0.18 ± 0.03 cm) was recorded in the SCI1 (MS + 0.0 mg/L IAA) treatment. There was a positive correlation between the concentration of IAA in the media and the root diameter of both types of shoots. However, it is notable that the observed differences were not statistically significant.

3.2.4. Survival Rates (%) of Rooted Shoots in Soil

The percentage of shoot survival in soil varied between 5.00 ± 2.88 and 82.50 ± 2.50% (Table 3). The survival rates of the hypocotyl and cotyledonary regenerants were higher for shoots rooted in I4 (MS + 0.75 mg/L) and I5 (MS + 100 mg/L), as compared to the other treatments. Specifically, the highest survival rate was observed in hypocotyl regenerants rooted in both I4 and I5, with a percentage of 82.50 ± 2.50%. The cotyledonary regenerants rooted in I4 and I5 exhibited a survival rate of 77.50 ± 2.50% and 77.50 ± 4.50%, respectively (Table 3). The present study also discovered an increased number and diameter of roots in response to the application of the two IAA concentrations (I4 and I5). These findings indicate that the survival rates of the regenerants in soil were influenced by the number and diameter of the roots. Previous studies have reported a varied survival rate of healthy rooted regenerants in soil. Yesmin et al. [43] reported a survival rate of 97% for rooted regenerated shoots, and Zaman et al. [29] documented a 100% survival rate. However, Jawad et al. [39] observed a lower success rate of 60% in the establishment of plantlets in soil.

Figure 7. Initiation of roots in shoots derived from cotyledon and hypocotyl explants of tomato cv. MT1. (**a**) Rooted hypocotyl shoots (SH) (scale bar: 4 cm); (**b**) rooted cotyledonary shoots (SC) (scale bar: 4 cm); (**c**) root length from shoots treated with different concentrations of IAA (scale bar: 4 cm); (**d**) root diameter from shoots treated with different concentrations of IAA; (**e**) thick root formed on medium I5 (MS + 1.0 mg/L IAA) (scale bar: 2 cm); (**f**) acclimatization of regenerated and rooted plants (scale bar: 4 cm); (**g**) surviving plants (scale bar: 4 cm). I1 = MS + 0 mg/L, I2 = MS + 0.25 mg/L, I3 = MS + 0.50 mg/L, I4 = MS + 0.75 mg/L, I5 = MS + 1.00 mg/L IAA.

3.3. Minimal Inhibitory Concentration of Kanamycin on Hypocotyl and Cotyledon Explants

3.3.1. Percentage of Explants Forming Shoots

The regeneration medium (MSB5 + 2 mg/L BAP + 0.5 mg/L IAA) without kanamycin resulted in the highest percentage of hypocotyl and cotyledon explants forming shoots, with a value of 98.25 + 1.75% and 98.44 + 1.56%, respectively (Figures 8 and 9, Table 4). The results also indicate that the addition of kanamycin at concentrations of 50 mg/L and 75 mg/L to the regeneration media led to direct shoot production in the hypocotyl explants at the rates of 48.44% and 13.88%, respectively (Figure 8 and Table 4). The percentage of hypocotyl explants forming shoots exhibited a significant decline in the medium with 75 mg/L kanamycin and was eventually suppressed at 100 mg/L. These findings were comparable with previous studies that used cv. MT1 for tomato transformation [13,44]. In a study conducted by Stavridou et al. [45], a concentration of 100 mg/L kanamycin was employed to inhibit the regeneration of non-transformed shoots in hybrid tomato cultivars Felina, Siena, and Don Jose. In another study, the screening of transformants of various tomato cultivars, including Moneymaker, Pusa Ruby, and Jinan, was performed using 50 mg/L kanamycin [32]. The inhibitory concentrations of kanamycin were observed to vary between 80–200 mg/L for different tomato cultivars, such as Zhongshu No. 5,

Daniela 144, Brillante 179, Annan 3017, Galina 3019, and Bernadine 5656 [44,46,47]. These findings suggest that the optimal concentration for selecting transformants depends on the genotypes of the tomato cultivars.

Table 4. Response of tomato cv. MT1 explants to different kanamycin concentrations.

Treatment	% Explants Forming Shoots	No. of Shoots per Explant
E1 = Hypocotyl	20.07 a ± 6.03	1.03 a ± 0.30
E2 = Cotyledon	12.30 b ± 5.85	0.81 b ± 0.39
Level of significance	**	**
MSD value	1.02	0.21
CV (%)	12.48	46.32
T1	98.34 a ± 1.09	5.63 a ± 0.50
T2	24.22 b ± 9.26	1.25 b ± 0.50
T3	6.94 c ± 2.70	0.50 c ± 0.19
T4	0 d ± 0	0 c ± 0
T5	0 d ±0	0 c ± 0
T6	0 d ±0	0 c ± 0
T7	0 d ±0	0 c ± 0
T8	0 d ±0	0 c ± 0
Level of significance	**	**
MSD value	3.2	0.67
CV (%)	12.48	46.32
E1T1	98.25 a ± 1.75	4.75 b ± 0.48
E1T2	48.44 b ± 2.99	2.50 c ± 0.29
E1T3	13.88 c ± 1.38	1.00 d ± 0
E1T4	0 d ± 0	0 d ± 0
E1T5	0 d ± 0	0 d ± 0
E1T6	0 d ± 0	0 d ± 0
E1T7	0 d ± 0	0 d ± 0
E1T8	0 d ± 0	0 d ± 0
E2T1	98.44 a ± 1.56	6.50 a ± 0.65
E2T2	0 d ± 0	0 d ± 0
E2T3	0 d ± 0	0 d ± 0
E2T4	0 d ± 0	0 d ± 0
E2T5	0 d ± 0	0 d ± 0
E2T6	0 d ± 0	0 d ± 0
E2T7	0 d ± 0	0 d ± 0
E2T8	0 d ± 0	0 d ± 0
MSD value	5.24	1.09
CV (%)	12.63	46.32
Interaction level (E × T) of significance	**	**

Means in a column that include the same letters are not statistically different at a 5% level using Tukey's HSD test. MSD = minimum significant difference, CV = coefficient of variation, T1 = 0, T2 = 50, T3 = 75, T4 = 100, T5 = 125, T6 = 150, T7 = 175, and T8 = 200 mg/L of kanamycin. Note: ns—not significant at $p > 0.05$, and **—significant at $p \leq 0.01$; ±—standard error (n = 4) (as ANOVA).

In contrast, the regeneration of shoots was not observed in the cotyledon explants, except in the regeneration medium without kanamycin, which served as a control (Figure 8 and Table 4). In addition, the occurrence of yellowing was observed on the cotyledon explants (Figure 9b). A similar observation was reported by Wang et al. [17] at a high concentration of kanamycin. Chen et al. [48] reported the observation of leaf yellowing in Arabidopsis explants cultured on kanamycin-containing media. According to Wang et al. (2015) [17], the yellowing and subsequent death of cells in green organs are caused by an interference of kanamycin with protein synthesis in the chloroplasts and mitochondria of plant cells that do not possess the detoxifying ability of the kanamycin resistance gene *nptII*. Duan et al. [49] also suggested that kanamycin affects the growth of cotyledon seedlings and leaves by preventing protein synthesis. Thus, the regeneration of shoots was also inhibited by the restriction of protein synthesis.

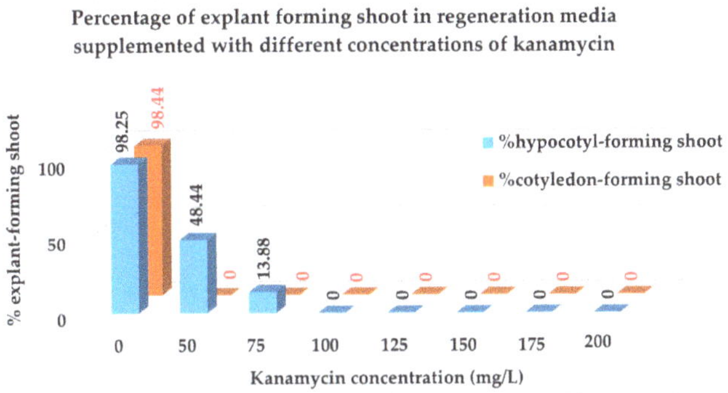

Figure 8. Percentage of explants forming shoots on regeneration media supplemented with different concentrations of kanamycin.

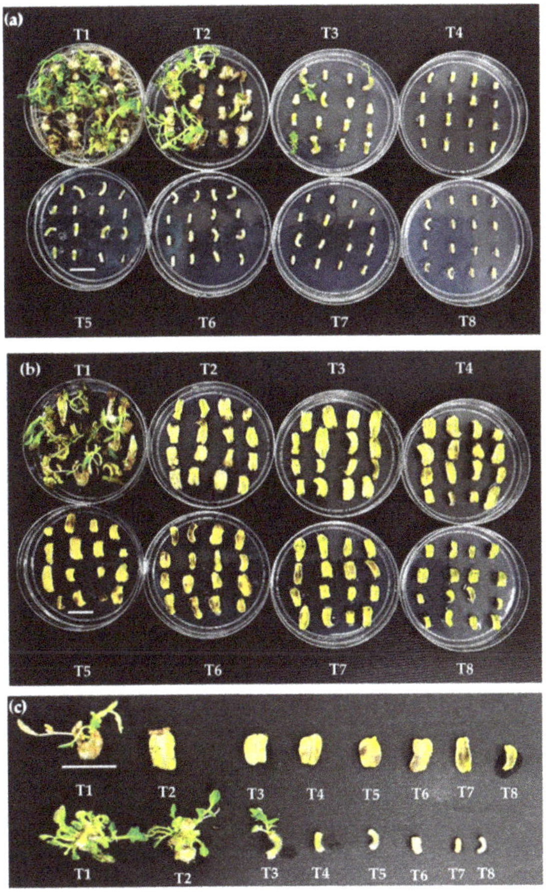

Figure 9. Response of tomato cv. MT1 explants to different kanamycin concentrations. (**a**) Response of hypocotyl explants; (**b**) response of cotyledon explants; (**c**) representative explants forming shoots. T1 = 0 mg/L, T2 = 50 mg/L, T3 = 75 mg/L, T4 = 100 mg/L, T5 = 125 mg/L, T6 = 150 mg/L, T7 = 175 mg/L, and T8 = 200 mg/L of kanamycin. Scale bar = 2 cm.

3.3.2. Number of Shoots per Explant

The results indicate that the cotyledon explants cultured on regeneration medium without kanamycin recorded the highest number of shoots (6.5 ± 0.65), followed by the hypocotyl explants (4.75 ± 0.48) on the same medium (Figure 10 and Table 4). The regeneration media containing 50 and 75 mg/L kanamycin yielded 2.5 ± 0.29 and 1 ± 0 shoots per explant, respectively, from the hypocotyl explants (Figure 10 and Table 4). Similar to the percentage of explants forming shoots, the number of shoots per explant also decreased with the increase in kanamycin concentration. Interestingly, no shoot was formed on the cotyledon explants on media with kanamycin. These results suggest that the toxicity of kanamycin is tissue-specific. Hung and Xie [50] observed the tissue-specific toxicity of kanamycin in *Astragalus racemosus*. They found that for the complete inhibition of callus and shoot formation in the cotyledon and hypocotyl, the concentration of kanamycin was the same, but a different concentration was required for root explants. In another study, Sharma et al. [51] found that 50 mg/L kanamycin did not completely inhibit the regeneration of shoots from tomato hypocotyl and cotyledon, but affected them slowly.

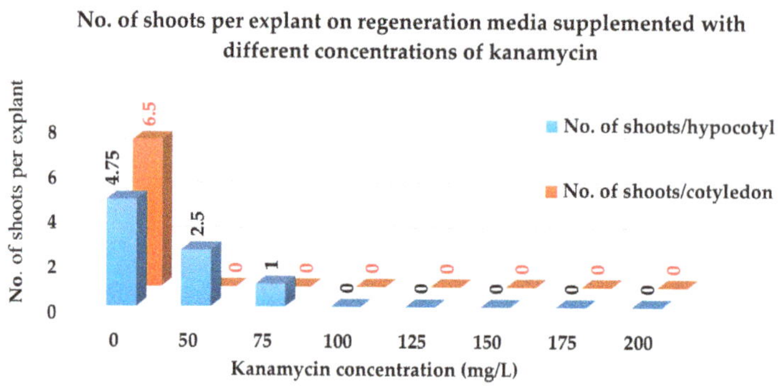

Figure 10. Number of shoots per explant on regeneration media supplemented with different concentrations of kanamycin.

3.3.3. Rooting and Acclimatization

The highest percentage of root induction was observed in shoots derived from the control regeneration medium (S1) (Figure 11 and Table 5). S1 showed a root formation rate of 72.5% when cultured on the MS medium. All the S1 shoots (100%) produced roots upon being cultured on MS + 0.75 mg/L IAA and MS + 1 mg/L IAA. The percentage of root formation was reduced for S2 (shoots from media containing 50 mg/L kanamycin) and S3 (shoots from media containing 75 mg/L kanamycin) shoots at all concentrations of IAA (Figure 11). The shoot type S3 exhibited the lowest percentage of rooting (10%) in T1, while a similar shoot type in T2 (MS + 0.75mg/L IAA) and T3 (MS + 1.0 mg/L IAA) displayed a rooting percentage of 40%.

The highest number of roots per shoot (32 ± 1.83) was recorded in T2 for S1 shoots, followed by T3 (27.25 ± 1.11) for the same type of shoot. The reduction in the root number followed a similar trend as the percentage of rooting. The number of roots decreased in the shoots derived from media with higher concentrations of kanamycin (Figure 12 and Table 5) and it was lowest for S3 in T1. These results indicate that shoots from media containing a higher concentration of kanamycin partially inhibited rooting in terms of the percentage of shoots inducing roots and the number of roots per shoot. Similarly, a reduction of root formation and root growth with an increase of the kanamycin concentration (>10 mg/L) was observed in Arabidopsis [49]. Of the rooted plants, 80% survived during the acclimatization stage.

Table 5. Rooting on shoots derived from kanamycin-treated explants.

Treatment	% Shoots Forming Roots	No. of Shoots per Explant
S1	90.83 a ± 4.17	20.67 a ± 3.92
S2	75.83 b ± 7.93	8.17 b ± 1.70
S3	30.00 c ± 4.77	5.33 c ± 1.16
Level of significance	**	**
MSD value	7.42	1.98
CV (%)	11.18	17.19
T1	40.83 b ± 8.02	1.42 c ± 0.36
T2	78.33 a ± 8.33	18.50 a ± 3.02
T3	77.50 a ± 8.27	14.25 b ± 2.84
Level of significance	**	**
MSD value	7.42	1.98
CV (%)	11.18	17.19
S1T1	72.50 b ± 4.79	2.75 ef ± 0.48
S1T2	100 a ± 0	32.00 a ± 1.83
S1T3	100 a ± 0	27.25 b ± 1.11
S2T1	40.00 c ± 4.08	1.00 f ± 0.41
S2T2	95.00 a ± 2.89	14.25 c ± 0.85
S2T3	92.50 a ± 4.79	9.25 d ± 1.11
S3T1	10.00 d ± 4.08	0.50 f ± 0.29
S3T2	40.00 c ± 4.08	9.25 d ± 1.11
S3T3	40.00 c ± 4.08	6.25 de ± 0.63
MSD value	18.39	4.39
CV (%)	11.67	16.03
Interaction level (E × G) significance	**	**

Means in a column that include the same letters are not statistically different at a 5% level using Tukey's HSD test. MSD = minimum significant difference, CV = coefficient of variation, S1 = shoots from from MSB5 + 2 mg/L BAP + 0.5 mg/L IAA, S2 = shoots from MSB5 + 2 mg/L BAP + 0.5 mg/L IAA + 50 mg/L kanamycin, and S3 = MSB5 + 2 mg/L BAP + 0.5 mg/L IAA + 75 mg/L kanamycin. T1 = MS, T2 = 0.75 mg/L IAA, T3 = 1 mg/L IAA. Note: ns—not significant at $p > 0.05$ and **—significant at $p \leq 0.01$; ±—standard error (n = 4); (as ANOVA).

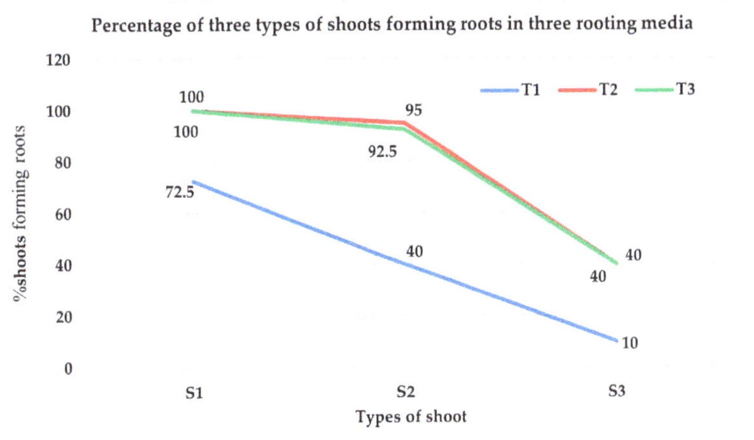

Figure 11. Percentage of three types of shoots forming roots on three rooting media.

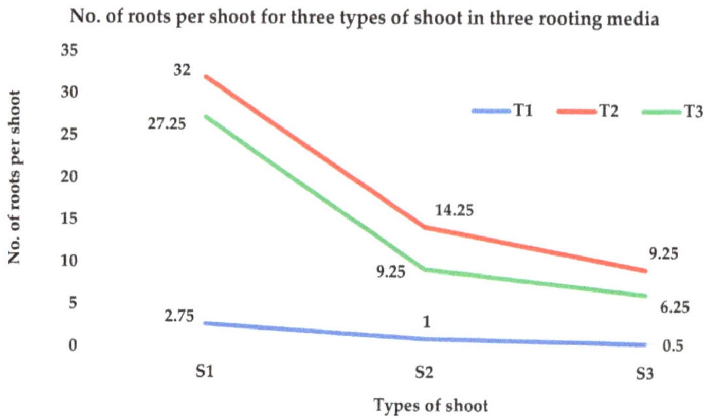

Figure 12. Number of roots per shoot for three types of shoots on three rooting media.

4. Conclusions

The best regeneration media for hypocotyl and cotyledon explants were found to be MSB5 + 2 mg/L BAP + 0.1 mg/L IAA (T14) and MSB5 + 2 mg/L BAP + 0.5 mg/L IAA (T18), respectively. The shoot elongation appeared to be optimal on MS media without PGRs. Based on the root induction and the survival rate of the regenerants, the optimal concentration of IAA in MS medium for root induction on regenerated shoots was found to be 0.75 mg/L. The findings also indicate that hypocotyl exhibited greater success in promoting a higher percentage of explant-forming shoots, while cotyledon was observed to be more effective in producing a higher number of direct shoot regeneration. The present study only employed BAP and IAA. The omission of other plant growth regulators (PGRs) precludes a comprehensive comparison of the effect of other growth regulators on the regeneration capacity. In future experiments regarding the in vitro regeneration of cv. MT1, it is recommended to include other PGRs and explants, such as the leaf, internode, and shoot tip. This will enable the identification of more optimal combinations of explant and growth regulators for shoot regeneration.

The results of the study indicate that a complete inhibition of shoot regeneration occurred on hypocotyl explants when exposed to medium containing 100 mg/L kanamycin. In contrast, the cotyledon explants exhibited total inhibition of regeneration when subjected to medium containing 50 mg/L kanamycin. The shoots derived from media supplemented with kanamycin showed a decrease in root formation, which was more pronounced in the shoots exposed to a greater concentration of kanamycin. The findings of the present study suggest that 100 mg/L kanamycin is suitable for the screening of transformants derived from hypocotyl of cv. MT1. At an initial concentration of 50 mg/L of kanamycin, the cotyledons were observed to have fully impeded regeneration, while the hypocotyls were still able to produce shoots. However, at a concentration of 75 mg/L, the hypocotyls were not completely inhibited, and a complete inhibiting effect was observed at a concentration of 100 mg/L kanamycin. It is possible that a more accurate concentration within the 0–50 mg/L range may be suitable for cotyledons, while a concentration range of 75–100 mg/L may be appropriate for hypocotyls. Therefore, it is recommended to evaluate smaller intervals of kanamycin concentrations, such as 10, 20, 30, 40, 50, 60, 70, 80, 90, and 100 mg/L, in future studies to obtain more precise outcomes.

Author Contributions: Conceptualization, S.A., W.A.S.W.A. and M.A.M.H.; methodology, S.A., W.A.S.W.A., M.A.Y.A. and M.A.M.H.; software, S.A. and R.A.; validation, M.A.M.H., A.S.J. and S.I.I.; formal analysis, S.A. and R.A.; writing—original draft preparation, S.A.; writing—review and editing, S.A., M.A.Y.A. and M.A.M.H.; supervision, M.A.M.H., M.A.Y.A., A.S.J. and S.I.I.; project administration, M.A.M.H., A.S.J. and S.I.I.; funding acquisition, M.A.M.H. and A.S.J. All authors have read and agreed to the published version of the manuscript.

Funding: This research was funded by the Ministry of Higher Education Malaysia (MoHE) through the Fundamental Research Grant Scheme (FRGS/1/2018/STG05/UPM/02/13) and the National Agricultural Technology Programme-Phase-II, Bangladesh (Project ID: 1100001758).

Data Availability Statement: The data presented in this study are available upon request from the corresponding author.

Acknowledgments: We thank all the staff members of the Agrobiotechnology Laboratory, Department of Agriculture Technology, Faculty of Agriculture, Universiti Putra Malaysia for their technical support during the research. We also extend our gratitude to Farahziatul Roshidah Nazri and Nur Afifah Raban for their help and support. Finally, we thank Maheran Abd Aziz for reviewing the manuscript.

Conflicts of Interest: The authors declare no conflict of interest.

References

1. FAOSTAT. Crops and Livestock Products. 2022. Available online: https://www.fao.org/faostat/en/#data/QCL/visualize (accessed on 14 May 2023).
2. Francois-Xavier, B. Worldwide (Total Fresh) Tomato Production Exceeds 187 Million Tonnes in 2020, in Tomato News Online Conference. 2022. Available online: https://www.tomatonews.com/en/worldwide-total-fresh-tomato-production-exceeds-187-million-tonnes-in-2020_2_1565.html (accessed on 14 May 2023).
3. Anwar, R.; Fatima, T.; Mattoo, A.K. *Tomatoes: A Model Crop of Solanaceous Plants*; Oxford University Press: Oxford, UK, 2019; pp. 1–50.
4. Liu, W.; Liu, K.; Chen, D.; Zhang, Z.; Li, B.; El-Mogy, M.M.; Tian, S.; Chen, T. *Solanum lycopersicum*, a Model Plant for the Studies in Developmental Biology, Stress Biology and Food Science. *Foods* **2022**, *11*, 2402. [CrossRef] [PubMed]
5. Van Eck, J.; Keen, P.; Tjahjadi, M. *Agrobacterium tumefaciens*-Mediated Transformation of Tomato. *Methods Mol. Biol.* **2019**, *1864*, 225–234. [PubMed]
6. Park, S.H.; Morris, J.L.; Park, J.E.; Hirschi, K.D.; Smith, R.H. Efficient and genotype-independent *Agrobacterium*-mediated tomato transformation. *J. Plant. Physiol.* **2003**, *160*, 1253–1257. [CrossRef]
7. Qiu, D.; Diretto, G.; Tavarza, R.; Giuliano, G. Improved protocol for *Agrobacterium*-mediated transformation of tomato and production of transgenic plants containing carotenoid biosynthetic gene *CsZCD*. *Sci. Hortic.* **2007**, *112*, 172–175. [CrossRef]
8. El-Shafey, N.; Hassan, N.; Khodary, S.; Badr, A. Differential In vitro Direct Regeneration of Tomato Genotypes on Various Combinations of Growth Regulators. *Biotechnology* **2017**, *16*, 155–164. [CrossRef]
9. Praveen, M.; Nanna, R.S. Effect of genotype, explant source and medium on in vitro regeneration of tomato. *Int. J. Genet. Mol. Biol.* **2011**, *3*, 45–50.
10. Bhatia, P.; Ashwath, N.; Senaratna, T.; Midmore, D. Tissue Culture Studies of Tomato (*Lycopersicon esculentum*). *Plant Cell Tissue Organ Cult.* **2004**, *78*, 1–21. [CrossRef]
11. Rahim, H.; Wahab, M.A.M.A.; Amin, M.Z.M.; Harun, A.; Haimid, M.T. Technological adoption evaluation of agricultural and food sectors towards modern agriculture: Tomato (Penilaian penggunaan teknologi bagi sektor pertanian dan makanan ke arah pertanian moden: Tomato. *Econ. Technol. Manag. Rev.* **2017**, *12*, 41–53.
12. Kamaladini, H.; Siti Nor, A.A.; Maheran, A.A. In vitro shoot regeneration for cotyledonary leaf explant of tomato variety MT1. In Proceedings of the Biodiversity and Biotechnology Symposium 2008, Kuching, Sarawak, Malaysia, 19–21 November 2008.
13. Behboodian, B.; Mohd Ali, Z.; Ismail, I.; Zainal, Z. Postharvest analysis of lowland transgenic tomato fruits harboring hpRNAi-ACO1 construct. *Sci. World J.* **2012**, *2012*, 439870. [CrossRef]
14. Raj, S.K.; Singh, R.; Pandey, S.; Singh, B. Agrobacterium-mediated tomato transformation and regeneration of transgenic lines expressing Tomato leaf curl virus coat protein gene for resistance against TLCV infection. *Curr. Sci.* **2005**, *88*, 1674–1679.
15. Plancarte-De la Torre, M.M.; Nunez-Palenius, H.G.; Gomez-Lim, M.A. Tomato Transformation with Genes Involved in Plant Immunity to Confer Broad Resistance Against Bacteria. *Rev. Fitotec. Mex.* **2016**, *39*, 349–358.
16. Gay, P.B.; Gillespie, S.H. Antibiotic resistance markers in genetically modified plants: A risk to human health? *Lancet Infect. Dis.* **2005**, *5*, 637–646. [CrossRef] [PubMed]
17. Wang, X.; Ryu, D.; Houtkooper, R.H.; Auwerx, J. Antibiotic use and abuse: A threat to mitochondria and chloroplasts with impact on research, health, and environment. *Bioessays* **2015**, *37*, 1045–1053. [CrossRef] [PubMed]
18. Farooq, N.; Nawaz, M.A.; Mukhtar, Z.; Ali, I.; Hundleby, P.; Ahmad, N. Investigating the In Vitro Regeneration Potential of Commercial Cultivars of Brassica. *Plants* **2019**, *8*, 558. [CrossRef]
19. Joersbo, M.; Okkels, F.T. A novel principle for selection of transgenic plant cells: Positive selection. *Plant Cell Rep.* **1996**, *16*, 219–221. [CrossRef] [PubMed]
20. Berry, M.; Gurung, A.; Easty, D.L. Toxicity of antibiotics and antifungals on cultured human corneal cells: Effect of mixing, exposure and concentration. *Eye* **1995**, *9 Pt 1*, 110–115. [CrossRef]

21. Gerszberg, A.; Grzegorczyk-Karolak, I. Influence of Selected Antibiotics on the Tomato Regeneration in in vitro cultures. *Not. Bot. Horti Agrobot. Cluj-Napoca* **2019**, *47*, 558–564. [CrossRef]
22. Andrews, J.M. Determination of minimum inhibitory concentrations. *J. Antimicrob. Chemother.* **2001**, *48* (Suppl. S1), 5–16. [CrossRef]
23. Saed Taha, R.; Ismail, I.; Zainal, Z.; Abdullah, S.N. The stearoyl-acyl-carrier-protein desaturase promoter (Des) from oil palm confers fruit-specific GUS expression in transgenic tomato. *J. Plant. Physiol.* **2012**, *169*, 1290–1300. [CrossRef]
24. Azzeme, A.M.; Abdullah, S.N.A.; Aziz, M.A.; Wahab, P.E.M. Oil palm drought inducible *DREB1* induced expression of DRE/CRT- and non-DRE/CRT-containing genes in lowland transgenic tomato under cold and PEG treatments. *Plant. Physiol. Biochem.* **2017**, *112*, 129–151. [CrossRef]
25. Murashige, T.; Skoog, F.K. A revised medium for rapid growth and bio assays with tobacco tissue cultures. *Physiol. Plant.* **1962**, *15*, 473–497. [CrossRef]
26. Gamborg, O.L.; Miller, R.A.; Ojima, K. Nutrient requirements of suspension cultures of soybean root cells. *Exp. Cell Res.* **1968**, *50*, 151–158. [CrossRef] [PubMed]
27. Cruz-Mendívil, A.; López, J.; Germán Báez, L.; Lopez-meyer, M.; Hernández-Verdugo, S.; López-Valenzuela, J.; Reyes Moreno, C.; Valdez, A. A Simple and Efficient Protocol for Plant Regeneration and Genetic Transformation of Tomato cv. Micro-Tom from Leaf Explants. *HortScience* **2011**, *46*, 1655–1660. [CrossRef]
28. Billah, M.; Banu, T.A.; Islam, M.; Banu, N.A.; Khan, S.; Akter, S.; Habib, A. In vitro regeneration and molecular characterization of some varieties of *Lycopersicon esculentum* Mill. in Bangladesh. *Bangladesh J. Sci. Ind. Res.* **2019**, *54*, 117–124. [CrossRef]
29. Zaman, M.R.U.; Bhuiyan, M.S.U.; Ahmed, T.; Haque, M.K.; Uddin, M. In Vitro Plant Regeneration of Tomato (*Lycopersicon esculentum*). *SF J. Agri Crop. Manag.* **2020**, *1*, 1001.
30. Sarker, R.; Islam, K.; Hoque, I. In vitro Regeneration and Agrobacterium-mediated Genetic Transformation of Tomato (*Lycopersicon esculentum* Mill.). *Plant Tissue Cult. Biotechnol.* **2010**, *19*, 101–111. [CrossRef]
31. Jamous, F.; Abu-Qaoud, H. In vitro regeneration of tomato (*Lycopersicon esculentum* Mill). *Plant Cell Biotechnol. Mol. Biol.* **2015**, *16*, 181–190.
32. Jawad, A.U.; Ay, A.-M.; Aj, K.; Al-Sadi, A. Enhanced somatic embryogenesis and *Agrobacterium*-mediated transformation of three cultivars of tomato by exogeneous application of putrescine. *Int. J. Agric. Biol.* **2014**, *16*, 81–88.
33. Ashakiran, K.; Sivankalyani, V.; Malaiy; Jayanthi, I.; Govindasamy, V.; Girija, S. Genotype specific shoots regeneration from different explants of tomato (*Solanum lycopersicum* L.) using TDZ. *Asian J. Plant. Sci. Res.* **2011**, *1*, 107–113.
34. Jehan, S.; Hassanein, A. Hormonal Requirements Trigger Different Organogenic Pathways on Tomato Nodal Explants. *Am. J. Plant Sci.* **2013**, *4*, 2118–2125. [CrossRef]
35. Osman, M.; Elhadi, E.; Khalafalla, M. Callus formation and organogenesis of tomato (*Lycopersicon esculentum* Mill, CV Omdurman) induced by thidiazuron. *Afr. J. Biotechnol.* **2010**, *9*, 4407–4413.
36. Gubis, J.; Lajchová, Z.; Faragó, J.; Jureková, Z. Effect of growth regulators on shoot induction and plant regeneration in tomato (*Lycopersicon esculentum* Mill.). *Biol.-Sect. Cell. Mol. Biol.* **2004**, *59*, 405–408.
37. Baye, E.; Matewos, T.; Belew, D. Effect of 6-Benzyl Amino Purine on In Vitro Multiplication of Tomato (*Lycopersicon esculentum* Mill.) Varieties using Shoot Explant. *J. Plant. Sci. Agric. Res.* **2020**, *4*, 32.
38. Mohamed, N.; Ismail, M.; Rahman, M. In vitro response from cotyledon and hypocotyls explants in tomato by inducing 6-benzylaminopurine. *Afr. J. Biotechnol.* **2010**, *9*, 4802–4807.
39. Jawad, Z.A.; Turker, M.; Ozdemir, F.A. Effect of different plant growth regulator on in vitro propagation of endangered plant; yellow tomato (*Lycopersicon esculentum* Mill.). *Int. J. Agric. For. Life Sci.* **2020**, *4*, 92–98.
40. Hanur, V.S.; Krishnareddy, B. In-Vitro Organogenesis in Tomato (*Solanum Lycopersicum*) Using Kinetin. *Adv. Plants Agric. Res.* **2016**, *4*, 397–401. [CrossRef]
41. Brassard, N.; Brissette, L.; Lord, D.; Laliberté, S. Elongation, rooting and acclimatization of micropropagated shoots from mature material of hybrid larch. *Plant Cell Tissue Organ Cult.* **1996**, *44*, 37–44. [CrossRef]
42. Arulananthu, G.; Bhat, S.G.; Ramesh, N. Callus induction and in vitro regeneration of tomato (*Lycopersicon esculentum* L.). *Res. J. Life Sci. Bioinform. Pharm. Chem. Sci.* **2019**, *5*, 491–503.
43. Yesmin, S.; Mollika, S.R.; Islam, M.N.; Nasrin, S. In vitro Regeneration of two BINA Tomato (*Lycopersicon esculentum* Mill.) Varieties of Bangladesh. *Plant Tissue Cult. Biotech.* **2022**, *32*, 43–51. [CrossRef]
44. Siti Suhaila, A.R.; Saleh, N.M. Inhibitory effect of Kanamycin on in vitro culture of *Lycopersicon esculentum* Mill cv. MT11. *J. Agrobiotech.* **2010**, *1*, 79–86.
45. Stavridou, E.; Tzioutziou, N.; Madesis, P.; Labrou, N.; Nianiou, I. Effect of different factors on regeneration and transformation efficiency of tomato (*Lycopersicum esculentum*) hybrids. *Czech J. Genet. Plant Breed.* **2019**, *55*, 120–127. [CrossRef]
46. Velcheva, M.; Faltin, Z.; Flaishman, M.; Eshdat, Y.; Perl, A. A liquid culture system for *Agrobacterium*-mediated transformation of tomato (*Lycopersicon esculentum* L. Mill.). *Plant Sci.* **2005**, *168*, 121–130. [CrossRef]
47. Ying, C.Y.; Huang, X.Q.; Guo, Y.Q.; Zhong, L.L.; Liu, Y.; Li, S.L.; Gu, X.M.; Zhou, X.H. Optimization of tomato genetic transformation, kanamycin-resistant screening and seed selection. *J. South. Med. Univ.* **2008**, *28*, 1117–1122.
48. Chen, C.; Xiaoyan, F.; Peng, R.; Tian, Y.; Yao, Q. Detoxifying processes during kanamycin-induced stress to Arabidopsis thaliana seedling growth. *Biotechnol. Biotechnol. Equip.* **2020**, *34*, 673–679. [CrossRef]

49. Duan, H.; Ding, X.; Song, J.; Zhikun, D.; Zhou, Y.; Zhou, C. Effects of kanamycin on growth and development of *Arabidopsis thaliana* seedling, cotyledon and leaf. *Pak. J. Bot.* **2009**, *41*, 1611–1618.
50. Hung, C.-Y.; Xie, J. Development of an Efficient Plant Regeneration System for the Selenium-hyperaccumulator *Astragalus racemosus* and the Nonaccumulator *Astragalus canadensis*. *Hortscience* **2008**, *43*, 2138–2142. [CrossRef]
51. Sharma, C.; Aggarwal, G.; Srivastava, D.K. Effect of antibiotic kanamycin on cultured cotyledon and hypocotyl tissues of Tomato (*Solanum lycopersicon* cv. Solan vajr). *Int. J. Agric. Environ. Biotechnol.* **2012**, *5*, 77–82.

horticulturae

MDPI

Article

Combined Study of Transcriptome and Metabolome Reveals Involvement of Metabolites and Candidate Genes in Flavonoid Biosynthesis in *Prunus avium* L.

Baochun Fu * and Yongqiang Tian

Pomology Institute, Shanxi Agricultural University, Taiyuan 030031, China
* Correspondence: byz_3525534@163.com

Abstract: Sweet cherry (*Prunus avium* L.) is a popular fruit tree grown for its juicy fruit and pleasing appearance. The fruit pf the sweet cherry contains active antioxidants and other chemical compounds essential for human health. For this study, we performed the transcriptomics and metabolomics analysis using young Green Peel (GP) and mature Red Peel (RP) from sweet cherries to understand the underlying genetic mechanism regulating fruit development and ripening. Using high-throughput RNA sequencing and ultra-performance liquid chromatography, with quadrupole time-of-flight tandem mass spectrometry, respectively, metabolic and transcript profiling was obtained. Relative to GP, there were equal quantities of pronouncedly varied metabolites in RP (n = 3564). Differentially expressed genes (DEGs, n = 3564), containing 45 transcription factor (TF) families, were recorded in RP. Meanwhile, 182 differentially expressed TF (DETF) members of 37 TF families, were displayed in abundance in RP compared to GP sweet cherries. The largest quantities of DETFs were members of the ERF (25) and basic helix–loop–helix (bHLH) (19) families, followed by the MYB (18), WRKY (18), and C2H2 (12) families. Interestingly, most ERF genes were down-regulated, whereas CCCH genes were mainly up-regulated in RP. Other DETFs exhibited significant variations. In addition, RT-QPCR results and metabolomics data together with transcriptomic data revealed that the abundance of catechin, epicatechin, rhoifolin, myricetin, keracyanin, and the other six glycosyltransferase genes was highly increased in RP when compared to GP sweet cherries. The relatively higher expression of DETFs, metabolite, and flavonoid biosynthesis in RP sweet cherries suggests the accumulation of distinct metabolites that cause red coloring during fruit development and ripening. Thus, the metabolomics and transcriptomic analysis of the current study are powerful tools for providing more valuable information for the metabolic engineering of flavonoids biosynthesis in sweet cherries. They are also helpful in understanding the relationship between genotype and phenotype.

Keywords: sweet cherry; metabolomics analysis; antioxidants; flavonoids; transcription factors; chemical compounds

1. Introduction

Sweet cherry (*Prunus avium* L.) is a non-climacteric, economically significant horticultural crop that is produced all over the world for its flesh fruit. The sweet cheery reproductive phase is short as it only takes two months from bloom to mature fruit [1]. Apart from its attractive color, the fruit of the sweet cherry is rich in antioxidants, vitamins, and minerals, and the tree is also suitable for cultivation in temperate regions. Sweet cherry fruits can have different colors, such as yellow, dark red, or blush [2]. Sweet cherry fruit goes through multiple developmental phases, from flower bud to fruit set. Following that, the ripening process begins, defining the quality of the final fruit. Various metabolic and genetic pathways control the fruit's development and ripening process. Some of the genetic and metabolic regulators of fruit development and ripening have been found in several other horticultural crops [3–8]. Anthocyanins is vital in the ripening process as it is involved in skin color pigmentation [9,10].

Flavonoids, the key secondary metabolites of plants, are sub-classed into anthocyanins, flavones, and flavanols [11]. Flavonoids have been found in numerous colored fruits and flowers and are involved in multiple plant functions. For example, they prevent plants from dormancy, ultraviolet light and phytopathogens-caused injury and modulate responses to other environmental stresses [12]. In humans, some reports highlighted flavonoid suppression in patients with chronic and cardiovascular diseases [13]. Furthermore, flavonoids and anthocyanins are the primary determinants of color in fruits and flowers. Six widely accepted anthocyanins (cyanidin, delphinidin, petunidin, pelargonin, peonidin, and malvidin) exist in plants. The peonidin is synthesized by the methylation of cyanidin, malvidin, and delphinium to variousdegrees [14]. In *Arabidopsis thaliana*, the structural genes in the biosynthetic pathway of anthocyanins can be classified as either "early" or "late" types [15]. Early biosynthetic genes, such as CHS and CHI, are mediated in a coordinated manner and encode enzymes that play a prominent role in the initial process of biosynthesis [16]. Late biosynthetic genes, such as DFR and ANS, participate in the late biosynthesis process [17].

In sweet cherry fruit, the role of flavonoids has previously been reported by study [18]. The study presented the involvement of flavonoids in postharvest sweet cherry fruits in response to UV-C light [18]. In *PavMADS7*-silenced plants, the level of anthocyanins decreased significantly, affecting fruit skin's reddening [19]. Several transcription factors (TFs) have been investigated for their role in the development and ripening of sweet cherry fruit. For instance, MIKC-type MADS-box TF *PavAGL15* regulates the size of sweet cherryfruits. Virus-induced gene silencing (VIGS) of *PavAGL15* significantly increased the size of sweet cherry fruits [20]. The *PavFUL* gene, when overexpressed, resulted in the formation of multi-silique and double fruits [21]. However, there is no body of research which relates the role of flavonoids and key TFs in regulating fruit development and ripening in sweet cherries.

The interpretation of metabolomics data offers an efficient method that may be used for the functional characterization of genes [22,23]. Metabolomics data can provide a plethora of information on the biochemical status of tissues. Metabolomics studies can also be used as a technical tool for association analysis. This can be used in conjunction with other data to examine the role of particular genes in a metabolic pathway of interest and to facilitate gene mining [24]. Systems biology research is now unable to function without the latest advancements and applications of high-throughput sequencing, high-resolution mass spectrometry, and information processing technologies [25,26]. Through the use of correlation and clustering analyses, transcript and metabolite datasets have been combined and can be visualized as connection networks between genes and metabolites [27]. These networks can show the response mechanism of rice to high nighttime temperatures [28], the regulation mechanism of delphinidin in flower color in grape hyacinths [29], the mechanism of potato pigmentation [30], the mechanism of blue flower formation in waterlilies [31], and catechin production in albino tea. The merging of transcriptomics and metabolomics offers substantial benefits for uncovering the biosynthetic mechanisms of important metabolic pathways [32–34]. The combined study of transcriptomics and metabolomics data has not yet been used to investigate the biosynthesis of phenolic chemicals.

In this study, we used a novel sweet cherry cultivar called "Hongmanao" which was developed by the Shanxi Academy of Agriculture Sciences' Institute of Pomology. For analysis, samples of the fruits' young green peel (GP) and mature red peel (RP) were taken. Young "GP" and mature "RP" sweet cherry fruit were used for the omics data analysis of extensively targeted metabolome and transcriptome. It was discovered that several significant TFs, including MYB, WRKY, and bHLH, control the biosynthesis of flavonoids. These findings will be beneficial for producing sweet and healthy plants, especially in the area of fruit quality.

2. Materials and Methods

2.1. Plant Materials

Sweet cherry trees (fifteen years old) were selected for this experiment. The orchard is located at the Fruit Research Institute of Shanxi Agricultural University (37°23′42″ N, 112°32′42″ E), Taigu District, Jinzhong City, Shanxi Province, China. The spacing between trees and rows was 2.5 m × 4.0 m. The same methods of cultivation and care were used on all the trees. Fruit samples were taken on 21 May 2021 (green peel), and 31 May 2021 (red peel) (RP). Eighteen tagged fruits were randomly selected, and three biological duplicates were selected at each stage. The fruit samples were cut into cross sections during sampling, frozen in liquid nitrogen, and kept at −80 °C for further studies. For the metabolome and transcriptome investigations, two developmental stages of the sweet cherry, C1 GP and C2 RP, were chosen and examined.

2.2. RNA Extraction, Library Preparation

Following the manufacturer's instructions, total RNA was extracted and purified using the TRIzol reagent (Invitrogen, Carlsbad, CA, USA). Using the NanoDrop ND-1000, each sample's RNA concentration and purity were measured (NanoDrop, Wilmington, DE, USA). The Bioanalyzer 2100 (Agilent, Santa Clara, CA, USA) used a RIN score of >7.0 to determine the integrity of the RNA, and denaturing agarose gel electrophoresis was used to corroborate the results. Using Dynabeads Oligo (dT)25-61005 (Thermo Fisher, Waltham, CA, USA), poly (A) RNA waspurified twice from 1 g of total RNA. Then, under 94 °C for 5–7 min, the poly(A) RNA was fragmented into tiny bits using a magnesium RNA fragmentation module (NEB, cat. e6150, San Francisco, CA, USA). Following the reverse transcription of the cleaved RNA fragments to produce the cDNA, *E. coli* DNA polymerase I (NEB, cat.m0209, USA), RNase H (NEB, cat.m0297, USA), and dUTP Solution (Thermo Fisher, cat. R0133, USA) were used to synthesize U-labeled second-stranded DNAs. The blunt ends of each strand were then given an A-base to help them ligate to the indexed adapters. Each adaptor had a T-base overhang that was used to ligate it to the DNA fragments with an A tail. The fragments were ligated to single- or dual-index adapters, and AMPureXP beads were used for size selection. After the heat-labile UDG enzyme (NEB, cat.m0280, USA) treatment of the U-labeled second-stranded DNAs, the ligated products were amplified with PCR underthe following conditions: initial denaturation at 95 °C for 3 min; 8 cycles of denaturation at 98 °C for 15 s, annealing at 60 °C for 15 s, and extension at 72 °C for 30 s; and then final ex-tension at 72 °C for 5 min. The resultant cDNA collection had an average insert size of 30,050 bp. Finally, using an illumine NovaseqTM 6000 (LC-Bio Technology Co., Ltd., Hangzhou, China), we carried out 2150bp paired-end sequencing (PE150) in accordance with the vendor's suggested procedure. Following RNA extraction, we used the Agilent 2100 device (Agilent Technologies, Beijing, China), Qubit v. 2.0 (Thermo Fisher Scientific), and the NanoDrop device (Thermo Fisher Scientific, Shanghai, China) to assess RNA purity, concentration, and integrity for the ensuing transcriptome sequencing. Using Qubit v. 2.0 and Agilent 2100 equipment, we evaluated and precisely measured the library concentration and insert size after library formation. Quantitative real-time PCR was used to check the quality of the library (qRT-PCR). With the help of a No-vaseq6000, high-throughput sequencing was carried out (Illumina, Beijing, China).

2.3. Bioinformatics Analysis of RNA-seq

Reads with adapter contamination, low quality bases, and indeterminate bases with default parameters were removed using the Fastp program (https://github.com/OpenGene/fastp accessed on 19 November 2021). Then, using fastp, the sequence quality was also confirmed. We utilized HISAT2 to map reads to the reference genome of Homo sapiens GRCh38 (https://ccb.jhu.edu/software/hisat2 accessed on 19 November 2021) (*Prunus avium* L.). The mapped readings of each sample were put together using StringTie (https://ccb.jhu.edu/software/stringtie accesse on 19 November 2021) using the default settings. Subsequently, gffcompare (https://github.com/gpertea/gffcompare/ accessed on 19 November 2021) was used to combine the transcriptomes from each sample to create a comprehensive transcriptome. Following the creation of the final transcriptome, all the expression levels of each transcript were estimated using StringTie. By using the formula FPKM (FPKM = [total exon fragments/mapped reads (millions) exon length (kB)]), StringTie was utilized to determine the expression level for mRNAs. With a parametric F-test comparing nested linear models (p value 0.05) and the R package edgeR (https://bioconductor.org/packages/release/bioc/html/edgeR.html accessed on 23 November 2021), the differentially expressed mRNAs were identified. Also, we assessed the read quality using FastQC (http://www.bioinformatics.babraham.ac.uk/projects/fastqc/ accessed on 23 November 2021), which is based on Q30 and GC levels. After acquiring high-quality sequencing data, Trinity [22], we constructed sequences. Functional annotation was aligned with the NR (RefSeq non-redundant proteins), and gene ontology GO and KEGG enrichment analysis were performed on the genes using the DAVID software (https://david.ncifcrf.gov accessed on 29 November 2021) [35–37]. Using fragments per kilobase of transcript per million mapped reads (FPKM) [38], we estimated gene expression.

2.4. Sample Preparation and Metabolite Extraction

The freeze-dried materials were ground in a mixer mill at 45 Hz for two minutes. Next, we added 100 mg of powder from each sample into a 5-mL EP tube and finished the extraction by adding 2000 L of a 3:1 *v/v* methanol and water solution. The samples were vortexed for 30 s, subjected to 15 min of ultrasonication in an ice bath, vibrated at 4 °C overnight, and then centrifuged for 15 min at 12,000 rpm and 4 °C. Afterwards, each supernatant (500 L) was dried while being gently blown by nitrogen. In 250 L of a 1:1 *v/v* solution of methanol and water, all residues were reconstituted using a 30-min vortexing and 15-min ultrasonication procedure in an ice bath. We preserved the supernatants in 2 mL glass vials at −80 °C for UHPLC-mass spectrometry (MS)/MS after centrifuging the samples for 15 min at 12,000 rpm and 4 °C. Finally, we made samples for quality control by combining equal aliquots of the supernatants from the sample.

2.5. Analyzing Wide-Target Metabolomics Data Qualitatively and Quantitatively

Following the implementation of qualitative metabolite assessment utilizing secondary spectrum data from the in-house metware database (MWDB) and metabolite data from a public database, metabolite quantification using multiple reaction monitoring (MRM) in triple quadrupole MS was carried out (TQMS). With the use of the TQMS, the required fragment ion was chosen, reducing interference from non-target ions and enhancing quantification accuracy and reproducibility. For inconsistent samples, metabolite MS data were collected, and the integration of MS peaks was put into practice to correct the same metabolite [39].

We also used TQMS to screen different ions from each drug. The signal intensities of these ions (measured in cps) were identified, and MULTIAQUANT software was used to open and read the sample MS file. The integrated data for the site were reserved, and the area of each chromatographic peak represented the relative concentrations of the respective metabolites. In order to ensure the accuracy of qualitative and quantitative analyses on the strength of the peak type and the reservation time of metabolites, contributing to the comparison of each metabolite level in disparate samples, correction of the mass spectrum peaks was later implemented for each metabolite in disparate samples.

As previously mentioned [40], we used principal component analysis (PCA) and partial least squares-discriminant analysis to identify cultivar-specific metabolite accumulation. To find the significantly different metabolites, fold change (FC) 2 or 0.5 and variable relevance in projection 1 were used as the selection criterion. Between DEGs and DEMs, Pearson correlation (PCA) analysis was carried out using R program.

Additionally, using the Roche LightCycler® 480 II Real-Time System, eight TFs in a 96-well plate were validated using qRT-PCR (Shanghai, China). The PCR amplification procedure used a thermal cycling schedule that consisted of 95 °C for 5 min, 95 °C for

10 s 40 cycles, and 60 °C for 30 s 40 cycles. All of the PCRs were carried out using the Hieff® qPCR SYBR Green Master Mix (No Rox) provided by Yeasen Biotech Co., Ltd., in accordance with the instructions' protocol (Shanghai, China). We established three technical and three biological replicates for each qRT-RCR.

2.6. Data Processing and Analysis

We utilized the 95% confidence interval ($p < 0.05$) and SPSS software (version 25.0, SPSS Inc., Chicago, IL, USA) to examine the data in this paper for statistical analysis (ANOVA) and statistical significance. For all measured parameters, the data are reported as the mean standard deviation (SD) of three biologically separate replicates; GraphPad Prism (version 8.0.2) (GraphPad Software, Inc., LA Jolla, CA, USA) was then used for graphical representations.

3. Results

3.1. Transcriptomic Analysis of Two Developmental Stages of Sweet Cheery Fruit

We implemented a transcriptomic analysis of the green and red peel cherries to confirm DEGs at two different developmental stages (Figure 1A). In agreement with FC $\leq$ 2.0 and FDR < 0.05, we filtered out 3812 DEGs in RP and GP sweet cherries, of which 1897 and 1915 genes showed an upward trend and a downward trend, respectively (Figure 1B,C). The GO was classified into molecular, biological, and cellular functions. The GO terms enriched in the cellular component classification were amino sugars and photosynthesis, while flavone and flavonol biosynthesis were the GO terms which were most enriched in the molecular function. DEGs regulating key molecular processes, such as ATP binding, RNA, and DNA binding, accumulated in abundance (Figure 2A). High numbers of DEGs reside in the nucleus and are key in the chloroplast thylakoid membrane, as shown in the cellular processes category (Figure 2A). Several key biological processes were also on display, such as flavonoid biosynthesis, response to abscisic acid, and auxin responsiveness. The KEGG pathway enrichment analysis highlighted that the majority of the DEGs are involved in photosynthesis–antenna proteins, circadian rhythm, amino sugars, photosynthesis, and flavone and flavonol biosynthesis (Figure 2B).

3.2. Metabolic Differences in Two Developmental Stages

We implemented LC-ESI-MS/MS sample analysis to compare the differences in two metabolite pathways. PCA analysis showed that GP (C1) and RP (C2) sweet cherry trees were distinct in the PC1 $\times$PC2 score plots (Figure 3A). In this work, we clarified 1942 up-regulated and 1622 down-regulated differentially expressed metabolites (DEMs) (3564 in total) in both groups (Figure 3B). As illustrated by the volcano plot (Figure 3C), the metabolite contents differed significantly between the two developmental stages. Based on the annotation information, the DEMs fell into 16 categories: pyrenes, flavonoids, prenol lipids, benzene and substituted derivatives, organ oxygen compounds, isobenzofurans, and indoles and their derivatives. From these, 25 flavonoids were screened, in 11 of which flavonoids accumulated in abundance, and14 were poorly enriched.

3.3. TFs Are Relevant to Differential Accumulation of Metabolites in "Red Peel" (C2) Sweet Cherries

The highly important TFs, including MYBs and bHLH, are instrumental in the biosynthesis and homeostasis of flavonoids [29,30]. Our data unveiled that 182 differentially expressed TFs (DETFs) from 37 TF families were discovered in the RP sweet cherries and this percentage is shown in Figure 4. The DETFs found in the largest quantities were members of the ERF (25) and bHLH (19) families, followed by the MYB (18), WRKY (18), and C2H2 (12) families. Interestingly, most ERF genes showed down-regulated expression, whereas CCCH genes were mainly up-regulated in C2. Apart from that, other DETFs displayed high expression variation among the tested samples.

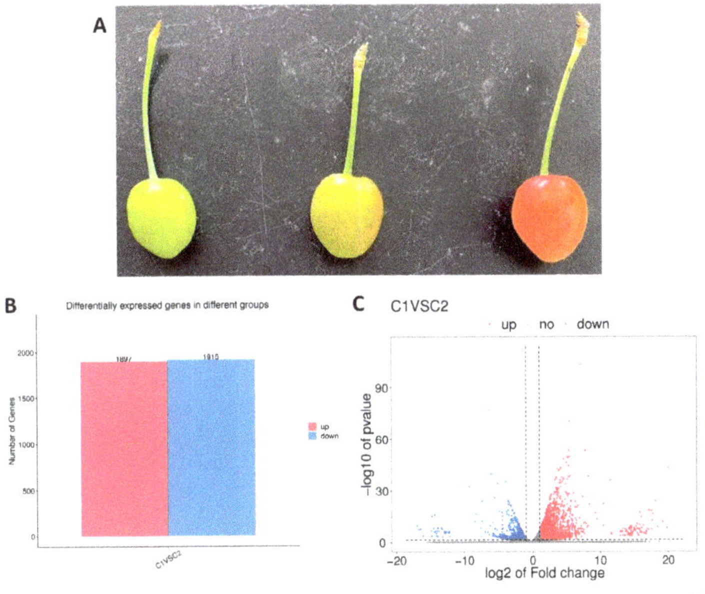

Figure 1. DEGs between young "Green Peel" and mature "Red Peel" sweet cherries. (**A**) Photograph of red peel, yellow peel and red peel sweet cherries; (**B**) Bar plot of DEGs; (**C**) As illustrated in the volcano plot, the difference in the expression level of genes displays statistical significance between the two groups of samples.

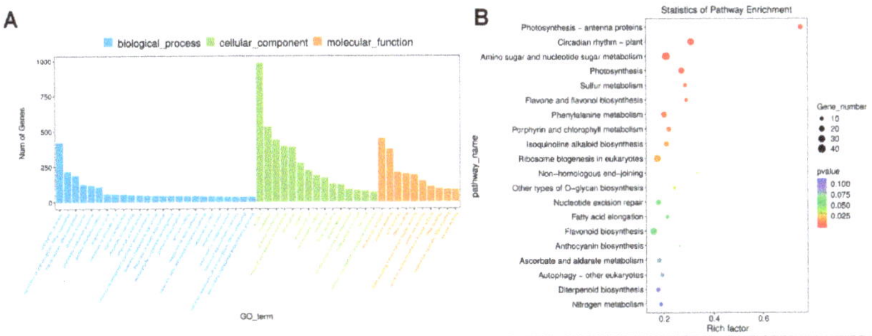

Figure 2. GO enrichment analysis (**A**) and KEGG enrichment analysis (**B**) of DEGs. The modules only illustrate the top 20 pathways showing the most pronounced enrichment.

3.4. Anthocyanin, Flavonoid, and Flavonol Biosynthesis Pathways

Omics data integration of transcriptomics and metabolomics analysis showed 44 DEGs genes (Table 1) and a total of 3564 DEMs and 7 genes which were involved in anthocyanin, flavonoid, flavone, and flavanol biosynthesis pathways are shown in (Table 2). Among the 7 DEMs, the levels of catechin, epicatechin, myricetin, rhoifolin, myricetin, and keracyanin were significantly higher in the C2, and only rutin was also up-regulated.

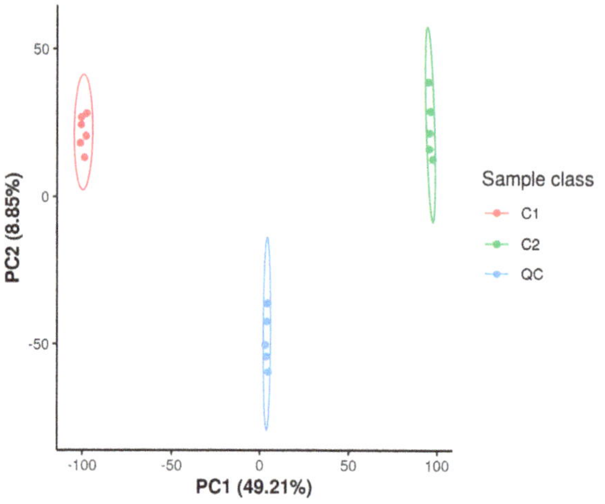

(**A**)

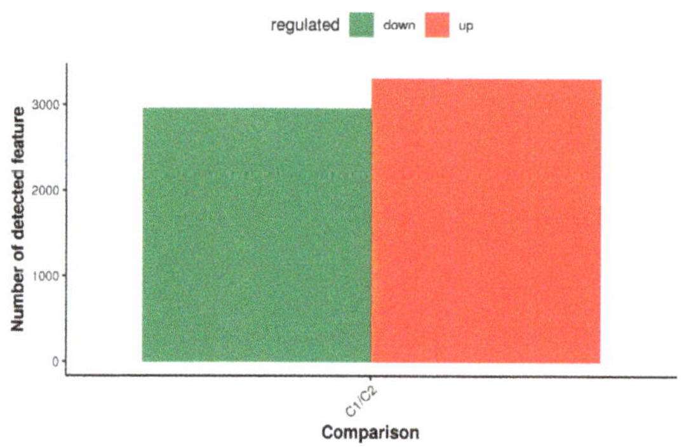

(**B**)

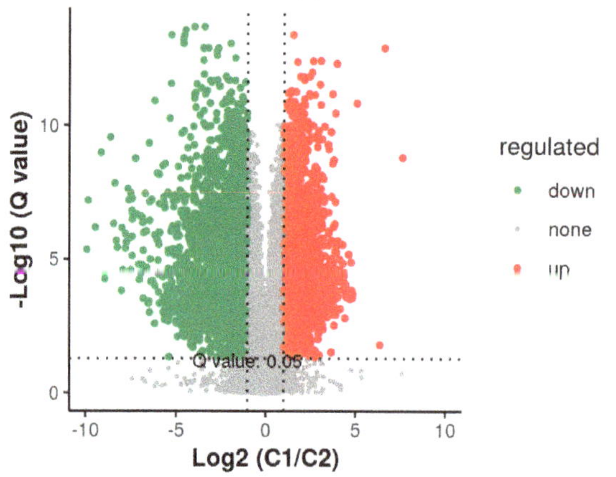

(**C**)

Figure 3. DEM profile. (**A**) PCA score map; (**B**) Comparison of DEMs between young "Green Peel" and mature "Red Peel" sweet cherry trees; (**C**) Volcano plot of DEMs.

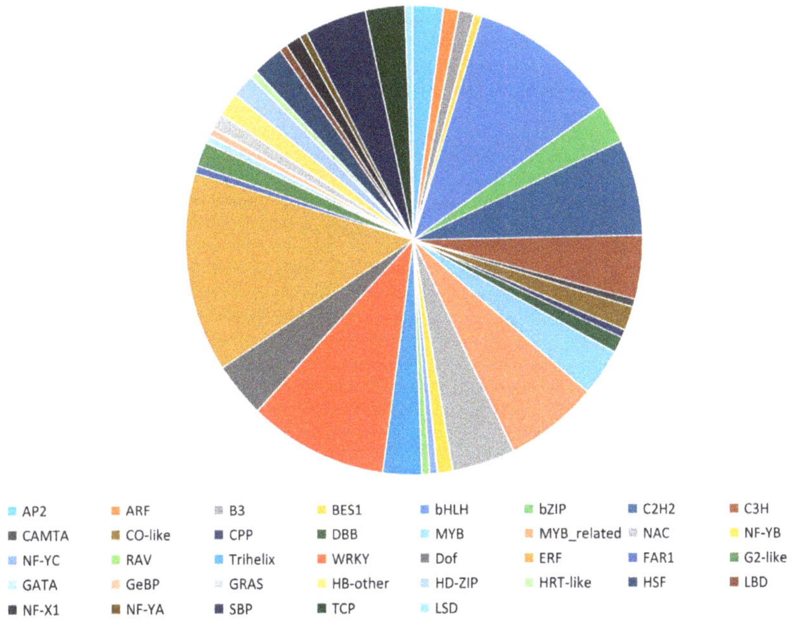

Figure 4. The pie chart shows the number of transcription factors.

Table 1. Omics data integration of transcriptome and metabolome yields 44 DEGs in anthocyanin, flavonoid, and flavone and flavonol biosynthesis pathways.

Pathway_id	Pathway_Name	Gene_ID	Description	Regulation
ko00942	Anthocyanin biosynthesis	Pav_sc0000138.1_g030.1.mk	anthocyanidin 3-O-glucosyltransferase 2-like [*Prunus avium*]	down
ko00942	Anthocyanin biosynthesis	Pav_sc0001323.1_g080.1.br	UDP-glycosyltransferase 88F5-like [*Prunus avium*]	down
ko00942	Anthocyanin biosynthesis	Pav_sc0001323.1_g100.1.br	UDP-glycosyltransferase 88F5-like [*Prunus avium*]	down
ko00942	Anthocyanin biosynthesis	Pav_sc0001323.1_g110.1.br	UDP-glycosyltransferase 88F5-like [*Prunus avium*]	down
ko00942	Anthocyanin biosynthesis	Pav_sc0001323.1_g140.1.mk	LOW QUALITY PROTEIN: UDP-glycosyltransferase 88F5-like [*Prunus avium*]	down
ko00941	Flavonoid biosynthesis	Pav_sc0000030.1_g1340.1.mk	flavonol synthase/flavanone 3-hydroxylase-like [*Prunus avium*]	up
ko00941	Flavonoid biosynthesis	Pav_sc0000044.1_g550.1.mk	naringenin,2-oxoglutarate 3-dioxygenase [*Prunus avium*]	down
ko00941	Flavonoid biosynthesis	Pav_sc0000045.1_g280.1.mk	chalcone synthase [*Prunus avium*]	down
ko00941	Flavonoid biosynthesis	Pav_sc0000091.1_g560.1.mk	ELMO domain-containing protein A [*Prunus avium*]	up
ko00941	Flavonoid biosynthesis	Pav_sc0000100.1_g580.1.mk	protein ECERIFERUM 26-like [*Prunus avium*]	up
ko00941	Flavonoid biosynthesis	Pav_sc0000107.1_g100.1.mk	leucoanthocyanidin dioxygenase [Prunus avium]	down
ko00941	Flavonoid biosynthesis	Pav_sc0000143.1_g510.1.mk	probable 2-oxoglutarate-dependent dioxygenase At5g05600 [*Prunus avium*]	up

Table 1. *Cont.*

Pathway_id	Pathway_Name	Gene_ID	Description	Regulation
ko00941	Flavonoid biosynthesis	Pav_sc0000206.1_g540.1.mk	trans-cinnamate 4-monooxygenase [*Prunus avium*]	down
ko00941	Flavonoid biosynthesis	Pav_sc0000358.1_g010.1.mk	flavonoid-3′-hydroxylase, partial [*Prunus avium*]	down
ko00941	Flavonoid biosynthesis	Pav_sc0000398.1_g280.1.mk	putative anthocyanidin reductase [*Prunus avium*]	down
ko00941	Flavonoid biosynthesis	Pav_sc0000465.1_g590.1.mk	codeine O-demethylase-like [*Prunus avium*]	up
ko00941	Flavonoid biosynthesis	Pav_sc0000465.1_g880.1.br	codeine O-demethylase-like [*Prunus avium*]	down
ko00941	Flavonoid biosynthesis	Pav_sc0000554.1_g2230.1.mk	probable LRR receptor-like serine/threonine-protein kinase MEE39 [*Prunus avium*]	down
ko00941	Flavonoid biosynthesis	Pav_sc0000791.1_g230.1.br	BAHD acyltransferase At5g47980-like [*Prunus avium*]	up
ko00941	Flavonoid biosynthesis	Pav_sc0000877.1_g1900.1.br	flavonoid 3′-monooxygenase [*Prunus avium*]	down
ko00941	Flavonoid biosynthesis	Pav_sc0001003.1_g370.1.mk	vacuolar-sorting receptor 7 isoform X2 [*Prunus avium*]	up
ko00941	Flavonoid biosynthesis	Pav_sc0001196.1_g1060.1.mk	cytochrome P450 98A2-like isoform X1 [*Prunus avium*]	down
ko00941	Flavonoid biosynthesis	Pav_sc0001196.1_g1070.1.br	cytochrome P450 98A2-like isoform X1 [*Prunus avium*]	down
ko00941	Flavonoid biosynthesis	Pav_sc0001196.1_g1090.1.mk	cytochrome P450 98A2-like [*Prunus avium*]	up
ko00941	Flavonoid biosynthesis	Pav_sc0001196.1_g1100.1.mk	cytochrome P450 98A2 [*Prunus persica*]	down
ko00941	Flavonoid biosynthesis	Pav_sc0001217.1_g200.1.mk	protein DMR6-LIKE OXYGENASE 2-like [*Prunus avium*]	down
ko00941	Flavonoid biosynthesis	Pav_sc0001345.1_g040.1.mk	probable 2-oxoglutarate-dependent dioxygenase At3g111800 [*Prunus avium*]	up
ko00941	Flavonoid biosynthesis	Pav_sc0001545.1_g080.1.mk	hypothetical protein PRUPE_4G029700 [*Prunus persica*]	down
ko00941	Flavonoid biosynthesis	Pav_sc0002208.1_g840.1.mk	bifunctional dihydroflavonol 4-reductase/flavanone 4-reductase [*Prunus avium*]	down
ko00941	Flavonoid biosynthesis	Pav_sc0002479.1_g160.1.br	vinorine synthase-like [*Prunus avium*]	down
ko00941	Flavonoid biosynthesis	Pav_sc0002479.1_g170.1.br	uncharacterized protein Pyn_04376 [*Prunus yedoensis* var. nudiflora]	up
ko00941	Flavonoid biosynthesis	Pav_sc0002479.1_g520.1.br	vinorine synthase-like [*Prunus avium*]	up
ko00941	Flavonoid biosynthesis	Pav_sc0003685.1_g130.1.mk	leucoanthocyanidin reductase [*Prunus avium*]	down
ko00941	Flavonoid biosynthesis	Pav_sc0005746.1_g030.1.mk	probable chalcone—flavonone isomerase 3 [*Prunus avium*]	down
ko00941	Flavonoid biosynthesis	Pav_sc0009537.1_g010.1.mk	chalcone synthase [*Prunus yedoensis* var. nudiflora]	down
ko00944	Flavone and flavonol biosynthesis	Pav_sc0000023.1_g010.1.br	phenolic glucoside malonyltransferase 1-like [*Prunus avium*]	up
ko00944	Flavone and flavonol biosynthesis	Pav_sc0000023.1_g130.1.br	phenolic glucoside malonyltransferase 1-like [*Prunus avium*]	up
ko00944	Flavone and flavonol biosynthesis	Pav_sc0000023.1_g140.1.br	phenolic glucoside malonyltransferase 1-like [*Prunus avium*]	up
ko00944	Flavone and flavonol biosynthesis	Pav_sc0000308.1_g350.1.mk	flavonoid 3-O-glucosyltransferase-like [*Prunus avium*]	up
ko00944	Flavone and flavonol biosynthesis	Pav_sc0000308.1_g360.1.mk	flavonoid 3-O-glucosyltransferase-like [*Prunus avium*]	down

Table 1. *Cont.*

Pathway_id	Pathway_Name	Gene_ID	Description	Regulation
ko00944	Flavone and flavonol biosynthesis	Pav_sc0000358.1_g010.1.mk	flavonoid-3′-hydroxylase, partial [*Prunus avium*]	down
ko00944	Flavone and flavonol biosynthesis	Pav_sc0000588.1_g030.1.mk	UDP-glycosyltransferase 89A2-like [*Prunus avium*]	down
ko00944	Flavone and flavonol biosynthesis	Pav_sc0000877.1_g1900.1.br	flavonoid 3′-monooxygenase [*Prunus avium*]	down
ko00944	Flavone and flavonol biosynthesis	Pav_sc0002358.1_g100.1.br	phenolic glucoside malonyltransferase 1-like [*Prunus avium*]	up

Table 2. Omics data integration of transcriptome and metabolome yielded DEMs in anthocyanin, flavonoid, and flavone and flavanol biosynthesis pathways.

Pathway ID	Regulated	MS2 Metabolite	MS2 Superclass	MS2 Class	MS2 KEGG
map00941	down	(+)-Catechin	Phenylpropanoids and polyketides	Flavonoids	C06562
map00941	down	Epicatechin	Phenylpropanoids and polyketides	Flavonoids	C09727
map00941	down	Myricetin	Phenylpropanoids and polyketides	Flavonoids	C10107
map00944	down	Rhoifolin	Phenylpropanoids and polyketides	Flavonoids	C12627
map00944	up	RUTIN	Phenylpropanoids and polyketides	Flavonoids	C05625
map00944	down	Myricetin	Phenylpropanoids and polyketides	Flavonoids	C10107
map00942	down	Keracyanin	Phenylpropanoids and polyketides	Flavonoids	C08620

The flavonoid biosynthesis pathway genes, such as anthocyanidin 3-*O*-glucosyltransferase 2 (3GT2) and four UDP-glycosyltransferase 88F5 (UGT 88f5), were all up-regulated in the C2 group. From the 30 DEGs in flavonoid biosynthesis, 11 DEGs were found to be highly expressed in C2, and 19 were down-regulated. In the flavone and flavanol biosynthesis pathway, four phenolic glucoside malonyl transferases (PMATs) and one flavonoid 3-*O*-glucosyltransferase (Pav_sc0000308.1_g350.1.mk) (3GT) were up-regulated. On the other hand, flavonoid 3-*O*-glucosyltransferase (Pav_sc0000308.1_g360.1.mk) (F3GT), flavonoid-3′-hydroxylase (F3′H), UDP-glycosyltransferase (UGT), and flavonoid 3′-monooxygenase were down-regulated in C2 vs. C1.

3.5. Correlation Analysis between DEGs and DEMs in Anthocyanin, Flavonoid, Flavone, and Flavanol Biosynthesis Pathways Reveals the Differential Regulatory Network

A correlation analysis was conducted between DEGs and DEMs in anthocyanin, flavonoid, flavone, and flavanol biosynthesis pathways. In total, 7 flavonoids and 44 genes underwent Pearson correlation analysis. As unveiled by the results, 20 genes were inextricably associated with 6 metabolites ($r2 > 0.9$, Figure 5A). The network analysis was divided into two clusters. Keracyanin was found to be abundant in cluster 1, and highly correlated with 5 genes. Myricetin, epicatechin, and catechin were more predominantly enriched in cluster 2, and highly connected to 15 genes (Figure 5B).

3.6. Verification of the Expression of Genes Relevant to the Route of Flavonoid Biosynthesis

We used qRT-PCR verification with eight TF genes to assess the precision and reproducibility of RNA-Seq data. All the genes' levels of expression were analyzed at every step, from development to ripening. Figure 6 displayed the expression level of specific genes. During the fruits' development through theirripening stages, the expression of the majority of genes was considerably altered. In light of these results, we can thus hypothesize that up-regulation of these genes may aid in fruit development up to the point of ripening.

(**A**)

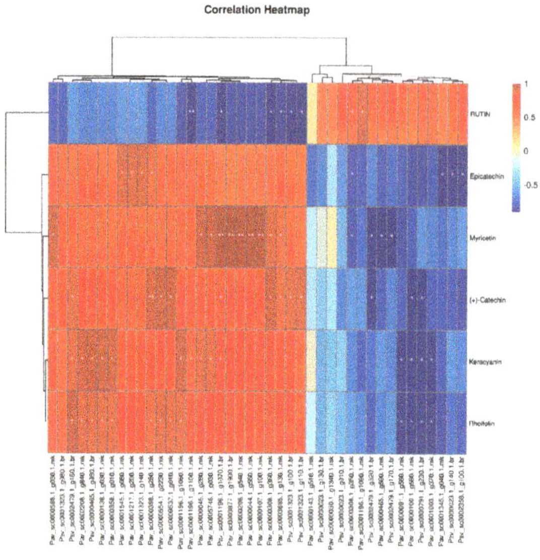

(**B**)

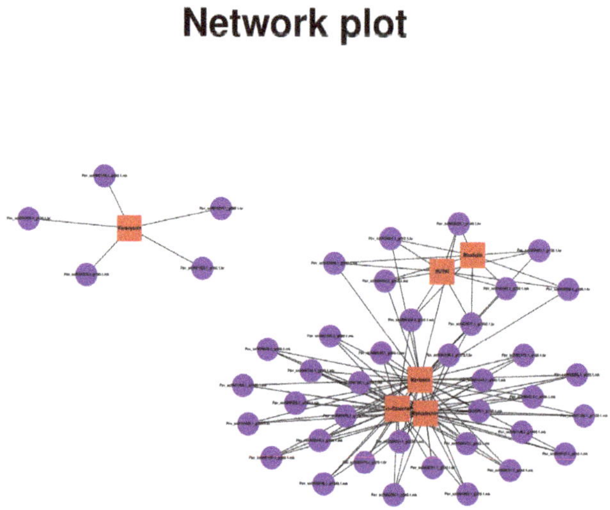

Figure 5. Principal component analysis (PCA) analysis of RNA−seq between the cultivars. Co-expression analysis of structural genes of anthocyanin, flavonoid, flavone, and flavanol biosynthesis pathways and transcription factors (TFs) in the RP and GP. (**A**) Clustering correlation heatmap with structural genes and metabolites using the OmicStudio tools; (**B**)The correlation networks of structural genes and metabolites anthocyanin, flavonoid, flavone, and flavanol biosynthesis.

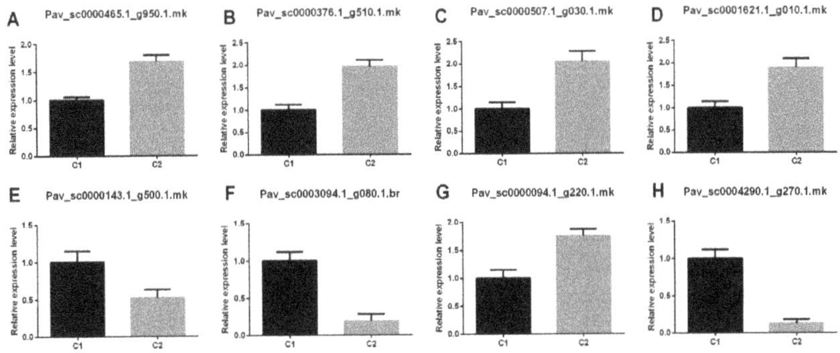

Figure 6. The qRT-PCR validation for 8 TFs. (**A**) Pav_sc0000465.1-g950.1.mk; (**B**) Pav_sc0000376.1-g510.1.mk; (**C**) Pav_sc0000507.1-g030.1.mk; (**D**) Pav_sc0001621.1-g010.1.mk; (**E**) Pav_sc0000143.1-g500.1.mk; (**F**) Pav_sc0003094.1-g080.1.mk; (**G**) Pav_sc0000094.1-g220.1.mk; (**H**) Pav_sc0004290.1-g270.1.mk. The histogram bars represent the means ± SE. All the data were calculated with three technical replicates, and the statistical significance was set at $p < 0.05$.

4. Discussion

Similar to genomics and proteomics, the new omics technique known as metabolomics can be used to identify and measure small-molecular-weight compounds that are present in an organism's cells [41]. In the post-genome era, a new area called plant metabolomics been extensively used for examining patterns of metabolite accumulation [42]. Furthermore, by identifying the genes involved in the metabolism, a subject of interest in contemporary plant biology, this approach was used to examine the underlying genetic basis of these patterns. Metabolites are the byproducts of acell's biological regulating process, and their concentrations can be viewed as the way a plant responds to genetic and environmental changes as it grows [43]. In this study, omics data assessment of metabolome and transcriptome provides extensive data about the transferred product profiles in secondary metabolism and the underlying variations in gene-expression networks. In the present study, 3812 DEGs were identified in RP sweet cherries, of which 1897 and 1915 genes were up-regulated and down-regulated, respectively. Meanwhile, in metabolome analysis, 3564 DEMs including 1942 up-regulated and 1622 down-regulatedmetabolites were also screened. The DEGs and DEMs are expected to be crucial in exploring the color change mechanism insweet cherries.

Flavonoids, which dominatesecondary metabolites, exist in heterogeneous plants [44,45]. They are determinants of the color of flowers, fruits, and leaves and are crucial for plant growth, development, and environmental adaptation. In particular, Flavonols, flavanones, and isoflavones, which are the distinct subclasses of flavonoids, are present in common vegetables and fruits [46]. For instance, catechin and epicatechin content was suppressed significantly in melatonin-treated bananas [47]. The bananas treated with melatonin showed delayed skin browning under cold temperature stress [48]. Augmented level of flavonoids and anthocyanins not only displayed the red color phenotype in mango, but they also enhanced resistance against cold stress and anthracnose disease [49]. A higher accumulation of carotenoid and flavonol was recorded in damaged-peel-colored apples [50]. In the present study, enrichment of 7 DEMs was unveiled in anthocyanin, flavonoid, and flavone and flavonol biosynthesis pathways. In particular, levels of catechin, epicatechin, rhoifolin, myricetin, and keracyanin were increased significantly in mature fruits, indicating the substances were responsible for the change of color.

The progression in fruit ripening initiates the colorful pathway which controls peel color. As a critical player, flavonoid regulates the pigmentation of peel color during fruit development and the ripening stage [51]. In the complex flavonoid pathway, anthocyanidin 3-*O*-galactosyltransferase is instrumental in catalyzing the galactosylation of anthocyanidin [52]. Higher transcriptional activity of galactosyltransferase and glucosyltransferase genes regulate the peel color in various horticultural crops [53,54]. In our study, one anthocyanidin 3GT2, four UGT 88f5, and one 3GT showed up-regulated expression in the RP sweet cherry group, indicating they are involvedt in the pigmentation of red peel sweet cherries (Table 1) [55].

TFs, namely transacting factors, initiate or impede gene expression to achieve regulations in response to environmental stresses, secondary metabolite biosynthesis, and plant growth and development [32,56]. In the present study, 182 DETFs, classified into 37 TF families, were found in RP indicating that the DETFs play a crucial role in the flavonoid-mediated peel pigmentation of sweet cherries. Previously, the MYB and bHLH (basic he-lix-loop-helix) TFs, the two largest TF families inplants, were intensively discussed in relation totheir role in flavonoid biosynthesis [57,58]. For instance, a bHLH3, when overexpressed, disturbed the flavonoid metabolic network by altering the levels of anthocyanins, flavonoids, and flavonols in mulberry fruits with discrepant colors [59]. In order to control the accumulation of plant flavonoids, MYB and bHLH collaborate with WDR protein to produce the ternary complex MBW [60,61]. The complex ternary MYB-113, a key regulator of flavonoid production in pepper [62], has no effect on how the MBW module functions. According to a study conducted on the plant *Arabidopsis thaliana*, the zinc finger protein TT1 (MYB TF) interacts with TT2 (MYB TF) to modify the expression of the genes involved in structural flavonoid biosynthesis [63].

In our research, bHLH (19) and MYB (18) account for high proportions of the TFs that have been discovered, including bHLH30, bHLH35, bHLH128, bHLH143, MYB8, MYB4, and MYB44 (Figure 4). Plants infiltrated with the MYB+bHLH combination construct showed specificity and significant accumulation of flavonoid pathway intermediates [64]. AtWRKY23 in *A. thaliana* impels flavonoid bio-synthesis by elevating enzyme-encoding genes in the flavonoid biosynthesis pathway [65]. In our study,18 WRKY genes significantly changed the expression levels in C2, which might play a significant role in the flavonoid synthesis that could induce fruit ripening. Over-expression of MdWRKY11 gears up the expressions of F3H, FLS, DFR, ANS, and UFGT, and facilitates anthocyanin and flavonoid accumulation in apple callus [66].

5. Conclusions

The present study involves combined metabolomics and transcriptomic assessments at two different stages, i.e., GP (young) and RP (mature), of fruit development and ripeningin sweet cherries. Our metabolome and transcriptome analyses provide a clear view of the participatory role of anthocyanin, flavonoid, flavone, and flavonol expression during the transition from young GP to mature RP sweet cherries. The candidate genes and metabolomics include six genes: catechin, epicatechin, rhoifolin, myricetin, keracyanin, and glycosyltransferase. These candidate genes give important information and useful references so we can better understand the regulation of the flavonoid biosynthesis pathway and the accumulation of metabolites in GP (young) and RP (mature) sweet cherries during distinct developmental stages. Even though these genes were predicted by the bioinformatics analysis, the validation of the transcriptome expression was verified by qRT-PCR. During fruit development and ripening, there is a greater accumulation of important TFs that control the pathway for secondary metabolites. However, the particular mechanism sill requires future research.

Author Contributions: The authors confirm their contributions to this work: B.F. in conceptualization, original draft preparation, and writing; and Y.T. in revising, and review. All authors have read and agreed to the published version of the manuscript.

Funding: This research was funded by Shanxi Agricultural University (YCX2020SJ10).

Data Availability Statement: The NCBI public database has the datasets produced for this investigation. The Bio Sample accession number PRJNA824997 in the Sequence Read Archive (SRA) contains all of the sequences.

Conflicts of Interest: The authors declare no conflict of interest.

References

1. Jin, W.; Wang, H.; Li, M.; Wang, J.; Yang, Y.; Zhang, X.; Yan, G.; Zhang, H.; Liu, J.; Zhang, K. The R2R3 MYB transcription factor PavMYB10.1 involves in anthocyanin biosynthesis and determines fruit skin colour in sweet cherry (*Prunus avium* L.). *Plant Biotechnol. J.* **2016**, *14*, 2120–2133. [CrossRef] [PubMed]
2. Wei, H.; Chen, X.; Zong, X.; Shu, H.; Gao, D.; Liu, Q. Comparative transcriptome analysis of genes involved in anthocyanin biosynthesis in the red and yellow fruits of sweet cherry (*Prunus avium* L.). *PLoS ONE* **2015**, *10*, e0121164. [CrossRef] [PubMed]
3. Khoso, M.A.; Hussain, A.; Ritonga, F.N.; Ali, Q.; Channa, M.M.; Alshegaihi, R.M.; Meng, Q.; Ali, M.; Zaman, W.; Brohi, R.D.; et al. WRKY transcription factors (TFs): Molecular switches to regulate drought, temperature, and salinity stresses in plants. *Front. Plant Sci.* **2022**, *13*, 1039329. [CrossRef]
4. Khan, M.; Ali, S.; Manghwar, H.; Saqib, S.; Ullah, F.; Ayaz, A.; Zaman, W. Melatonin Function and Crosstalk with Other Phytohormones under Normal and Stressful Conditions. *Genes* **2022**, *13*, 1699. [CrossRef]
5. Su, L.; Rahat, S.; Ren, N.; Kojima, M.; Takebayashi, Y.; Sakakibara, H.; Wang, M.; Chen, X.; Qi, X. Cytokinin and auxin modulate cucumber parthenocarpy fruit development. *Sci. Hortic.* **2021**, *282*, 110026. [CrossRef]
6. Kong, M.; Sheng, T.; Liang, J.; Ali, Q.; Gu, Q.; Wu, H.; Chen, J.; Liu, J.; Gao, X. Melatonin and Its Homologs Induce Immune Responses via Receptors trP47363-trP13076 in *Nicotiana benthamiana*. *Front. Plant Sci.* **2021**, *12*, 691835. [CrossRef]
7. Raza, A.; Tabassum, J.; Fakhar, A.Z.; Sharif, R.; Chen, H.; Zhang, C.; Ju, L.; Fotopoulos, V.; Siddique, K.H.M.; Singh, R.K.; et al. Smart reprograming of plants against salinity stress using modern biotechnological tools. *Crit. Rev. Biotechnol.* **2022**, *42*, 1–28. [CrossRef]
8. Sharif, R.; Su, L.; Chen, X.; Qi, X. Hormonal interactions underlying parthenocarpic fruit formation in horticultural crops. *Hortic. Res.* **2022**, *9*, uhab024. [CrossRef]
9. Michailidis, M.; Karagiannis, E.; Tanou, G.; Karamanoli, K.; Lazaridou, A.; Matsi, T.; Molassiotis, A. Metabolomic and physico-chemical approach unravel dynamic regulation of calcium in sweet cherry fruit physiology. *Plant Physiol. Biochem. PPB* **2017**, *116*, 68–79. [CrossRef]
10. Martínez-Esplá, A.; Zapata, P.J.; Valero, D.; García-Viguera, C.; Castillo, S.; Serrano, M. Preharvest application of oxalic acid increased fruit size, bioactive compounds, and antioxidant capacity in sweet cherry cultivars (*Prunus avium* L.). *J. Agric. Food Chem.* **2014**, *62*, 3432–3437. [CrossRef]
11. Mohotti, S.; Rajendran, S.; Muhammad, T.; Strömstedt, A.A.; Adhikari, A.; Burman, R.; de Silva, E.D.; Göransson, U.; Hettiarachchi, C.M.; Gunasekera, S. Screening for bioactive secondary metabolites in Sri Lankan medicinal plants by microfractionation and targeted isolation of antimicrobial flavonoids from Derris scandens. *J. Ethnopharmacol.* **2020**, *246*, 112158. [CrossRef] [PubMed]
12. Wang, A.; Li, R.; Ren, L.; Gao, X.; Zhang, Y.; Ma, Z.; Ma, D.; Luo, Y. A comparative metabolomics study of flavonoids in sweet potato with different flesh colors (*Ipomoea batatas* (L.) Lam). *Food Chem.* **2018**, *260*, 124–134. [CrossRef] [PubMed]
13. Casili, G.; Lanza, M.; Campolo, M.; Messina, S.; Scuderi, S.; Ardizzone, A.; Filippone, A.; Paterniti, I.; Cuzzocrea, S.; Esposito, E. Therapeutic potential of flavonoids in the treatment of chronic venous insufficiency. *Vasc. Pharmacol.* **2021**, *137*, 106825. [CrossRef] [PubMed]
14. Tanaka, Y.; Brugliera, F.; Chandler, S. Recent progress of flower colour modification by biotechnology. *Int. J. Mol. Sci.* **2009**, *10*, 5350–5369. [CrossRef]
15. Pelletier, M.K.; Murrell, J.R.; Shirley, B.W. Characterization of flavonol synthase and leucoanthocyanidin dioxygenase genes in Arabidopsis. Further evidence for differential regulation of "early" and "late" genes. *Plant Physiol.* **1997**, *113*, 1437–1445. [CrossRef]
16. Li, X.W.; Li, J.W.; Zhai, Y.; Zhao, Y.; Zhao, X.; Zhang, H.J.; Su, L.T.; Wang, Y.; Wang, Q.Y. A R2R3-MYB transcription factor, GmMYB12B2, affects the expression levels of flavonoid biosynthesis genes encoding key enzymes in transgenic Arabidopsis plants. *Gene* **2013**, *532*, 72–79. [CrossRef]
17. Wei, J.; Yang, J.; Jiang, W.; Pang, Y. Stacking triple genes increased proanthocyanidins level in Arabidopsis thaliana. *PLoS ONE* **2020**, *15*, e0234799. [CrossRef]
18. Zhang, Q.; Yang, W.; Liu, J.; Liu, H.; Lv, Z.; Zhang, C.; Chen, D.; Jiao, Z. Postharvest UV-C irradiation increased the flavonoids and anthocyanins accumulation, phenylpropanoid pathway gene expression, and antioxidant activity in sweet cherries (*Prunus avium* L.). *Postharvest Biol. Technol.* **2021**, *175*, 111490. [CrossRef]
19. Qi, X.; Liu, C.; Song, L.; Li, M. PaMADS7, a MADS-box transcription factor, regulates sweet cherry fruit ripening and softening. *Plant Sci.* **2020**, *301*, 110634. [CrossRef]
20. Dong, Y.; Qi, X.; Liu, C.; Song, L.; Ming, L. A sweet cherry AGAMOUS-LIKE transcription factor PavAGL15 affects fruit size by directly repressing the PavCYP78A9 expression. *Sci. Hortic.* **2022**, *297*, 110947. [CrossRef]
21. Wang, J.; Sun, W.; Wang, L.; Liu, X.; Xu, Y.; Sabir, I.A.; Jiu, S.; Wang, S.; Zhang, C. FRUITFULL is involved in double fruit formation at high temperature in sweet cherry. *Environ. Exp. Bot.* **2022**, *201*, 104986. [CrossRef]

22. Grabherr, M.G.; Haas, B.J.; Yassour, M.; Levin, J.Z.; Thompson, D.A.; Amit, I.; Adiconis, X.; Fan, L.; Raychowdhury, R.; Zeng, Q.; et al. Full-length transcriptome assembly from RNA-Seq data without a reference genome. *Nat. Biotechnol.* **2011**, *29*, 644–652. [CrossRef] [PubMed]
23. Chen, C.; Chen, H.; Yang, W.; Li, J.; Tang, W.; Gong, R. Transcriptomic and Metabolomic Analysis of Quality Changes during Sweet Cherry Fruit Development and Mining of Related Genes. *Int. J. Mol. Sci.* **2022**, *23*, 7402. [CrossRef] [PubMed]
24. Fukushima, A.; Kusano, M.; Redestig, H. Integrated omics approaches in plant systems biology. *Curr. Opin. Chem. Biol.* **2009**, *13*, 532–538. [CrossRef] [PubMed]
25. Kuhn, N.; Maldonado, J.; Ponce, C.; Arellano, M.; Time, A.; Multari, S.; Martens, S.; Carrera, E.; Donoso, J.M.; Sagredo, B.; et al. RNAseq reveals different transcriptomic responses to GA_3 in early and midseason varieties before ripening initiation in sweet cherry fruits. *Sci. Rep.* **2021**, *11*, 13075. [CrossRef]
26. Li, Y.K.; Fang, J.B.; Qi, X.J.; Lin, M.M.; Zhong, Y.P.; Sun, L.M.; Cui, W. Combined analysis of the fruit metabolome and transcriptome reveals candidate genes involved in flavonoid biosynthesis in Actinidia arguta. *Int. J. Mol. Sci.* **2018**, *19*, 1471. [CrossRef]
27. Kleessen, S.; Irgang, S.; Klie, S.; Giavalisco, P.; Nikoloski, Z. Integration of transcriptomics and metabolomics data specifies Chlamydomonas' metabolic response to rapamycin tratment. *Plant J.* **2015**, *81*, 822–835. [CrossRef]
28. Glaubitz, U.; Li, X.; Schaedel, S.; Erban, A.; Sulpice, R.; Kopka, J.; Hincha, D.K.; Zuther, E. Integrated analysis of rice transcriptomic and metabolomic responses to elevated night temperatures identifies sensitivity- and tolerance-related profiles: Integrated profiling analysis of rice under HNT. *Plant Cell Environ.* **2017**, *40*, 121–137. [CrossRef]
29. Hunter, D.A.; Napier, N.J.; Erridge, Z.A.; Saei, A.; Chen, R.K.Y.; McKenzie, M.J.; O'Donoghue, E.M.; Hunt, M.; Favre, L.; Lill, R.E.; et al. Transcriptome Responses of Ripe Cherry Tomato Fruit Exposed to Chilling and Rewarming Identify Reversible and Irreversible Gene Expression Changes. *Front. Plant Sci.* **2021**, *12*, 685416. [CrossRef]
30. Cho, K.; Cho, K.S.; Sohn, H.B.; Ha, I.J.; Hong, S.Y.; Lee, H.; Kim, Y.M.; Nam, M.H. Network analysis of the metabolome and transcriptome reveals novel regulation of potato pigmentation. *J. Exp. Bot.* **2016**, *67*, 1519. [CrossRef]
31. Wu, Q.; Wu, J.; Li, S.S.; Zhang, H.J.; Feng, C.Y.; Yin, D.D.; Wu, R.Y.; Wang, L.S. Transcriptome sequencing and metabolite analysis for revealing the blue flower formation in waterlily. *BMC Genom.* **2016**, *17*, 897. [CrossRef]
32. Liu, G.F.; Han, Z.X.; Feng, L.; Gao, L.P.; Gao, M.J.; Gruber, M.Y.; Zhang, Z.L.; Xia, T.; Wan, X.C.; Wei, S. Metabolic flux redirection and transcriptomic reprogramming in the albino tea cultivar 'Yu-Jin-Xiang' with an emphasis on catechin production. *Sci. Rep.* **2017**, *7*, 45062. [CrossRef]
33. Zhu, G.T.; Wang, S.C.; Huang, Z.J.; Zhang, S.B.; Liao, Q.G.; Zhang, C.Z.; Lin, T.; Qin, M.; Peng, M.; Yang, C.K.; et al. Rewiring of the fruit metabolome in tomato breeding. *Cell* **2018**, *172*, 249–261.e12. [CrossRef]
34. Islam, M.A.U.; Nupur, J.A.; Khalid, M.H.B.; Din, A.M.U.; Shafiq, M.; Alshegaihi, R.M.; Ali, Q.; Ali, Q.; Kamran, Z.; Manzoor, M.; et al. Genome-Wide Identification and In Silico Analysis of ZF-HD Transcription Factor Genes in *Zea mays* L. *Genes* **2022**, *13*, 2112. [CrossRef]
35. Poux, S.; Arighi, C.N.; Magrane, M.; Bateman, A.; Wei, C.H.; Lu, Z.; Boutet, E.; Bye, A.J.H.; Famiglietti, M.L.; Roechert, B.; et al. On expert curation and scalability: UniProtKB/Swiss-Prot as a case study. *Bioinformatics* **2017**, *33*, 3454–3460. [CrossRef]
36. Ashburner, M.; Ball, C.A.; Blake, J.A.; Botstein, D.; Butler, H.; Cherry, J.M.; Davis, A.P.; Dolinski, K.; Dwight, S.S.; Eppig, J.T.; et al. Gene ontology: Tool for the unification of biology. The Gene Ontology Consortium. *Nat. Genet.* **2000**, *25*, 25–29. [CrossRef]
37. Kanehisa, M.; Goto, S.; Kawashima, S.; Okuno, Y.; Hattori, M. The KEGG resource for deciphering the genome. *Nucleic Acids Res.* **2004**, *32*, D277–D280. [CrossRef]
38. Trapnell, C.; Williams, B.A.; Pertea, G.; Mortazavi, A.; Kwan, G.; van Baren, M.J.; Salzberg, S.L.; Wold, B.J.; Pachter, L. Transcript assembly and quantification by RNA-Seq reveals unannotated transcripts and isoform switching during cell differentiation. *Nat. Biotechnol.* **2010**, *28*, 511–515. [CrossRef]
39. Fraga, C.G.; Clowers, B.H.; Moore, R.J.; Zink, E.M. Signature-discovery approach for sample matching of a nerve-agent precursor using liquid chromatography-mass spectrometry, XCMS, and chemometrics. *Anal. Chem.* **2010**, *82*, 4165–4173. [CrossRef]
40. Wang, M.; Avula, B.; Wang, Y.H.; Zhao, J.; Avonto, C.; Parcher, J.F.; Raman, V.; Zweigenbaum, J.A.; Wylie, P.L.; Khan, I.A. An integrated approach utilising chemometrics and GC/MS for classification of chamomile flowers, essential oils and commercial products. *Food Chem.* **2014**, *152*, 391–398. [CrossRef]
41. Guo, H.H.; Guo, H.X.; Zhang, L.; Tang, Z.M.; Yu, X.M.; Wu, J.F.; Zeng, F.C. Metabolome and transcriptome association analysis reveals dynamic regulation of purine metabolism and flavonoid synthesis in transdifferentiation during somatic embryogenesis in cotton. *Int. J. Mol. Sci.* **2019**, *20*, 2070. [CrossRef]
42. Hall, R.; Beale, M.; Fiehn, O.; Hardy, N.; Sumner, L.; Bino, R. Plant metabolomics: The missing link in functional genomics strategies. *Plant Cell* **2002**, *14*, 1437–1440. [CrossRef]
43. Jalal, A.; Ali, Q.; Manghwar, H.; Zhu, D. Identification, Phylogeny, Divergence, Structure, and Expression Analysis of A20/AN1 Zinc Finger Domain Containing *Stress-Associated Proteins* (*SAPs*) Genes in *Jatropha curcas* L. *Genes* **2022**, *13*, 1766. [CrossRef]
44. Mehrtens, F.; Kranz, H.; Bednarek, P.; Weisshaar, B. The Arabidopsis transcription factor MYB12 is a flavonol-specific regulator of phenylpropanoid biosynthesis. *Plant Physiol.* **2005**, *138*, 1083–1096. [CrossRef]
45. Liu, J.; Osbourn, A.; Ma, P. MYB Transcription Factors as Regulators of Phenylpropanoid Metabolism in Plants. *Mol. Plant* **2015**, *8*, 689–708. [CrossRef]

46. Xu, W.; Dubos, C.; Lepiniec, L. Transcriptional control of flavonoid biosynthesis by MYB-bHLH-WDR complexes. *Trends Plant Sci.* **2015**, *20*, 176–185. [CrossRef] [PubMed]
47. Wang, Z.; Yu, Q.; Shen, W.; El Mohtar, C.A.; Zhao, X.; Gmitter, F.G., Jr. Functional study of CHS gene family members in citrus revealed a novel CHS gene affecting the production of flavonoids. *BMC Plant Biol.* **2018**, *18*, 189. [CrossRef]
48. Chen, C.; Zhou, G.; Chen, J.; Liu, X.; Lu, X.; Chen, H.; Tian, Y. Integrated Metabolome and Transcriptome Analysis Unveils Novel Pathway Involved in the Formation of Yellow Peel in Cucumber. *Int. J. Mol. Sci.* **2021**, *22*, 1494. [CrossRef] [PubMed]
49. Wang, Z.; Pu, H.; Shan, S.; Zhang, P.; Li, J.; Song, H.; Xu, X. Melatonin enhanced chilling tolerance and alleviated peel browning of banana fruit under low temperature storage. *Postharvest Biol. Technol.* **2021**, *179*, 111571. [CrossRef]
50. Sivankalyani, V.; Feygenberg, O.; Diskin, S.; Wright, B.; Alkan, N. Increased anthocyanin and flavonoids in mango fruit peel are associated with cold and pathogen resistance. *Postharvest Biol. Technol.* **2016**, *111*, 132–139. [CrossRef]
51. Naeem, M.; Shahzad, K.; Saqib, S.; Shahzad, A.; Nasrullah; Younas, M.; Afridi, M.I. The *Solanum melongena COP1LIKE* manipulates fruit ripening and flowering time in tomato (*Solanum lycopersicum*). *Plant Growth Regul.* **2022**, *96*, 369–382. [CrossRef]
52. Shen, N.; Wang, T.; Gan, Q.; Liu, S.; Wang, L.; Jin, B. Plant flavonoids: Classification, distribution, biosynthesis, and antioxidant activity. *Food Chem.* **2022**, *383*, 132531. [CrossRef] [PubMed]
53. He, X.; Huang, R.; Liu, L.; Li, Y.; Wang, W.; Xu, Q.; Yu, Y.; Zhou, T. CsUGT78A15 catalyzes the anthocyanidin 3-O-galactoside biosynthesis in tea plants. *Plant Physiol. Biochem. PPB* **2021**, *166*, 738–749. [CrossRef] [PubMed]
54. Ranganath, K.G.; Shivashankara, K.S.; Roy, T.K.; Dinesh, M.R.; Geetha, G.A.; Pavithra, K.C.; Ravishankar, K.V. Profiling of anthocyanins and carotenoids in fruit peel of different colored mango cultivars. *J. Food Sci. Technol.* **2018**, *55*, 4566–4577. [CrossRef]
55. Liu, Y.; Tikunov, Y.; Schouten, R.E.; Marcelis, L.F.M.; Visser, R.G.F.; Bovy, A. Anthocyanin Biosynthesis and Degradation Mechanisms in Solanaceous Vegetables: A Review. *Front. Chem.* **2018**, *6*, 52. [CrossRef]
56. Zhang, Q.; Wang, L.; Liu, Z.; Zhao, Z.; Zhao, J.; Wang, Z.; Zhou, G.; Liu, P.; Liu, M. Transcriptome and metabolome profiling unveil the mechanisms of Ziziphus jujuba Mill. peel coloration. *Food Chem.* **2020**, *312*, 125903. [CrossRef]
57. Feller, A.; Machemer, K.; Braun, E.L.; Grotewold, E. Evolutionary and comparative analysis of MYB and bHLH plant transcription factors. *Plant J. Cell Mol. Biol.* **2011**, *66*, 94–116. [CrossRef]
58. Wang, X.-Y.; Tian, L.; Feng, S.-J.; Wei, A.-Z. Identifying potential flavonoid biosynthesis regulator in Zanthoxylum bungeanum Maxim. by genome-wide characterization of the MYB transcription factor gene family. *J. Integr. Agric.* **2022**, *21*, 1997–2018. [CrossRef]
59. Zhu, L.; Guan, Y.; Zhang, Z.; Song, A.; Chen, S.; Jiang, J.; Chen, F. CmMYB8 encodes an R2R3 MYB transcription factor which represses lignin and flavonoid synthesis in chrysanthemum. *Plant Physiol. Biochem.* **2020**, *149*, 217–224. [CrossRef]
60. Zhu, J.-H.; Xia, D.-N.; Xu, J.; Guo, D.; Li, H.-L.; Wang, Y.; Mei, W.-L.; Peng, S.-Q. Identification of the bHLH gene family in Dracaena cambodiana reveals candidate genes involved in flavonoid biosynthesis. *Ind. Crops Prod.* **2020**, *150*, 112407. [CrossRef]
61. Arlotta, C.; Puglia, G.D.; Genovese, C.; Toscano, V.; Karlova, R.; Beekwilder, J.; De Vos, R.C.H.; Raccuia, S.A. MYB5-like and bHLH influence flavonoid composition in pomegranate. *Plant Sci.* **2020**, *298*, 110563. [CrossRef] [PubMed]
62. Li, H.; Yang, Z.; Zeng, Q.; Wang, S.; Luo, Y.; Huang, Y.; Xin, Y.; He, N. Abnormal expression of bHLH3 disrupts a flavonoid homeostasis network, causing differences in pigment composition among mulberry fruits. *Hortic. Res.* **2020**, *7*, 83. [CrossRef] [PubMed]
63. Naik, J.; Misra, P.; Trivedi, P.K.; Pandey, A. Molecular components associated with the regulation of flavonoid biosynthesis. *Plant Sci.* **2022**, *317*, 111196. [CrossRef] [PubMed]
64. Fiehn, O.; Kopka, J.; Dormann, P.; Altmann, T.; Trethewey, R.N.; Willmitzer, L. Metabolite profiling for plant functional genomics. *Nat. Biotechnol.* **2000**, *18*, 1157–1161. [CrossRef]
65. Liu, Y.; Lv, J.; Liu, Z.; Wang, J.; Yang, B.; Chen, W.; Ou, L.; Dai, X.; Zhang, Z.; Zou, X. Integrative analysis of metabolome and transcriptome reveals the mechanism of color formation in pepper fruit (*Capsicum annuum* L.). *Food Chem.* **2020**, *306*, 125629. [CrossRef]
66. Appelhagen, I.; Lu, G.H.; Huep, G.; Schmelzer, E.; Weisshaar, B.; Sagasser, M. TRANSPARENT TESTA1 interacts with R2R3-MYB factors and affects early and late steps of flavonoid biosynthesis in the endothelium of Arabidopsis thaliana seeds. *Plant J. Cell Mol. Biol.* **2011**, *67*, 406–419. [CrossRef]

horticulturae

MDPI

Article

Differential Responses of Cherry Tomatoes (*Solanum lycopersicum*) to Long-Term Heat Stress

Bo-Mi Park [1,†], Hyo-Bong Jeong [2,†], Eun-Young Yang [3], Min-Kyoung Kim [1], Ji-Won Kim [1], Wonbyoung Chae [1], Oak-Jin Lee [2], Sang Gyu Kim [2] and Sumin Kim [1,*]

[1] Department of Environmental Horticulture and Landscape Architecture, Environmental Horticulture, Dankook University, Cheonan 31116, Republic of Korea
[2] Vegetable Research Division, National Institute of Horticultural & Herbal Science, RDA, Wanju 55365, Republic of Korea
[3] Department of Horticulture, Korea National College of Agriculture and Fisheries, Jeonju 54874, Republic of Korea
* Correspondence: sumin.kim@dankook.ac.kr; Tel.: +82-042-550-3644
† These authors contributed equally to this work.

Abstract: As global warming increases day/night temperatures as well as frequencies of heat waves, studying physiological responses in long-term heat stress is required for sustainable yield production in the future. In this study, effects of long-term heat stress on photosynthetic, morphological, and yield parameters of three cherry tomato accessions, HR17, HR22, and HR24, were evaluated. The experiment was conducted under two temperature greenhouse conditions, where temperature set-point for ventilation was 30 °C and 35 °C during the day for 57 days, respectively. Plants were harvested on the 35th days and 57th days after heat treatments, and their physiological and morphological characteristics and yield traits were measured. Under control conditions, HR17 and HR22 had 0.5–0.6 harvest index, while HR24 had 0.3 harvest index. On 35th days after heat treatment, although yield loss percentage of HR17 was high (43%), it produced the highest fruit yield among all three accessions. However, after longer heat treatment, HR24 produced the highest fruit yields among all accessions with the smallest yield loss (34%). Furthermore, yield loss was highly associated with reductions in nitrogen use efficiency and water content in plant body under heat stress. The results of this study will provide breeders with a new insight into selecting heat-tolerant genotypes in cherry tomatoes.

Keywords: harvest index; heat-tolerant genotype; greenhouse; nitrogen use efficiency; photosynthesis

1. Introduction

Heat stress is caused by a combination of several environmental factors including temperature, relative humidity, air movement, solar radiation, and precipitation, with negative impacts on both plant growth and its productivity. In South Korea, the average air temperature has continuously increased since the early 21st century [1]. According to the report by Greenpeace [1], the number of extreme hot days has doubled over the past ten years. This heat warning has increasingly attracted the attention of many scientists and farmers, and several reports regarding heat stress in greenhouse vegetables were published [2,3]. Cherry tomato (*Solanum lycopersicum*) is an important vegetable, which is mostly grown in a greenhouse, with 30–40 million tons of average annual total production in South Korea [4]. The optimal day temperature for the growth of cherry tomato is in the range of 22 to 26 °C [5], and night optimal temperature is in the range of 15 to 20 °C [6]. Temperatures higher than the plants' optimal temperatures can cause heat stress, which negatively affects their growth, quality of fruits, and productivity.

Many previous studies have found that heat stress is highly related to reduction in the tomato fruit quantity or quality considerably caused by starch depletion [7], decrease in chlorophyll-carotenoid [8,9], abortion of the male gametophyte [10], decreases in root growth [11], etc. While immediate response or short-term heat stress response (<7 days) are relatively well studied [12–14], the physiological and photosynthetic processes underlying long-term (>45 days) heat stress in cherry tomato plants are still not well understood. Saidi et al. [15] reported that both short and long terms of heat exposures critically affected membrane transport and increased the damage to membrane fluidity and permeability of cells. In addition, long-term high temperature can have a greater negative impact on nutrient uptake. For example, the prolonged exposure to extreme heat can cause lower oxygen availability which leads to root browning [16,17]. As healthy portions of root turn brown, the plant may not absorb the nutrients as much as it needs, resulting in yield reduction. To have a better understanding on the effects of longer heat stress exposures, more experimentation evaluating physiological responses to long-term heat stress exposures is needed.

Heat stress responses, including physiological processes, growth and development, and yields, vary with cultivars [18] as well as species because they have different strategies to adapt the heat stress conditions [19]. Under stress conditions, plants either grow slowly to adapt stress conditions, or sacrifice their growth to respond to heat stress. The growth speed under stress is achieved through stress-triggered cell signaling [19]. The stress tolerance, also known as a relative ability to grow under stress condition, is usually determined by evaluating decreases in growth rate, fruit production, or biomass accumulation [19]. Some plants can increase the growth of certain plant organs, such as roots [20] or stems [21], as a response to stress exposures, which can result in higher biomass accumulation under stress. This higher biomass accumulation may reflect better stress tolerance. Although some plants can recover after short-term heat stress exposure (<1 day) [22], most plants have more sensitive stress-response programs under long-term heat stress. A comparison of various stress responses between different genotypes will help breeders to have a better understanding of different adaptation strategies or defense mechanisms, which will provide useful information for selection in breeding programs.

The main aim of this study is to investigate the heat stress responses of three different cherry tomato accessions that have different growth patterns and stress tolerance and are grown in a controlled greenhouse condition. Under heat stress, plants develop certain efficient strategies to avoid or tolerate the heat stress which allows them to adapt to and defend themselves from heat stress [23]. The adaptation strategies or stress resistances of three accessions were evaluated through the investigation of changes in physiological characteristics, yields, and nutrient uptake under long-term heat stress conditions.

2. Materials and Methods

2.1. Plant Material and Experimental Design

Three cherry tomato (*Solanum lycopersicum*) accessions, including HR17 (moderate heat-tolerant), HR22 (heat-sensitive), and HR24 (heat-tolerant commercial cultivar, 'Joeungyeo', Farm Hannong, Seoul, Republic of Korea), were compared. Both HR17 and HR22 have round shaped fruits, while HR24 has oval shaped fruits (Figure 1). The fruit weights for all cultivars ranged between 15 and 25 g.

Seeds of HR17 and HR22 were sown on 16 March 2022 in plastic trays (54 × 28 cm in size, 5 × 10 cm cells with pot volume 3.7 L) containing commercial bed soil ('Bio Sangto'; Seoul, Republic of Korea), containing cocopeat (67.5%), peat moss (17.0%), zeolite (5.0%), perlite (10.0%), pH adjuster (0.3%), humectant (0.014%), and fertilizers (0.185%) containing 270 mg kg^{-1} of each of N, P, and K, respectively. The seedlings were grown to seven to nine fully expanded mature leaf stage (25–30 cm height) in a glasshouse (26 °C/18 °C in day/night (16/8 h) with relative humidity within 65–70%) at the National Institute of Horticultural and Herbal Science (Wanju, Republic of Korea, 35°83′ N, 127°03′ E). Tomato seedlings were transferred into greenhouses on 3 May 2022 (48 days after sowing). Black plastic mulch film was applied to the test beds. Plants were regularly watered with a drip irrigation system and fertigated weekly with nutrient solution A (N 5.5%, K 4.5%, Ca 4.5%, B 0.00014%, Fe 0.05%, Zn 0.0001%, and Mo 0.0002%) and B (N 6%, P 2%, K 4%, Mg 1%, B 0.05%, Mn 0.01%, Zn 0.005%, and Cu 0.0015%) (Mulpure, Daeyu, Seoul, Republic of Korea). Before heat treatment, the seedling plants were grown for two weeks in greenhouse conditions with a temperature set-point for ventilation of 30 °C during the day for adaptation in new environmental conditions. After two weeks, the temperature set-point of a greenhouse was reprogrammed to 35 °C during the day, while the other greenhouse for control was kept unchanged. The average air temperatures in a control greenhouse were relatively stable at 25–35 °C, and average air temperature for heat stress was 2–5 °C higher than that in the control greenhouse (Figure 2). The total number of days of heat treatments were shown in Figure 2. The number of days when the maximum temperatures were over 40 °C were 28 days and 10 days in a heat-treated and control greenhouse, respectively (Figure 2). The relative humidity was kept between 50 and 85% in both greenhouses.

Figure 1. Fruit phenotypes of HR17, HR22, and HR24 grown in control-condition greenhouse on 57th day after treatment.

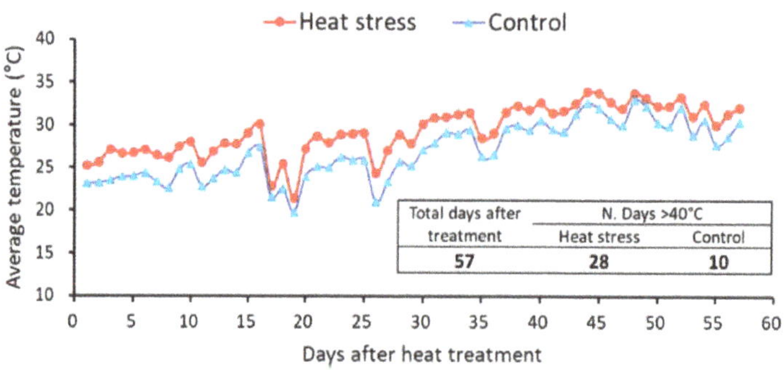

Figure 2. Average temperatures changes in heat-treated greenhouse and control greenhouse (Wanju, Republic of Korea). In table, the number of total treatment days, and the number of days when the daily maximum temperatures were over 40 °C in both greenhouses were presented. The heat stress was exposed to plants when plants were grown for about 62 days after sowing.

The experiment plots were laid out as a split design with three replicates. The main plot was a temperature treatment (heat stress or control), while the sub-plot was genotype in three accessions, including HR17, HR22, and HR24. The main plots were carried in two greenhouses. In each greenhouse, the sub-plot was laid out in randomized completed design with 1.5 m long single row plots consisting of five transplants (30 cm apart) and three replicates. The distance between single-row plots was 140 cm.

2.2. Evaluation of Physiological Characteristics and Yields

Plant height and stem thickness of three tomato cultivars were measured 1, 2, 4, 12, 25, 35, 43, and 57 days after heat treatment. For each replicate, at least three plants were measured. The plant height (cm) was measured from the base of plant to the tip. The plant stem thickness (mm) was measured using a digital caliper (CD-20APX, Mitutoyo Co., Ltd., Kanagawa, Japan). Plants were harvested on the 35th days and 57th days after heat treatment. At harvest, fresh weights (g) of fruits, leaves, and stems, and total leaf area (LA) were measured. The leaf area of each plant per m^2 covered by the crop was measured using an integrator of LA (LICOR-300, Lincoln, NE, USA). The harvested samples were dried at 70 °C, and dry weights (g) of stem, leaves, and fruits were measured. The moisture contents of each plant were measured using fresh and dry weights. The dry matter distribution was calculated using the dry weight data for each plant part.

2.3. Leaf Gas Exchange and Chlorophyll Fluorescence

On days of 1, 2, 4, 12, 25, 35, 43, and 57 days after heat treatment, a LI-COR LI-6800 (LI-COR Inc., Lincoln, NE, USA) gas exchange instrument was used to measure a net photosynthesis rate (A, μmol $m^{-2}s^{-1}$). On the 35th day, the readings were not stable; thus, the data from the 35th day were excluded from the data analysis. The measurements were taken on a newly fully expanded leaf between 10:00 and 14:00. The temperatures in Li-Cor chamber were set at 25 °C and 35 °C for control and heat treatment, respectively. The light intensity was 600 μmol $m^{-2}s^{-1}$, and the CO_2 concentration was set to 400 μmol mol^{-1} CO_2 with 60% relative humidity for both greenhouse conditions. We exposed the selected leaves to various levels of irradiation for 4–5 min until the CO_2 uptake curve was stabilized, and then data were collected.

On the same days as the measurements of net photosynthesis rates, chlorophyll fluorescence was also collected on the same days as gas exchange measurements. Photon system Instrument (FluorPen, FFP 110, PSI, Drasov, Czech Republic) was used to measure PSII photochemical efficiency (Qy, Fv/Fm). The measurements were taken on a newly fully expanded leaf after 15 min of dark adaptation. After dark adaptation, saturating light was given at 3000 μmol (photon) $m^{-2}s^{-1}$, actinic light 1000 μmol (photon) $m^{-2}s^{-1}$, and measuring at 3000 μmol (photon) $m^{-2}s^{-1}$.

2.4. Determination of Electrolyte Leakage Potential in Seedlings Leaves under Heat Stress

A newly fully expanded leaf for each replicate was collected on days 2, 4, 12, 25, 35, 43, and 57 after heat treatment to measure electrolyte leakage. Leaves were sampled from three different plants for each accession by using a cork bore as a punch. The punched leaf disks were 5.5 mm in diameter. The punched samples were placed in a 15 mL tube containing 10 mL of deionized water and then incubated on a shaker at 25 °C for 30 min. The conductivity (EC1) of water was measured using a STARA-HB conductivity meter (Thermo Orion, Waltham, MA, USA). The tube was heated in a boiling water bath (100 °C) for 20 min and cooled at room temperature for 20 min, and then the conductivity (EC2) was measured. Final EC content was calculated as the percentage of EC1/EC2.

2.5. Calculation Nitrogen Use Efficiency

After harvesting plants on the 30th and 58th days after heat treatment, plants were dried and ground for nitrogen analysis. The total N contents in the dry matters of the fruits, stems, and leaves were analyzed based on the Kjedahl method (PanReacAplliChem, 2018) [24]. Approximately 1 g of each ground sample was placed into 300 mL glass tubes. The samples were digested in 15 mL concentrated H_2SO_4 using the Kjeldahl digestion system (SH420F, Hanon, Jinan, China). The digested samples were distilled with a small amount of NaOH using the distillation (K9840, Shandong Haineng Technology Instrument Co., Ltd., Shandong, China). After distillation, 0.1 N HCl was slowly added to the samples to determine total N contents. A more detailed protocol can be found in reports from

PanReacApliChem [24]. To find the total N (g/kg), it can be calculated with the following equation [24]:

$$\frac{\left(\text{ml HCl}_{\text{sample}} - \text{ml HCl}_{\text{bland}}\right) \times [\text{HCl}_{\text{con}}] \times 14.01 \times 100)}{1000 \times \text{weight of samples(g)}} \quad (1)$$

Nitrogen use efficiencies (NUE) of fruit (f) and biomass (b) for all varieties grown in both control and heat treatment greenhouses were calculated with the following equation:

$$\text{NUE}_\text{f}(\%) = \frac{\text{Total N yield in fruit}}{\text{Total N accumulation (soil + fertilizer)}} \times 100 \quad (2)$$

$$\text{NUE}_\text{b}(\%) = \frac{\text{Total N yield in biomass(stem + leaves)}}{\text{Total N accumulation (soil + fertilizer)}} \times 100 \quad (3)$$

2.6. Statistical Analysis

The effects of heat treatment and accession were accessed by means of analysis of variance (ANOVA). The heat treatment and accession were treated as fixed factors. Pearson correlation procedures were conducted to analyze the relationships between the measured traits in control and heat treatment greenhouses in two different heat stress periods. For plant height, net photosynthesis rate, photosynthesis efficiency, and leaf leakage rates were also statistically tested by means of ANOVA. Treatment, days, accession, and interactions between them were tested. The effects of treatment, accession, organ parts, and interactions between them on NUE were also statistically tested. All statistical analyses were performed using the SAS program (SAS 9.4, Cary, NC, USA).

3. Results

3.1. Physiological Characteristic Measurements and Analysis Simple Correlation Factors

Physiological characteristics and yields of three different cherry tomato accessions, including HR17, HR22, and HR24, were determined during various periods of exposure to heat stress in both greenhouses. Fresh weights, fruit weight, and harvest index of all three accessions were summarized in Table 1. According to the results of statistical analysis (Table 2), fresh weights on the 35th days were significantly affected by heat treatment (p = 0.0032). Under heat stress, the fruit yields of three accessions were reduced. Although there were no significant differences found on the 57th day, except for in HR24, all accessions experienced a yield reduction under heat stress. Fruit weights significantly differed by accession (p = 0.004) as HR22 had the highest fruit yield among all accessions. There was a significant interaction between accession and treatment (p = 0.04). Under heat stress, HR17 showed the highest fruit yield on the 35th day after treatment. Although there were no significant differences found on the 58th day after treatment, under heat stress, HR24 produced the highest fruit yields among accessions. There were large fruit yield reductions observed in HR17 and HR22. On the 35th day, harvest index significantly differed by accession (p = 0.001) as HR24 had the lowest harvest index of 0.25. On the 58th day, the harvest index of HR24 was still lower than others, but its harvest index increased from 0.25 to 0.4 under control conditions. The harvest index of other accessions decreased from 0.6 to 0.5. The largest yield losses were observed in HR22, with 52–57% of total yield and fruit yield losses at 35 days of treatment and 58–67% yield losses at 57 days of treatment. Based on these results, significant yield losses in fresh weight and fruits were also observed in HR17 (Table 1), showing fruit losses of 43% and 64% on the 35th and 57th days, respectively; however, they showed better yield than HR22 (Table 1).

Under control conditions, in general, three accessions had different growth patterns. HR17 and HR22 typically had higher harvest index and produced larger fruit yields than biomass (stem + leaves) at both harvest dates (Table 1). On the 35th day, for example, the HR17 had the highest harvest index of 0.61, approximately 2.44-fold higher than HR24. On the 57th day, HR17 still had the highest fruit yields among three accessions. HR24 had the smallest harvest index on both harvest dates, but its fresh weight was higher or the same as HR17 under control conditions. HR22 had a similar harvest index and yields to HR17 up to the 35th day, with the smallest fresh weights on the 57th day, but its harvest index was kept in the range 0.54–0.62. According to the yield data in a control greenhouse, HR17 and HR22 tended to produce more fruits than plant biomass, while HR24 produced more plant biomass than fruits at the early growth stage. The growth patterns of the three accessions were similar at elevated temperatures. However, significant negative effects of long-term heat stress were found in plant biomass and fruit yields of two accessions, HR17 and HR22. HR24 showed the most tolerance to heat stress among all accessions. At 35 days of treatment, heat stress had no effect on fruit yields (300 g) of HR24, compared to the control (293 g). HR24 had the smallest harvest index value compared to other accessions over two periods, but it produced the highest fruit yield compared to other accessions on the 57th day. These results indicated that the growth pattern of HR24, which produced more plant biomass than fruits, had less heat stress effects for longer periods than the accessions that had harvest index values of around 0.5–0.6.

Table 1. Effects of long-term heat treatments on fresh weight (g), fresh fruit weight (g), harvest index of three cherry tomato accessions, including HR17, HR22, and HR24, in heat treatment and control conditions during two different time periods (35 days and 57 days). Yield differences of total weight and fruits were calculated for each accession in each heat-treated period. n.a. data are not available.

	Control			Heat Stress			Yield Difference (%)	
Tomato Varieties	Total Fresh Weight (g)	Fresh Fruit Weight (g)	Harvest Index	Total Fresh Weight (g)	Fresh Fruit Weight (g)	Harvest Index	Total Weight	Fruit
After 35 days								
HR17	1123 ± 192	687 ± 129	0.61	640 ± 14	395 ± 57	0.62	−43	−43
HR22	1325 ± n.a.	800 ± n.a.	0.6	640 ± 14	348 ± 25	0.55	−52	−57
HR24	1193 ± 131	293 ± 74	0.25	893 ± 272	300 ± 7	0.36	−25	2
After 57 days								
HR17	1594 ± 1071	1078 ± 721	0.50	742 ± 366	390 ± 227	0.50	−53	−64
HR22	1453 ± 787	821 ± 463	0.56	617 ± 293	268 ± 118	0.44	−58	−67
HR24	1559 ± 764	664 ± 299	0.43	1430 ± 97	438 ± 146	0.30	−8	−34

Table 2. ANOVA of the effects of heat treatment (Trt.), accession (AC), and their interactions (Trt*AC) on morphological characteristics at alpha = 0.05. n.s. data are not available.

Heat Stress Days	Factors	Fresh Weight	Fruit Weight	Harvest Index	Moisture Content	Leaf Area Index	Height	Stem
35	Trt.	0.0032	0.004	n.s.	0.0002	n.s.	n.s.	n.s.
	AC	n.s.	0.008	0.001	0.003	0.04	0.004	n.s.
	Trt*AC	n.s.	0.04	n.s.	n.s.	n.s.	n.s.	n.s.
57	Trt.	n.s.	n.s.	0.0005	0.04	n.s.	n.s.	n.s.
	AC	n.s.	n.s.	0.0004	n.s.	n.s.	0.08	n.s.
	Trt*AC	n.s.	n.s.	n.s.	n.s.	n.s.	n.s.	n.s.

In addition to yield investigations, several physiological characteristics were investigated in three tomato accessions (Table 3). The moisture contents were affected by treatment (p = 0.0002) and accessions (p = 0.003) on the 35th day, while moisture content was only significantly impacted by treatment (p = 0.04). In general, the moisture contents were significantly reduced under heat stress on both the 35th and 57th days. On the 35th day, HR22 had higher moisture content than others under control and heat stress, while HR24 had the lowest moisture content among all accessions. Leaf area index significantly differed by accession, as HR24 had the highest leaf area index (1.69) among all accessions on the 35th day. The plant height of all accessions increased as temperature increased; however, it significantly differed by accession. HR22 was the shortest (162–172 cm) among all accessions, while both HR17 and HR24 were around 180–196 cm. There were no significant effects found on stem thickness. As shown in Table 1, HR17 and HR22 had the most significant heat stress effects on yields, showing significant reductions in fruit yield and leaf area index of HR17 and HR22 in heat stress conditions (Table 1).

Table 3. Effects of long-term heat treatment on moisture content, leaf area index (LAI), plant height, and stem thickness of three cherry tomato accessions, in HR17, HR22, and HR24, grown in two different time periods, 35 days and 57 days, in heat-treated and control greenhouses. n.a. data are not available.

Effect	Tomato Varieties	35 Days				57 Days			
		Moisture Content (%)	LAI	Plant Height (cm)	Stem Thickness (mm)	Moisture Content (%)	LAI	Plant Height (cm)	Stem Thickness (mm)
Control	HR17	89 ± 1.1	0.87 ± 0.05	180 ± 4	14.05 ± 0.33	87 ± 3.6	1.82 ± 1.40	186 ± 17	14.89 ± 1.55
	HR22	90 ± n.a.	1.28 ± n.a.	162 ± 10	13.95 ± 0.7	85 ± 4.6	1.62 ± 1.32	171 ± 50	14.54 ± 2.15
	HR24	87 ± 0.1	1.69 ± 0.37	185 ± 5	14.72 ± 0.47	86 ± 4.3	2.69 ± 1.96	189 ± 21	15.91 ± 3.6
Heat	HR17	86 ± 0.8	0.62 ± 0.32	189 ± 6	15.25 ± 1.80	78 ± 6.4	1.16 ± 0.63	198 ± 7	15.11 ± 1.69
	HR22	87 ± 0.3	0.67 ± 0.09	166 ± 24	14.82 ± 1.47	73 ± 12.2	1.48 ± 0.30	173 ± 37	15.24 ± 1.88
	HR24	83 ± 0.2	1.18 ± 0.54	188 ± 5	14.92 ± 0.86	84 ± 1.8	2.52 ± 0.19	190 ± 11	15.31 ± 2.33

Under control conditions at 35 days after treatment, there were significant positive correlations between fruit yields with harvest index and moisture content, while a significant positive correlation with harvest index was observed under the heat stress condition (Figure 3). At 57 days after treatment, fruit yield had significant positive correlations with total yield under control conditions. Under the heat stress condition, fruit yields had a significant negative correlation with stem thickness, while there were significant positive correlations with total weight and moisture content (Figure 3).

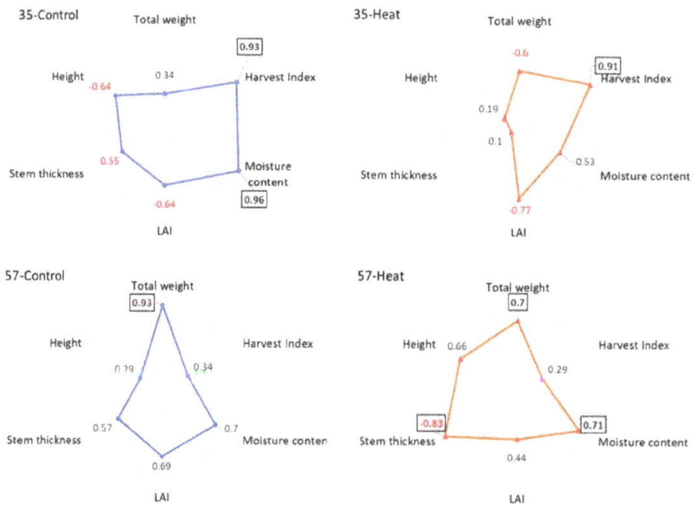

Figure 3. Correlation spider charts for showing correlation between fresh fruit weight and other morphological characteristics, including total weight, harvest index, moisture content, leaf area index (LAI), stem thickness, and plant height of all accessions treated with heat (orange line) for 35 days and 57 days and grown in control conditions (blue line). Red number indicates the negative relationship between accessions. Numbers in black box indicate significant correlation at alpha = 0.05.

3.2. Periodic Growth Responses in Morphological and Photoperiodic Parameters to Prolonged Heat Exposures

Plant height, the PSII efficiency (Qy, Fv/Fm), and net photosynthesis rate (A) of three cherry tomato accessions were measured on days 2, 4, 12, 25, 35, 43, and 57 after heat treatment (Figure 4). According to statistical analysis, shown in Table 4, plant heights were significantly affected by heat treatment and heat exposure days, both at $p < 0.0001$. Additionally, there was a significant interaction between treatment and heat stress exposure days ($p < 0.0001$). In both HR17 and HR22, there was increased sensitivity to heat stress 5–10 days after treatments, coincident with decreased plant growth. Although there were no significant differences among accessions, HR24 maintained its maximum growth under long-term heat stress. Photosynthesis was significantly affected by interaction between treatment and exposure days at $p = 0.0049$ and $p = 0.006$, respectively. Under heat conditions, Qy values of HR22 continuously decreased after four days following treatment, while it started to decrease after 25 days under control conditions (Figure 4). HR17 showed slight decreases in Qy after 10 days in control conditions, while it maintained its photosynthesis efficiency under the heat stress condition. The HR24 maintained its photosynthesis efficiency under both control and heat stress conditions during the experiment periods. All accessions started to decrease their net photosynthesis rates (A) 25 days after treatments under heat stress conditions (Figure 4).

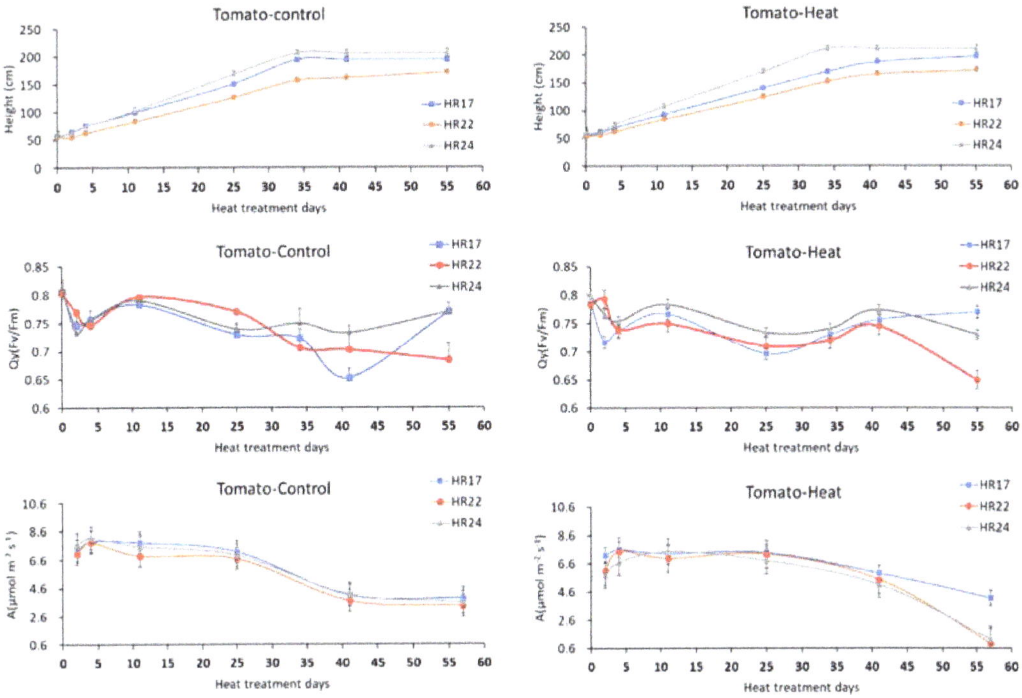

Figure 4. Effects of long-term heat treatment on plant height, the PSII efficiency (Qy, Fv/Fm), net photosynthesis rate (A) of three cherry tomato accessions, HR17, HR22, and HR24, grown in heat-treated and control greenhouses.

Table 4. ANOVA of effects of treatment days, heat stress, accessions, and their interactions on plant height, PSII efficiency, photosynthesis rates, and leaf heat damage levels. n.s. indicates no significant difference. * indicates the interaction between variables.

Factors	Height	Qy	A	EC
Trt. Days (D)	<0.0001	<0.0001	<0.0001	<0.0001
Treatment (Trt)	<0.0001	n.s.	0.058	n.s.
D*Trt	<0.0001	0.0049	0.006	0.0345
Accession (AC)	n.s.	n.s.	n.s.	n.s.
D*AC	n.s.	0.0234	n.s.	n.s.
Trt*AC	n.s.	n.s.	n.s.	n.s.
D*Trt*AC	n.s.	n.s.	n.s.	n.s.

3.3. *Effects of Long-Term Heat Stress on Leaf Heat Damage Levels and Nutrient Use Efficiency*

Leaf damage levels were investigated, with electrolyte leakages from leaf discs as an indicator of heat injury on days 2, 4, 12, 25, 35, 43, and 57 after heat treatment (Figure 5). According to statistical analysis, in Table 4, there was a significant interaction between growth days and treatment ($p = 0.0345$). Under control conditions, the injury was decreased as plants grew, while the leaf damage level tended to increase as heat stress exposure time increased under the heat stress condition (Figure 5). Although there was no significant difference detected among accessions, HR24 had the lowest damage level under the heat stress condition across the experimental periods (Figure 5), which reflected the NUE results. Although there were significant reductions in NUE for all accessions under the heat stress condition, NUE in HR24 vegetative organs was higher than other accessions at both 35 days (15.9%) and 57 days (18%) after treatments. Under control conditions, HR17 and HR22 had higher NUE in fruit organs at both 35 days and 57 days than HR24. However, all NUE in fruits at 57 days decreased to almost one-third of values at 35 days, under both control and heat stress conditions ($p < 0.0001$, Table 5).

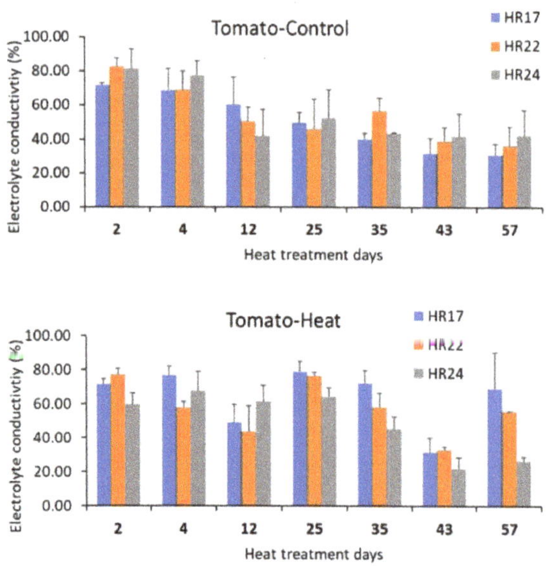

Figure 5. Effects of heat stress on leaf electrolyte leakage rate of three cherry tomato accessions for a long-term period in heat stress treatment and control greenhouses.

Table 5. Nitrogen Use Efficiency (NUE, %) and relative ratios of fruit and biomass NUE of three cherry tomato accessions grown in heat-treated and control greenhouses for 35 days and 57 days. ANOVA of effects of treatment days, heat stress, accessions, organs (biomass and fruit), and their interactions on NUE were presented. n.s. indicates no significant difference. * indicates the interaction between variables.

Treatment	Accessions	NUE (%)				Fruit/Biomass	
		35D		57D			
		Fruit	Biomass	Fruit	Biomass	35D	57D
Control	HR17	49.2	49.9	14.1	33.1	1.0	0.4
	HR22	40.7	40.6	13.3	26.9	1.0	0.5
	HR24	32.2	75.9	10.1	53.1	0.4	0.2
Heat	HR17	6.9	6.4	2.2	8.5	1.1	0.3
	HR22	3.0	7.3	1.1	9.2	0.4	0.1
	HR24	8.0	15.9	2.8	18.0	0.5	0.2

Effect	*p*-value	Effect	*p*-value
Days	<0.0001	Organ	<0.0001
Treatment (Trt)	0.045	Days*Organ	<0.0001
Days*Trt	0.016	Trt*Organ	n.s.
Accessions (AC)	<0.0001	Days*Trt*Organ	0.045
Days*AC	n.s.	AC*Organ	<0.0001
Trt*AC	n.s.	Days*AC*Organ	n.s.
Days*Trt*AC	n.s.	Trt*AC*Organ	n.s.
		Days*Trt*AC*Organ	n.s.

4. Discussion

The physiological and yield responses in heat stress conditions were significantly varied by accessions and growth stages. According to the previous studies [13,25], traits associated with heat stress can vary during the vegetative and reproductive growth stages. Thus, it is crucial to monitor changes in vegetative and reproductive traits from multiple genotypes during both vegetative and reproductive stages to adapt to elevated temperatures in the present and future [1]. In addition, the identification of key traits associated with heat tolerance will enhance the speed of the tomato breeding program by the early selection of heat-tolerant genotypes.

This study had monitored the growth, photosynthesis, and yield changes during different growth stages in three cherry tomato accessions. These tomato accessions had different growth patterns under control conditions. For example, HR17 (moderate heat-tolerant accession) and HR22 (heat-sensitive accession) tended to have high harvest index around 0.54–0.6, which means that it produced slightly more fruits than biomass under control conditions. In contrast, HR24 (heat-tolerant accession) produced less fruit yield compared to its biomass, resulting in lower harvest index values (0.25–0.46). As shown in Figure 2, the number of days when the maximum temperatures reached over 40 °C in a heat treatment greenhouse was 28 days, which was three times greater than in control conditions. As plants had exposure to more extreme heats for a long-term period, significant yield reductions were mostly observed in two accessions, HR17 and HR22. When plants produced more fruits in earlier stages, they were more sensitive to heat stress, which resulted in 43–67% fruit yield losses as in HR17 and HR22. When plants produced more biomass than fruits (e.g., HR24), they tended to be more tolerant to heat stress than others. Although HR24 produced less fruit yield than HR17 on the 34th day, its fruit yield was slightly increased by 2% under the heat stress condition. At the 57th day, the fruit yield of HR24 was only decreased by 34% and produced the highest fruit yields among all accessions.

Based on the results from correlation analysis, under heat stress conditions, fruit yields were still strongly influenced by harvest index up to the 35th day, but total fresh weight, including biomass and fruits, was strongly associated with fruit yields at the 57th day. These results indicate that plants with more leaves will be more tolerant to long-term heat stress. HR24 increased height and leaf area index as temperature increased, which resulted in the highest fresh total weight among all accessions under the heat stress condition. Since the leaf area index was high in HR24, its photosynthesis efficiency was maintained higher than other accessions for a long-term period. Many previous studies have reported that heat stress significantly affected vegetative parameters, including leaf fresh and dry weight, leaf area, plant height, stem thickness, etc. [26]. However, there are conflicting and contradictory results regarding correlation between heat stress tolerant ability and growth responses. In many cases, greater elongations of stem and leaves were observed as temperature increased [27]. This elongation response may help plants to avoid the heat dissipation through raising their leaves and meristematic tissues towards a cooling breeze [28]. In addition, heat-tolerant plants are characterized by high photosynthesis rates [29,30], and sustain gas exchange rate under heat stress [31]. Unlike our results, some previous studies had different results, showing that vegetative growth was not strongly associated with fruit yields [32,33]. Moreover, Abdelmageed and Gruda [34] reported that vegetative growth parameters, including fresh leaf weight and leaf area, of heat-tolerant tomato plants were smaller at a high temperature. This suggests that comparative studies among multiple heat-tolerant tomato accessions that have different growth patterns are needed to understand interactions between genotype and abiotic stress.

The results of NUE reflected the growth pattern of each accession. HR17 had high NUE in fruits at 35 days of growth under control conditions, and resulted in higher fruit yield production. In contrast, HR24 had high NUE in biomass because it produced more stem and leaves than fruit parts at 35 days of plant growth. Under heat stress, the overall NUE reduced significantly for all accessions. Many studies have reported that the heat stress reduced the NUE due to reduced photosynthetic leaf area [35–37] and root growth [11]. Under heat stress, fresh yield of all accessions was reduced, which resulted in a large reduction in NUE. HR24 had greater NUE than others because its fresh weight was only reduced by 34% under long-term heat stress. NUE is highly correlated with water use efficiency (WUE) [38]. According to Elio et al. [38], when cherry tomato WUE decreased, the plant absorbed less nitrogen, resulting in yield reduction. A similar result was also observed in this study. The moisture contents were significantly affected by treatment, as seen on the 35th and 57th days after treatment. Under heat stress, water contents were significantly reduced in comparison to control conditions. A greater damage level observed under heat stress also explained why NUE and water content decreased as temperature increased. A reduction in the number of healthy leaf cells caused water stress-induced leaf area reduction, which was also observed in many other species (e.g., *Sesbania aculeate*, *Phaseolus vulgaris* [39], *Sesamum indicum* [40], *Pennisctum glaucum* L. [41], and *Solanum lycopersicum* [42]).

5. Conclusions

In this study, the responses of three cherry tomato accessions, that have different growth patterns, to long-term heat stress were evaluated. The results of this study suggested some key characteristics that made the accessions more tolerant to long-term heat stress. For comparatively short-term heat stress, genotypes with a high harvest index (for example, HR17) were more favorable. However, for long-term heat stress, genotypes that had greater stem elongation and leaf area were more tolerant to heat stress. In addition, according to our results, NUE was correlated with water stress, and all accessions had lower NUE and water contents under heat stress. This suggests that a sufficient amount of water irrigation might help plants to survive under long-term heat stress. Further investigations of combination effects of water stress and heat stress on the growth of these accessions are needed to find the optimal cropping management under long-term heat stress conditions.

Author Contributions: Conceptualization, E.-Y.Y., W.C. and S.K.; data curation, B.-M.P., H.-B.J., M.-K.K. and J.-W.K.; funding acquisition, W.C.; investigation, B.-M.P., H.-B.J., E.-Y.Y., M.-K.K., J.-W.K., O.-J.L. and S.G.K.; methodology, B.-M.P., H.-B.J., E.-Y.Y., M.-K.K., J.-W.K., W.C. and S.K.; project administration, W.C.; supervision, E.-Y.Y. and S.K.; visualization, S.K.; writing—original draft, B.-M.P., H.-B.J. and S.K.; writing—review and editing, H.-B.J., W.C. and S.K. All authors have read and agreed to the published version of the manuscript.

Funding: This research was funded by (Project No: PJ01669601 "A study on the high temperature tolerance mechanism of hot pepper and tomato using an evaluation population") from the National Institute of Horticultural and Herbal Science, Rural Development Administration.

Institutional Review Board Statement: Not applicable.

Data Availability Statement: The datasets presented in this study are available upon request to the corresponding author.

Conflicts of Interest: The authors declare no conflict of interest.

References

1. Greenpeace. Extreme Temperature Rise in East Asia Cities. 2021. Available online: https://www.greenpeace.org/eastasia/press/6727/greenpeace-maps-growing-climate-risk-from-extreme-weather-in-chinas-major-cities/ (accessed on 7 September 2022).
2. Nie, W.F.; Xing, E.; Wang, J.; Mao, Y.; Ding, X.; Guo, J. Emerging Strategies Mold Plasticity of Vegetable Plants in Response to High Temperature Stress. *Plants* **2022**, *11*, 959. [CrossRef]
3. Formisano, L.; El-Nakhel, C.; Corrado, G.; Pascale, S.; Rouphael, Y. Biochemical, Physiological, and Productive Response of Greenhouse Vegetables to Suboptimal Growth Environment Induced by Insect Nets. *Biology* **2020**, *9*, 432. [CrossRef]
4. Kostat. Tomato Reports. 2022. Available online: https://sri.kostat.go.kr/board.es?mid=a90102010100&bid=11918 (accessed on 7 September 2022).
5. Sato, S.; Peet, M.; Thomas, J. Physiological factors limit fruit set of tomato (*Lycopersicon esculentum* Mill.) under chronic, mild heat stress. *Plant Cell Environ.* **2000**, *23*, 719–726. [CrossRef]
6. Charles, W.B.; Harris, R. Tomato fruit-set at high and low temperatures. *Can. J. Plant Sci.* **1972**, *52*, 497–506. [CrossRef]
7. Dinar, M.; Rudich, J.; Zamski, E. Effects of heat stress on carbon transport from tomato leaves. *Ann. Bot.* **1983**, *51*, 97–103. [CrossRef]
8. Camejo, D.; Rodríguez, P.; Morales, M.A.; Dell'amico, J.M.; Torrecillas, A.; Alarcón, J.J. High temperature effects on photosynthetic activity of two tomato cultivars with different heat susceptibility. *J. Plant Physiol.* **2005**, *162*, 281–289. [CrossRef]
9. Saeed, A.; Hayat, K.; Khan, A.A.; Iqbal, S. Heat tolerance studies in tomato (*Lycopersicon esculentum* Mill.). *Intl. J. Agr. Biol.* **2007**, *9*, 649–652.
10. Reddy, K.R.; Kakani, V.G. Screening Capsicum species of different origins for high temperature tolerance by in vitro pollen germination and pollen tube length. *Sci. Hortic.* **2007**, *112*, 130–135. [CrossRef]
11. Giri, A.; Heckathorn, S.; Mishra, S.; Krause, C. Heat Stress Decreases Levels of Nutrient-Uptake and-Assimilation Proteins in Tomato Roots. *Plants* **2017**, *6*, 6. [CrossRef]
12. Bita, C.E.; Zenoni, S.; Vriezen, W.H.; Mariani, C.; Pezzotti, M.; Gerats, T. Temperature stress differentially modulates transcription in meiotic anthers of heat-tolerant and heat-sensitive tomato plants. *BMC Genom.* **2011**, *12*, 384. [CrossRef]
13. Rajametov, S.N.; Yang, E.Y.; Jeong, H.B.; Cho, M.C.; Chae, S.Y.; Paudel, N. Heat Treatment in Two Tomato Cultivars: A Study of the Effect on Physiological and Growth Recovery. *Horticulturae* **2021**, *7*, 119. [CrossRef]
14. Qu, G.Q.; Liu, X.; Zhang, Y.L.; Yao, D.; Ma, Q.M.; Yang, M.Y.; Luo, Y.B. Evidence for programmed cell death and activation of specific caspa se-like enzymes in the tomato fruit heat stress response. *Planta* **2009**, *229*, 1269–1279. [CrossRef]
15. Saidi, Y.; Peter, M.; Finka, A.; Cicekli, C.; Vigh, L.; Goloubinoff, P. Membrane lipid composition affects plant heat sensing and modulates Ca2+-dependent heat shock response. *Plant Signal. Behav.* **2010**, *5*, 1530–1533. [CrossRef]
16. Fukuoka, N.; Enomoto, T. The occurrence of internal browning induced by high soil temperature treatment and its physiological function in Raphanus root. *Plant Sci.* **2001**, *161*, 117–124. [CrossRef]
17. Wells, C.E.; Eissenstat, D.M. Beyond the roots of young seedlings: The influence of age and order on fine root physiology. *J. Plant Growth Regul.* **2002**, *21*, 324–334. [CrossRef]
18. Guo, T.; Gull, S.; Ali, M.M.; Yousef, A.F.; Ercisli, S.; Kalaji, H.M.; Telesiński, A.; Auriga, A.; Wróbel, J.; Radwan, N.S.; et al. Heat stress mitigation in tomato (*Solanum lycopersicum* L.) through foliar application of gibberellic acid. *Sci. Rep.* **2022**, *12*, 11324. [CrossRef]
19. Zhang, H.; Zhao, Y.; Zhu, J.K. Thriving udner stress: How plants balance growht and the tresess resonse. *Dev. Cell.* **2020**, *55*, 529–543. [CrossRef]
20. Hsiao, T.C.; Acevedo, E.; Fereres, E.; Henderson, D.W. Water stress, growth and osmotic adjustment. *Philos. Trans. R. Soc. Lond. B Biol. Sci.* **1976**, *26*, 479–500. [CrossRef]
21. Zhao, Y.; Xing, L.; Wang, X.; Hou, Y.J.; Gao, J.; Wang, P.; Duan, C.G.; Zhu, X.; Zhu, J.K. The ABA receptor PYL8 promotes lateral root growth by enhancing MYB77-dependent transcription of auxin-responsive genes. *Sci. Signal.* **2014**, *3*, 53. [CrossRef]

22. Song, Y.; Chen, Q.; Ci, D.; Shao, X.; Zhang, D. Effects of high temperature on photosynthesis and related gene expression in poplar. *BMC Plant Biol.* **2014**, *14*, 111. [CrossRef]
23. Elangbam, L.D.; Sudhir, K.; Basanta, S.; Susheel, K.S. Adaptation Strategies and Defense Mechanisms of Plants during Environmental Stress. *Med. Plants* **2017**, *20*, 359–413. [CrossRef]
24. PanReacAplliChem. Nitrogen Determination by Kjeldahl Method. 2018. Available online: https://byjus.com/chemistry/kjeldahl-method/ (accessed on 21 January 2023).
25. Sun, M.; Jiang, F.; Zhang, C.; Shen, M.; Wu, Z. A new comprehensive evaluation system for thermo-tolerance in tomato at different growth stage. *J. Agric. Sci. Technol. B* **2016**, *6*, 152–168. [CrossRef]
26. Aldubai, A.A.; Alsadon, A.A.; Migdadi, H.H.; Alghamdi, S.S.; Al-Faifi, S.A.; Afzal, M. Response of Tomato (*Solanum lycopersicum* L.) Genotypes to Heat Stress Using Morphological and Expression Study. *Plants* **2022**, *11*, 615. [CrossRef] [PubMed]
27. Myster, J.; Moe, R. Effect of diurnal temperature alterations on plant morphology in some greenhouse crops—A mini review. *Sci. Hort.* **1995**, *62*, 205–215. [CrossRef]
28. Gray, W.M.; Ostin, A.; Sandberg, G.; Romano, C.P.; Estelle, M. High temperature promotes auxin-mediated hypocotyl elongation in Arabidopsis. *Proc. Natl. Acad. Sci. USA* **1998**, *95*, 7197–7202. [CrossRef] [PubMed]
29. Scafaro, A.P.; Haynes, P.A.; Atwell, B.J. Physiological and molecular changes in Oryza meridionalis Ng, a heat-tolerant species of wild rice. *J. Exp. Bot.* **2010**, *61*, 191–202. [CrossRef]
30. Nagarajan, S.; Jagadish, S.V.K.; Prasad, A.H.; Thomar, A.K.; Anand, A.; Pal, M.; Agarwal, P.K. Local climate affects growth, yield and grain quality of aromatic and non-aromatic rice in northwestern India. *Agric. Ecosyst. Environ.* **2010**, *138*, 274–281. [CrossRef]
31. Craita, E.B.; Tom, G. Plant tolerance to high temperature in a changing environment: Scientific fundamentals and production of heat stress-tolerant crops. *Front. Plant Sci.* **2013**, *4*, 273. [CrossRef]
32. Zhou, R.; Yu, X.; Ottosen, C.O.; Rosenqvist, E.; Zhao, L.; Wang, Y.; Yu, W.; Zhao, T.; Wu, Z. Drought stress had a predominant effect over heat stress on three tomato cultivars subjected to combined stress. *BMC Plant Biol.* **2017**, *17*, 24. [CrossRef]
33. Xu, J.; Wolters-Arts, M.; Mariani, C.; Huber, H.; Rieu, I. Heat stress affects vegetative and reproductive performance and trait correlations in tomato (*Solanum lycopersicum*). *Euphytica* **2017**, *213*, 156. [CrossRef]
34. Abdelmageed, A.H.A.; Gruda, N. Influence of heat shock pretreatment on growth and development of tomatoes under controlled heat stress conditions. *J. Appl. Bot. Food Qual.* **2007**, *81*, 26–28. [CrossRef]
35. Farooq, M.; Bramley, H.; Palta, J.A.; Siddique, K.H.M. Heat stress in wheat during reproductive and grain-filling phases. *Crit. Rev. Plant Sci.* **2011**, *30*, 491–507. [CrossRef]
36. Mu, Q.; Cai, H.J.; Sun, S.K.; Wen, S.S.; Xu, J.T.; Dong, M.Q.; Saddique, Q. The physiological response of winter wheat under short-term drought conditions and the sensitivity of different indices to soil water changes. *Agric. Water Manag.* **2021**, *243*, 106475. [CrossRef]
37. Alsamir, M.; Mahmood, T.; Trethowan, R.; Ahmad, N. An overview of heat stress in tomato (*Solanum lycopersicum* L.). *Saudi J. Biol. Sci.* **2021**, *28*, 1654–1663. [CrossRef]
38. Elio, E.T.M.; Chao, W.; Ashutus, S.; Elke, B.; Peiman, Z.; Beata, B.K.; Aminu, D.; Yaosheng, W. Reduced nitrogen proportion during the vegetative growth stage improved fruit yield and nitrogen uptake of cherry tomato plants under sufficient soil water regime. *Acta Agric. Scand. B Soil Plant Sci.* **2022**, *72*, 700–708. [CrossRef]
39. Ashraf, M.; Iram, A.I. Drought stress induced changes in some organic substances in nodules and other plant parts of two potential legumes differing in salt tolerance. *Flora* **2005**, *200*, 535–546. [CrossRef]
40. Hassanzadeh, M.; Asghari, A.; Jamaati-E-Somarin, S.H.; Saeidi, M.; Zabihi-E-Mahmoodabad, R.; Hokmalipour, S. Effects of water deficit on drought tolerance indices of sesame (*Sesamum indicum* L.) genotypes in Moghan region. *J. Environ. Sci.* **2009**, *3*, 116–121. [CrossRef]
41. Kusaka, M.; Lalusin, A.G.; Fujimura, T. The maintenance of growth and turgor in pearl millet (*Pennisetum glaucum* [L.] Leeke) cultivars with different root structures and osmo-regulation under drought stress. *Plant Sci.* **2005**, *168*, 1–14. [CrossRef]
42. Medyouni, I.; Zouaoui, R.; Rubio, E.; Serino, S.; Ahmed, H.B.; Bertin, N. Effects of water deficit on leaves and fruit quality during the development period in tomato plant. *Food Sci. Nutr.* **2021**, *9*, 1949–1960. [CrossRef]

horticulturae

MDPI

Article

Ascophyllum nodosum and Silicon-Based Biostimulants Differentially Affect the Physiology and Growth of Watermelon Transplants under Abiotic Stress Factors: The Case of Drought

Filippos Bantis * and Athanasios Koukounaras

Department of Horticulture, Aristotle University, 54124 Thessaloniki, Greece
* Correspondence: fbantis@agro.auth.gr

Abstract: Climate change is an inevitable process characterized by an abrupt increase in global temperature and a decrease in precipitations leading to drought incidents. Biostimulants could be a valuable tool for mitigating these harsh conditions. The objective of our study was to test the efficiency of two biostimulants, a silicon-based seaweed and the seaweed *Ascophyllum nodosum*, to mitigate the drought stress endured by watermelon transplants during the first few weeks after transplanting. In order to achieve this, three water treatments (100%, 75%, and 50% of field capacity) were applied in pots. Important growth parameters (leaf number, fresh weight, and plant area) deteriorated depending on water availability. This was also the case for the root system development displayed by root dry weight, total length, and surface area. It is the first time the OJIP transient has been evaluated after the application of *A. nodosum* for drought-stressed plants. Chlorophyll fluorescence parameters showed that the photosynthetic apparatus was more stressed when *A. nodosum* was applied, especially in the harshest conditions (i.e., 50% field capacity). Overall, the silicon-based biostimulant failed to demonstrate drought-mitigating potential compared to the non-treated counterparts. On the other hand, *A. nodosum* alleviated the negative effects of water deficit, especially in the harshest conditions.

Keywords: climate change; water deficit; photosynthetic apparatus; OJIP transient; seaweed; root system architecture; grafted seedlings; *Citrullus lanatus*

1. Introduction

Climate change is an inevitable human-induced process. According to the Sixth Assessment Report of the Intergovernmental Panel on Climate Change, global surface temperature rose by 0.85 °C between 1880 and 2021, and is expected to rise further by 0.3–0.7 °C between 2016 and 2035. In addition, heat waves and heavy rainfall have increased in frequency over large parts of Europe [1]. The effects of global warming are already being observed on a global level and are expected to further worsen in the coming years. In addition to the expected increase in average seasonal temperatures, crops are expected to suffer from another key stressor: extreme summer drought that is largely associated with extreme heat [2]. Such conditions prevailed in Central Europe in 2003 [3] and 2018, in Italy in 2017 [4], and an exceptional heatwave also occurred in Europe in early summer of 2019 [5] and 2021.

Drought is nowadays a common but very important stress factor responsible for declining crop yields. This abiotic stressor leads to morphological and physiological alterations in plants by affecting their metabolism. Estimations of water deficiency effects on plants highlighted drought's negative impact on crop quality [6] and production, which exhibited 50% losses [7], while the electron transport chain within the photosynthetic apparatus is also severely impaired [8].

Biostimulants could be a valuable tool for improving plant production. Biostimulants consist of many subclasses (humic and fulvic acids, hydrolyzed proteins, pollen grains, mycorrhiza fungi, etc.) with different modes of action showing positive results in the treatment of specific issues such as drought and salinity implications [9–13]. A silicon-based biostimulant mitigated the effects of water stress on lettuce [9]. Furthermore, brown seaweed (*Ascophyllum nodosum*) is nowadays accepted as an effective agronomic input in crop production to alleviate the effects of drought and salinity [10,11]. For example, in a study with tomatoes, *A. nodosum* alleviated the negative effects of saline irrigation [12].

Watermelon (*Citrullus lanatus*) is native to West Africa, cultivated for its fruits and established in the soil mainly through transplanted seedlings. While it is considered a berry, the crop's fruit is large and contains over 90% water. According to the FAOSTAT database, watermelon is an economically important crop throughout the world, producing over 100 million tons per year (FAOSTAT database), and is grown mainly in East Asian and Mediterranean/South European countries [14,15]. In 2021, the value of watermelon exports worldwide amounted to EUR 2.02 billion, while European countries accounted for 47% (EUR 949 million) of the global export value.

The first few weeks after transplantation are critical for plant development due to the occurrence of transplanting shock. The objective of our study was to test the efficiency of two biostimulants to mitigate the water deficiency conditions endured by watermelon transplants until flower blooming. The biostimulants used were a silicon-based seaweed and the seaweed *A. nodosum*, which were applied on the foliar, while three water (including the control) treatments were applied. Environmental stress factors can damage the photosynthetic mechanism. This damage can be evaluated by the chlorophyll fluorescence OJIP transients which correspond to the reduction stages of the electron transport chain. The photosynthetic performance of rice using silicon has already been tested with the OJIP transient [16]. However, our study is the first attempt to evaluate the plant photosynthetic apparatus under drought stress using the OJIP transient after the application of the seaweed *A. nodosum*.

2. Materials and Methods

2.1. Plant Growth

The experiment was performed at the greenhouse of the Laboratory of Vegetable Crops (N 40.536; E 22.995) of the Aristotle University of Thessaloniki, Greece, in May 2021. The grafted seedlings were provided by a professional nursery (Agris S.A., Kleidi, Imathia, Greece) at the stage of 3–4 leaves. Specifically, the plant material consisted of watermelon "Celine F1" scions grafted onto interspecific squash (*Cucurbita maxima* × *C. moschata*) "TZ-148" hybrid rootstocks. The root system was completely cut-off during grafting, as suggested by Lee and Oda [17] for rapid root growth.

The seedlings were immediately transplanted in plastic pots (volume of 1L) filled with a 2:1 mixture of peat and perlite. Each pot contained one seedling. The transplanted seedlings were irrigated with plenty of water in order for the substrate to be fully saturated. Following this, the pots were transferred into a plastic greenhouse and placed on a large bench at a randomized complete block (RCBD) with six replicates (pots).

2.2. Water Deficiency and Biostimulant Treatments

The drought experiment was initiated two days after transplanting (DAT). At DAT 2 and every two days onwards, the plants were irrigated with different amounts of nutrient solution (Hoagland pH 6.5; electric conductivity 2.6 mS cm^{-1}). Specifically, 18 pots were irrigated with 200 mL which was designated as 100% of field capacity (Control treatment) after watering, 18 pots were irrigated with 150 mL, which was designated as 75% of field capacity after watering, and 18 pots were irrigated with 100 mL, which was designated as 50% of field capacity after watering. The 100% of field capacity (200 mL) was designated upon irrigating several pots (not involved in this experiment) until a runoff of 15–20% water. In the 75% and 50% treatments, there was no runoff water from the pots. A preliminary experiment was conducted in order to determine the proper field water capacity treatments.

Moreover, at DAT 2, six plants per water deficiency treatment (18 pots in total) were sprayed with a silicon-based biostimulant (30 kg/ha) hereby labeled as "Si", while six different plants per water deficiency treatment (18 pots in total) were sprayed with an *A. nodosum* seaweed biostimulant (4 L/ha) hereby labeled as "Asc". The Si biostimulant is comprised of >85% SiO_2 (*w*/*v*). The Asc is comprised of seaweed extract of *A. nodosum* 19.5% (*w*/*v*), P_2O_5 (9.8%), and K_2O (13.7%). The amounts of foliar application for each biostimulant were in accordance with the instructions on the label for watermelon crops. Biostimulant applications were only performed once throughout the experiment, at DAT 2.

2.3. Determinations and Analysis

The plants were grown until the blooming of the first few flowers at DAT 20. Upon blooming initiation, we determined the leaf number, the stem diameter (using a digital caliper), the female and male flowers, and the relative chlorophyll content using a CCM-200 plus chlorophyll meter (Opti-Sciences, Hudson, NH, USA) which provides dimensionless values. Plant area was determined from images using WinRHIZO Pro software (Regent Instruments Inc., Québec, QC, Canada).

Chlorophyll fluorescence parameters (after 20 min dark adaptation) were determined using a Pocket PEA chlorophyll fluorometer (Hansatech, King's Lynn, UK). Specifically, PI_{abs} (performance index), φ_{P0} (maximum quantum yield for primary photochemistry), ψ_{E0} (probability that an electron moves further than Q_A), RC/ABS (Q_A reducing reaction centers per PSII antenna), V_J (the relative fluorescence at the J-step), and ΔV_{IP} (relative fluorescence increase between the intersystem carriers and electron end acceptors of PSI [18,19].

Upon plant destruction, we measured the shoot fresh weight, as well as the relative water content (RWC). RWC was measured on three fully developed leaves (4th, 5th, and 6th leaf from the lateral bud) per sample using the equation "RWC [%] = [(FW − DW) / (TW − DW)] × 100" with FW = fresh weight, TW = turgor weight, DW = dry weight. TW was measured after the leaves were moisturized in plastic bags filled with water for 24 h. The root dry weight was measured after drying the rinsed roots in an oven (72 °C for 3 days). In addition, the root architecture parameters, such as total root length and root surface area, were determined using WinRHIZO Pro software (Regent Instruments Inc., Québec, QC, Canada).

Statistical analysis was performed with SPSS software (SPSS 23.0, IBM Corp., Armonk, NY, USA) using analysis of variance (ANOVA). Post-hoc analysis was performed with the Tukey method at $\alpha = 0.05$.

3. Results and Discussion

Climate change is nowadays associated with elevated heat and less precipitations leading to increased frequency and severity of drought incidents. Among the risks arising from global warming, its impact on crop physiological and developmental processes will be devastating.

In our research, we studied the potential of silicon-based and *A. nodosum* seaweed biostimulants to alleviate the effects of water deficiency imposed on watermelon transplants. Among the evaluated morphometric parameters, the stem diameter and female flower number did not show significant differences regardless of the water and biostimulant treatments (Table 1). Watermelon is a monoecious plant where male flowers start to bloom a few days earlier than female ones. Since the experiment lasted until the blooming of the first few flowers, female flowers did not manage to exhibit potentially significant differences among the different treatments, which, in any case, was not the aim of the study. In addition, stem diameter was also similar in all treatments, which can be attributed to the transplanting shock which ensues the first few weeks after transplantation. However, male flowers were significantly more at 100% compared to 100% Asc, and 50% Si (Table 1). Following this, *A. nodosum* foliar application alleviated the effects of water deficiency stress on several important growth parameters. Specifically, the leaf number was significantly greater at 100% Asc compared to 75%, 75% Si, and all the 50% water treatments (Figure 1A).

The shoot fresh weight was significantly greater at 100% Asc compared to all the 75% and 50% treatments (Figure 1B). Both in the leaf area and in the shoot fresh weight, the 100% treatments showed greater values compared to 75% and 50%, irrespective of the biostimulants, while 50% showed the lowest values. In addition, the plant area was significantly greater at 100% Si and 100% Asc than 50% (Figure 1C). It is obvious that plants irrigated with greater amounts of nutrient solution responded better compared to the water-deficient ones, an effect which was visible even after a rather short period of 20 days after transplantation.

Table 1. Developmental and physiological parameters of watermelon plants 20 days after transplantation treated with two biostimulants (Si: silicon-based, and Asc: *Ascophyllum nodosum* seaweed) and additionally irrigated with different amounts of nutrient solution. Mean values (n = 6; ± SE) within a column followed by different letters are significantly different (a < 0.05). RWC: relative water content; RCC: relative chlorophyll content.

Treatments	Stem Diameter (mm)	Female Flowers	Male Flowers	RWC (%)	RCC
100%	6.95 ± 0.53 a	0.67 ± 0.42 a	2.33 ± 0.67 a	73.66 ± 2.91 a	81.68 ± 6.60 a
100% Si	6.85 ± 0.20 a	0.00 ± 0.00 a	2.17 ± 0.87 ab	73.11 ± 2.88 a	105.45 ± 8.57 a
100% Asc	6.38 ± 0.21 a	0.33 ± 0.21a	0.17 ± 0.17 bc	71.47 ± 0.67 a	78.18 ± 9.13 a
75%	6.78 ± 0.19 a	0.00 ± 0.00 a	1.00 ± 0.37 abc	69.62 ± 2.40 a	81.17 ± 12.98 a
75% Si	6.09 ± 0.17 a	0.00 ± 0.00 a	0.50 ± 0.34 abc	69.83 ± 3.36 a	75.67 ± 9.70 a
75% Asc	6.09 ± 0.10 a	0.17 ± 0.17a	1.17 ± 0.40 abc	67.74 ± 0.64 a	78.03 ± 10.97 a
50%	6.14 ± 0.17 a	0.00 ± 0.00 a	1.00 ± 0.52 abc	63.65 ± 1.27 a	69.52 ± 4.17 a
50% Si	6.35 ± 0.35 a	0.00 ± 0.00 a	0.00 ± 0.00 c	63.52 ± 0.69 a	72.05 ± 10.51 a
50% Asc	6.08 ± 0.11 a	0.00 ± 0.00 a	0.20 ± 0.16 abc	64.22 ± 3.73 a	89.53 ± 8.93 a

Lettuce plants treated with a silicon-based biostimulant under water deficit showed greater shoot biomass and increased antioxidant content compared to the non-treated counterpart [9], an effect that was not observed in our case. In a study with water-stressed tomatoes, a product including seaweeds (*A. nodosum* and *Laminaria digitata*) and yeast resulted in greater shoot fresh weight during flowering (similar to our experiment) compared to non-treated plants [20]. Upon seaweed extract application for plant growth, phytohormones such as auxins and cytokinins or macro- and micronutrients might impose significant positive effects [21,22]. Overall, the plant growth-promoting factors found in *A. nodosum* induce cell division, subsequently affecting leaf formation and general biomass accumulation [23].

RWC has been shown to be an important indicator of water stress in Brassica [24]. In our case, the relative water content was unaffected by the different water treatments, but 50% water led to significantly lower values regardless of biostimulant application (Table 1). It seems that watermelon has developed a drought avoidance strategy through its evolution by which it forms a dense root system allowing the roots to exploit more water, while the leaves are structured in a way to avoid water losses through transpiration [25]. Moreover, other strategies include modification of photosynthetic proteins to protect photosystem II from photoinhibition, and additionally, alteration in the proteome and transcriptome to regulate tolerance [26]. Drought stress is known to reduce stomatal conductance leading to lower RWC values [27]. Silicon has been shown to increase RWC in wheat [28] but this was not evident in our case where no differences were observed within each water treatment. The difference could be attributed to the shorter time of our experiment, which was possibly insufficient for the exhibition of variable RWC values. Quite similarly, Salvi et al. [29] did not observe stem water potential differences in *Vitis vinifera* treated with *A. nodosum* extract compared to non-treated plants. Moreover, a study with *Brassica juncea* showed greater RWC and related water parameters (i.e., water and osmotic potential) when plants were foliarly sprayed with silicon-based and *A. nodosum* biostimulants [22]. Galvao et al. [30] performed an experiment including the application of *Bacillus amyloliquefaciens* (rhizobacteria) and *A. nodosum* to alleviate drought stress in common beans, and they reported that it was not possible to correlate the use of biostimulants with the mitigation of water deficit effect. It is possible that the stress imposed by the water deficit is very difficult to strongly mitigate only by applying the known biostimulants, especially in crops already equipped with drought resistance mechanisms such as watermelon. In another study, protein hydrolysates derived from casein and soybean showed a great ability to reduce water stress in *Vitis vinifera* cv. Corvina [31].

The root system is probably the first plant tissue affected by water deficit. The root dry weight was significantly enhanced at the 100% water treatments compared to the 75% and 50% treatments, irrespective of the biostimulants. Specifically, 100% Asc showed significantly greater root dry mass than all the 75% and 50% treatments (Figure 1D). Regarding root architecture parameters, the total root length and root surface area were significantly enhanced at the 100% water treatments compared to the 75% and 50% treatments, irrespective of the biostimulants. Specifically, 100% Asc showed significantly greater total root length and root surface area than all the 75% and 50% treatments (except for 75% Asc root surface area) (Figure 1E,F). Tobacco plants treated with silicon-based biostimulants were shown to develop mechanisms such as greater root surface area and root biomass in order to absorb water in a more sufficient manner [27]. However, this was not the case for watermelon plants where silicon did not affect their root system development compared to 100%. After Si application, a thin layer of silicon was evenly spread on top of the substrate particles, which possibly reduced water transpiration, allowing the plants to absorb more water and limiting the need to develop a substantial root system. On the other hand, *A. nodosum* showed a strong tendency for increased root system development displayed by root dry biomass, total root length and surface area, which nonetheless was not significant compared to the other treatments. In an experiment with sorghum, the authors reported that silicon treatments mitigated the effects of reduced water on the root system attributes leading to higher drought tolerance, while silicon treatments did not increase the relative expression of aquaporin genes [32]. In addition, humic and fulvic acids in combination with maleic hydrazide showed a clear desmutagenic action in the root tips of *Vicia faba* [33].

The OJIP transient is a means to display the efficiency of primary photochemistry as well as the structure of the photosystem II (PSII) photosynthetic structure [34]. Even though chlorophyll fluorescence measurements revealed significant differences between certain treatments, all plants developed well, which was also evident by the absolute values of each chlorophyll fluorescence parameter. In our case, PI_{abs}, a parameter that sums up the effects of φ_{P0}, ψ_{E0}, and RC/ABS, was greater at the non-biostimulant and Si treatments compared to the Asc treatments. Specifically, 75% had greater values than 100%, 75% Asc and 50% Asc (Figure 2A). Similarly, light trapping (i.e., RC/ABS) was significantly greater at 100% Asc, 75%, 75% Si, 50%, and 50% Si compared to 50% Asc (Figure 2B). The latter followed a similar trend with PI_{abs} showing that differences in PI_{abs} can mostly be attributed to the RC/ABS component. In a study with rice, Wang et al. [16] found that the PI_{abs} of drought-stressed plants diminished, while the addition of silicon enhanced the activity of the electron transfer chain and improved their adaptability. The authors concluded that silicon alters the components of the thylakoid membrane protein, ultimately playing an important part in the transfer of light energy. However, drought-imposed watermelon transplants did not significantly alter their photosynthetic reactions after silicon application.

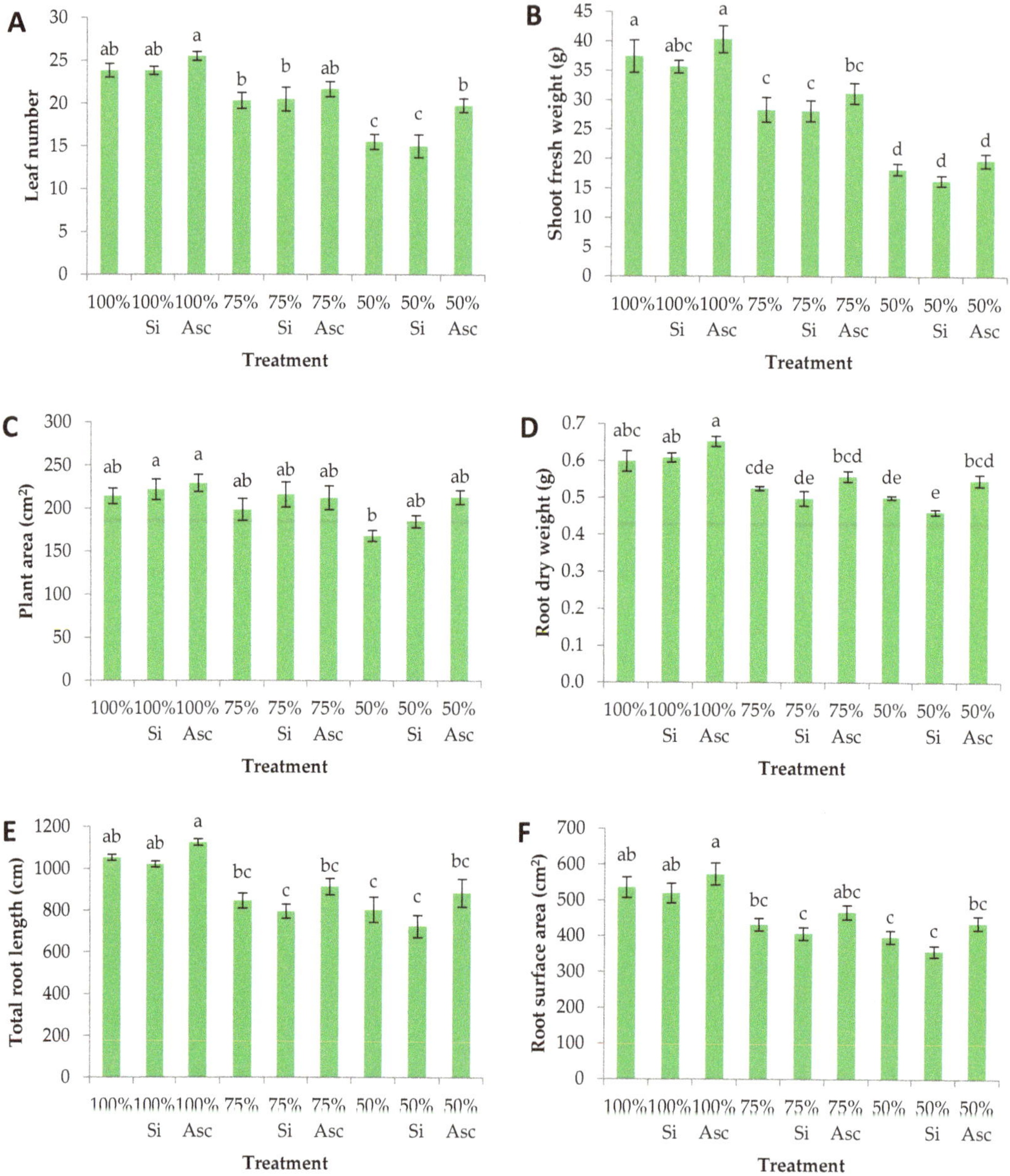

Figure 1. (**A**) Leaf number, (**B**) shoot fresh weight, (**C**) plant area, (**D**) root dry weight, (**E**) total root length, and (**F**) root surface area of watermelon plants 20 days after transplantation treated with two biostimulants (Si: silicon-based, and Asc: *Ascophyllum nodosum* seaweed) and additionally irrigated with different amounts of nutrient solution. Mean values (n = 6; ± SE) within a row followed by different letters are significantly different (a < 0.05).

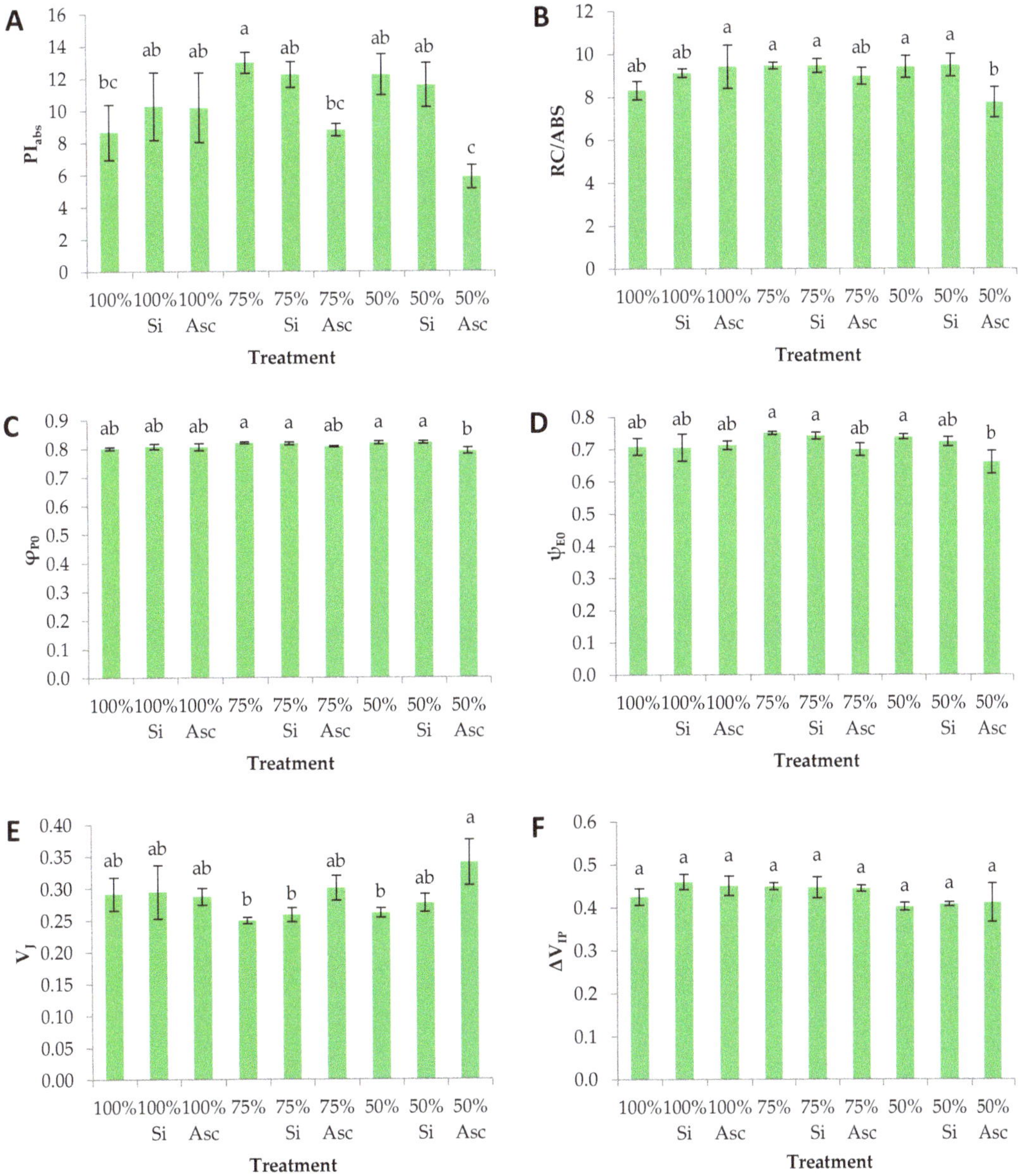

Figure 2. (**A**) PI$_{abs}$, (**B**) RC/ABS, (**C**) φ_{P0}, (**D**) ψ_{E0}, (**E**) V_J, and (**F**) ΔV_{IP} of watermelon plants 20 days after transplantation treated with two biostimulants (Si: silicon-based, and Asc: *Ascophyllum nodosum* seaweed) and additionally irrigated with different amounts of nutrient solution. Mean values (n = 6; ± SE) within a row followed by different letters are significantly different (a < 0.05).

Both the quantum efficiency of Q_A reduction (φ_{P0}) and the probability of electron transport from Q_A to the intersystem carriers of the photosynthetic apparatus (ψ_{E0}) showed greater values at 75%, 75% Si, 50%, and 50% Si compared to 50% Asc (Figure 2C,D). φ_{P0} was greater at the Si than the Asc treatments, while ψ_{E0} was greater at the non-biostimulant than the Asc treatments, irrespective of the water amount. Nonetheless, both parameters showed high values in all treatments. In particular, φ_{P0} (also known as Fv/Fm) has been reported to reach values of 0.78–0.86 in non-stressed plants [35]. In drought-stressed lettuce plants irrigated at 75% field capacity, PI_{abs} and φ_{P0} (stated as Fv/Fm) were greater after the input of a silicon-based biostimulant [27]. Nonetheless, PI_{abs} is considered a more sensitive parameter for the detection of abiotic stress factors compared to φ_{P0} [36–38]. The relative fluorescence at the J-step of OJIP transient (V_J) was significantly greater at 50% Asc than at 50%, 75%, and 75% Si (Figure 2E). In the same parameter, the Si and Asc water treatments showed greater values compared to the non-biostimulant treatments, irrespective of the water amount. The increased V_J values in the Asc treatments mentioned above indicate inhibition of the photosystem II through diminished electron flow to Q_A. This could be due to damage caused by water deficit or by controlled silencing to protect the photosynthetic apparatus from over-reduction.

Watermelon is a very sturdy species with great ability to withstand prolonged drought which is displayed by the vast root system and thick leaves. In our study, the artificial reduction of field water capacity did not manage to considerably deteriorate the crops' photosynthetic performance. However, foliar application of *A. nodosum* seaweed during the first few days after transplanting led to (statistically significant) diminished photosynthetic activity, as displayed by several OJIP parameters. This discomfort might have been perceived by the plants as a threat; an alarm provoking them to orientate their efforts to form an extensive root system, to accumulate biomass, and to form plenty of leaves, hence the relatively reduced number of flowers (c.f. Table 1). The above leads to the conclusion that a positive response, a eustress, ensues with the inclusion of the seaweed *A. nodosum* during a strong water deficit imposed on watermelon transplants.

Moreover, ΔV_{IP}, which is the relative fluorescence increase between the intersystem carriers and the PSI electron end acceptors, exhibited insignificant differences among the water deficiency treatments (Figure 2F). This was also the case for the relative chlorophyll content, which showed similar values in all treatments, even though 50% Asc showed a strong tendency for reduced values (Table 1). Nonetheless, studies [39] showed that drought stress causes chlorophyll decomposition in plants, an effect that can be alleviated in rice by the addition of silicon [16]. On the other hand, Xu and Leskovar [40] reported that under 100% field capacity the seaweed *A. nodosum* imposed no effects on the chlorophyll content and chlorophyll fluorescence of spinach. Chlorophyll decomposition is a protective mechanism of the photosynthetic apparatus during extreme conditions [41]. The insignificant differences in relative chlorophyll content enhance the scenario by which watermelon crops have the capacity to withstand extreme water deficit without considerably damaging their photosynthetic apparatus and ultimately diminishing their development. Chlorophyll degradation is also known to be a result of reduced nitrogen content, which is rather used by plants for active growth [42]. Moreover, the photosynthetic capacity is often positively correlated with the nitrogen concentration in the leaves, thus, chlorophyll and nitrogen content can be used to determine the plant's photosynthetic activity [43].

4. Conclusions

Two biostimulants, a silicon-based seaweed and the seaweed *A. nodosum*, were tested as means to mitigate the water deficit effects on watermelon transplants. As expected, 100% field water capacity led to greater aboveground and underground development compared to water-deficient counterparts of 75% and 50% field capacity. Silicon neither deteriorated nor improved the situation. However, *A. nodosum* significantly enhanced important parameters such as leaf number and plant area, and showed a strong tendency for extensive root system development. Moreover, our study is the first attempt to evaluate the plant photosynthetic apparatus under drought stress using the OJIP transient after the application of the seaweed *A. nodosum*. Watermelon is a very sturdy species with great ability to withstand prolonged drought, thus photosynthetic parameters were not considerably depleted when water was deficient. In addition, silicon application did not significantly attenuate the impact of reduced water on the photosynthetic apparatus.

However, *A. nodosum* application triggered a slight stress which provoked a positive response during a strong water deficit. Therefore, watermelon transplants concentrated their efforts to form a vast root system, to accumulate biomass, and to form plenty of leaves. Overall, water reduction depleted watermelon growth attributes, but supplementation with *A. nodosum* mitigated the impact of stress.

Author Contributions: Conceptualization, methodology, and data analysis, F.B. and A.K.; experimental measurements, F.B.; writing—original draft preparation, F.B.; writing—review and editing, F.B. and A.K.; supervision and project administration, A.K. All authors have read and agreed to the published version of the manuscript.

Funding: This research has been co-financed by the European Union and Greek national funds through the Operational Program Competitiveness, Entrepreneurship and Innovation, under the call RESEARCH—CREATE—INNOVATE (project code: T1EDK-00960, LEDWAR.gr).

Acknowledgments: The authors would like to express their gratitude to Christodoulos Dangitsis and Agris S.A. for their valuable suggestions regarding the seedlings, as well as Eleni Papoui, Emmanouil Kokolakis, and Anna Gkotzamani, Agronomists, for their assistance with the cultivation practices.

Conflicts of Interest: The authors declare no conflict of interest.

References

1. IPCC. Climate Change 2021: The Physical Science Basis. In *Contribution of Working Group I to the Sixth Assessment Report of the Intergovernmental Panel on Climate Change*; Masson-Delmotte, V., Zhai, P., Pirani, A., Connors, S.L., Péan, C., Berger, S., Caud, N., Chen, Y., Goldfarb, L., Gomis, M.I., et al., Eds.; Cambridge University Press: Cambridge, UK; New York, NY, USA, 2021; 2391p.
2. Carnicer, J.; Coll, M.; Ninyerola, M.; Pons, X.; Sanchez, G.; Penuelas, J. Widespread crown condition decline, food web disruption, and amplified tree mortality with increased climate change-type drought. *Proc. Natl. Acad. Sci. USA* **2011**, *108*, 1474–1478. [CrossRef] [PubMed]
3. Breda, N.; Huc, R.; Granier, A.; Dreyer, E. Temperate forest trees and stands under severe drought: A review of ecophysiological responses, adaptation processes and long-term consequences. *Ann. For. Sci.* **2006**, *63*, 625–644. [CrossRef]
4. Puletti, N.; Mattioli, W.; Bussotti, F.; Pollastrini, M. Monitoring the effects of extreme drought events on forest health by Sentinel-2 imagery. *J. Appl. Remote Sens.* **2019**, *13*, 020501. [CrossRef]
5. Bantis, F.; Graap, J.; Fruchtenicht, E.; Bussotti, F.; Radoglou, K.; Bruggemann, W. Field Performances of Mediterranean Oaks in Replicate Common Gardens for Future Reforestation under Climate Change in Central and Southern Europe: First Results from a Four-Year Study. *Forests* **2021**, *12*, 678. [CrossRef]
6. Osakabe, Y.; Osakabe, K.; Shinozaki, K.; Tran, L.-S.P. Response of plants to water stress. *Front. Plant Sci.* **2014**, *5*, 86. [CrossRef]
7. Machado, R.M.A.; Serralheiro, R.P.; Machado, R.M.A.; Serralheiro, R.P. Soil Salinity: Effect on Vegetable Crop Growth. Management Practices to Prevent and Mitigate Soil Salinization. *Horticulturae* **2017**, *3*, 30. [CrossRef]
8. Hayat, S.; Hasan, S.A.; Fariduddin, Q.; Ahmad, A. Growth of tomato (*Lycopersicon esculentum*) in response to salicylic acid under water stress. *J. Plant Interact.* **2008**, *3*, 297–304. [CrossRef]
9. Hidalgo-Santiago, L.; Navarro-León, E.; López-Moreno, F.J.; Arjó, G.; González, L.M.; Ruiz, J.M.; Blasco, B. The application of the silicon-based biostimulant Codasil® offset water deficit of lettuce plants. *Sci. Hortic.* **2021**, *285*, 110177. [CrossRef]
10. Paul, K.; Sorrentino, M.; Lucini, L.; Rouphael, Y.; Cardarelli, M.; Bonini, P.; Miras Moreno, M.B.; Reynaud, H.; Canaguier, R.; Trtílek, M.; et al. A combined phenotypic and metabolomic approach for elucidating the biostimulant action of a plant-derived protein hydrolysate on tomato grown under limited water availability. *Front. Plant Sci.* **2019**, *10*, 493. [CrossRef]
11. Carillo, P.; Ciarmiello, L.F.; Woodrow, P.; Corrado, G.; Chiaiese, P.; Rouphael, Y. Enhancing sustainability by improving plant salt tolerance through macro-and micro-algal biostimulants. *Biology* **2020**, *9*, 253. [CrossRef]
12. Ikuyinminu, E.; Goñi, O.; O'Connell, S. Enhancing Irrigation Salinity Stress Tolerance and Increasing Yield in Tomato Using a Precision Engineered Protein Hydrolysate and *Ascophyllum nodosum*-Derived Biostimulant. *Agronomy* **2022**, *12*, 809. [CrossRef]
13. Papoui, E.; Bantis, F.; Kapoulas, N.; Ipsilantis, I.; Koukounaras, A. A Sustainable Intercropping System for Organically Produced Lettuce and Green Onion with the Use of Arbuscular Mycorrhizal Inocula. *Horticulturae* **2022**, *8*, 466. [CrossRef]
14. Lee, J.M.; Kubota, C.; Tsao, S.J.; Bie, Z.; Hoyos Echevarria, P.; Morra, L.; Oda, M. Current status of vegetable grafting: Diffusion, grafting techniques, automation. *Sci. Hortic.* **2010**, *127*, 93–105. [CrossRef]
15. Bantis, F.; Panteris, E.; Dangitsis, C.; Carrera, E.; Koukounaras, A. Blue light promotes hormonal induced vascular reconnection, while red light boosts the physiological response and quality of grafted watermelon seedlings. *Sci. Rep.* **2021**, *11*, 21754. [CrossRef] [PubMed]
16. Wang, Y.; Zhang, B.; Jiang, D.; Chen, G. Silicon improves photosynthetic performance by optimizing thylakoid membrane protein components in rice under drought stress. *Environ. Exp. Bot.* **2019**, *158*, 117–124. [CrossRef]
17. Lee, J.M.; Oda, M. Grafting of Herbaceous Vegetable and Ornamental Crops. In *Horticultural Review*; Janick, J., Ed.; John Wiley & Sons: New York, NY, USA, 2003; pp. 61–124.

18. Strasser, R.J.; Tsimilli-Michael, M.; Srivastava, A. Analysis of the Chlorophyll a Fluorescence Transient. In *Chlorophyll a Fluorescence: A Signature of Photosynthesis*; Advances in Photosynthesis and Respiration; Papageorgiou, G.C., Govindjee, Eds.; Springer: Dordrecht, The Netherlands, 2004; pp. 321–362.
19. Bantis, F.; Fruchtenicht, E.; Graap, J.; Stroll, S.; Reininger, N.; Schafer, L.; Pollastrini, M.; Holland, V.; Bussotti, F.; Radoglou, K.; et al. The JIP-test as a tool for forestry in times of climate change. *Photosynthetica* **2020**, *58*, 224–236. [CrossRef]
20. Campobenedetto, C.; Agliassa, C.; Mannino, G.; Vigliante, I.; Contartese, V.; Secchi, F.; Bertea, C.M. A Biostimulant Based on Seaweed (*Ascophyllum nodosum* and *Laminaria digitata*) and Yeast Extracts Mitigates Water Stress Effects on Tomato (*Solanum lycopersicum* L.). *Agriculture* **2021**, *11*, 557. [CrossRef]
21. Mansori, M.; Chernane, H.; Latique, S.; Benaliat, A.; Hsissou, D.; El Kaoua, M. Seaweed extract effect on water deficit and antioxidative mechanisms in bean plants (*Phaseolus vulgaris* L.). *J. Appl. Phycol.* **2015**, *27*, 1689–1698. [CrossRef]
22. Goyal, V.; Baliyan, V.; Avtar, R.; Mehrotra, S. Alleviating Drought Stress in *Brassica juncea* (L.) Czern & Coss. by Foliar Application of Biostimulants—Orthosilicic Acid and Seaweed Extract. *Appl. Biochem. Biotechnol.* **2022**, 1–29.
23. Jacomassi, L.M.; Viveiros, J.O.; Oliveira, M.P.; Momesso, L.; de Siqueira, G.F.; Crusciol, C.A.C. A seaweed extract-based biostimulant mitigates drought stress in sugarcane. *Front. Plant Sci.* **2022**, *13*, 865291. [CrossRef]
24. Chaghakaboodi, Z.; Kakaei, M.; Zebarjadi, A. Study of relationship between some agrophysiological traits with drought tolerance in rapeseed (*Brassica napus* L.) genotypes. *Cent. Asian J. Plant Sci. Innov.* **2021**, *1*, 1–9.
25. Yoshimura, K.; Masuda, A.; Kuwano, M.; Yokota, A.; Akashi, K. Programmed Proteome Response for Drought Avoidance/Tolerance in the Root of a C3 Xerophyte (Wild Watermelon) Under Water Deficit. *Plant Cell Physiol.* **2008**, *49*, 226–241. [CrossRef] [PubMed]
26. Li, H.; Mo, Y.; Cui, Q.; Yang, X.; Guo, Y.; Wei, C.; Yang, J.; Zhang, Y.; Ma, Y.; Zhang, X. Transcriptomic and physiological analyses reveal drought adaptation strategies in drought-tolerant and -susceptible watermelon genotypes. *Plant Sci.* **2019**, *278*, 32–43. [CrossRef] [PubMed]
27. Hajiboland, R.; Cheraghvareh, L.; Poschenrieder, C. Improvement of drought tolerance in tobacco (*Nicotiana rustica* L.) plants by silicon. *J. Plant Nutr.* **2017**, *40*, 1661–1676. [CrossRef]
28. Gong, H.; Chen, K. The regulatory role of silicon on water relations, photosynthetic gas exchange, and carboxylation activities of wheat leaves in field drought conditions. *Acta Physiol. Plant.* **2012**, *34*, 1589–1594. [CrossRef]
29. Salvi, L.; Brunetti, C.; Cataldo, E.; Niccolai, A.; Centritto, M.; Ferrini, F.; Mattii, G.B. Effects of *Ascophyllum nodosum* extract on *Vitis vinifera*: Consequences on plant physiology, grape quality and secondary metabolism. *Plant Physiol. Biochem.* **2019**, *139*, 21–32. [CrossRef] [PubMed]
30. Galvao, I.M.; dos Santos, O.F.; de Souza, M.L.C.; Guimaraes, J.J.; Kuhn, I.E.; Broetto, F. Biostimulants action in common bean crop submitted to water deficit. *Agric. Water Manag.* **2019**, *225*, 105762. [CrossRef]
31. Boselli, M.; Bahouaoui, M.A.; Lachhab, N.; Sanzani, S.M.; Ferrara, G.; Ippolito, A. Protein hydrolysates effects on grapevine (*Vitis vinifera* L., cv. Corvina) performance and water stress tolerance. *Sci. Hortic.* **2019**, *258*, 108784. [CrossRef]
32. Avila, R.G.; Magalhães, P.C.; da Silva, E.M.; Júnior, C.C.G.; Lana, U.G.D.P.; de Alvarenga, A.A.; de Souza, T.C. Silicon Supplementation Improves Tolerance to Water Deficiency in Sorghum Plants by Increasing Root System Growth and Improving Photosynthesis. *Silicon* **2020**, *12*, 2545–2554. [CrossRef]
33. Ferrara, G.; Loffredo, E.; Simeone, R.; Senesi, N. Evaluation of antimutagenic and desmutagenic effects of humic and fulvic acids on root tips of *Vicia faba*. *Environ. Toxicol.* **2000**, *15*, 513–517. [CrossRef]
34. Lazár, D. The polyphasic chlorophyll a fluorescence rise measured under high intensity of exciting light. *Funct. Plant Biol.* **2006**, *33*, 9–30. [CrossRef] [PubMed]
35. Björkman, O.; Demmig, B. Photon yield of O_2 evolution and chlorophyll fluorescence characteristics at 77 K among vascular plants of diverse origin. *Planta* **1987**, *170*, 489–504. [CrossRef] [PubMed]
36. Strasser, R.; Srivastava, A.; Tsimilli-Michael, M. The Fluorescence Transient as a Tool to Characterize and Screen Photosynthetic Samples. In *Probing Photosynthesis: Mechanisms, Regulation and Adaptation*; Yunus, M., Pathre, U., Mohanty, P., Eds.; Taylor & Francis Publishers: London, UK, 2000; pp. 445–483.
37. Christen, D.; Schönmann, S.; Jermini, M.; Strasser, R.J.; Dèfago, G. Characterization and early detection of grapevine (*Vitis vinifera*) stress responses to esca disease by in situ chlorophyll fluorescence and comparison with drought stress. *Environ. Exp. Bot.* **2007**, *60*, 504–514. [CrossRef]
38. Bantis, F.; Fotelli, M.; Ilic, Z.S.; Koukounaras, A. Physiological and Phytochemical Responses of Spinach Baby Leaves Grown in a PFAL System with LEDs and Saline Nutrient Solution. *Agriculture* **2020**, *10*, 574. [CrossRef]
39. Wu, F.Z.; Bao, W.K.; Li, F.L.; Wu, N. Effects of drought stress and N supply on the growth, biomass partitioning and water-use efficiency of *Sophora davidii* seedlings. *Environ. Exp. Bot.* **2008**, *63*, 248–255. [CrossRef]
40. Xu, C.; Leskovar, D.I. Effects of *A. nodosum* seaweed extracts on spinach growth, physiology and nutrition value under drought stress. *Sci. Hortic.* **2015**, *183*, 39–47. [CrossRef]
41. Bantis, F.; Radoglou, K.; Brüggemann, W. Differential ecophysiological responses to seasonal drought of three co-existing oak species in northern Greece. *Plant Biosyst. Int. J. Deal. All Asp. Plant Biol.* **2018**, *153*, 378–384. [CrossRef]

42. Furlani Júnior, E.; Nakagawa, J.; Bulhões, L.J.; Moreira, J.A.A.; Grassi Filho, H. Correlation between chlorophyll readings and levels of nitrogen applied in bean. *Bragantia* **1996**, *55*, 171–175.
43. Pal, P.K.; Prasad, R.; Singh, R.D. Evaluating the non-destructive method for determining the chlorophyll and nitrogen content in *Stevia rebaudiana* (Bertoni) leaf. *Plant Biosyst.* **2015**, *149*, 131–135. [CrossRef]

horticulturae

Article

Mitigation of Powdery Mildew Disease by Integrating Biocontrol Agents and Shikimic Acid with Modulation of Antioxidant Defense System, Anatomical Characterization, and Improvement of Squash Plant Productivity

Nour El-Houda A. Reyad [1], Samah N. Azoz [2], Ayat M. Ali [3] and Eman G. Sayed [4,*]

[1] Plant Pathology Department, Faculty of Agriculture, Cairo University, Giza 12613, Egypt
[2] Agricultural Botany Department, Faculty of Agriculture, Cairo University, Giza 12613, Egypt
[3] Central Lab of Organic Agriculture, Agricultural Research Center, Giza 12619, Egypt
[4] Vegetable Crops Department, Faculty of Agriculture, Cairo University, Giza 12613, Egypt
* Correspondence: eman.ali@agr.cu.edu.eg; Tel.: +20-100-636-2604

Abstract: Squash (*Cucurbita pepo* L.) is a globally important vegetable, the production of which is severely constrained by powdery mildew caused by *Podosphaera xanthii*. In this study, we examined the effects of *Trichoderma asperellum* (MW965676), *Streptomyces rochei* (MN700192), and a mixture of the two foliar sprays with or without shikimic acid seed priming treatment on powdery mildew severity, plant growth, and total yield during the 2020–2021 and 2021–2022 growing seasons. We also studied their immune eliciting properties by examining their enzymatic, phenolic, and hormonal functions. The combination of *Trichoderma asperellum, Streptomyces rochei,* and shikimic acid triggered plant defense responses, which elicited enzyme activities such as peroxidase (POD), superoxide dismutase (SOD), and catalase (CAT), phenolic compound accumulation, and increased salicylic acid (SA) and jasmonic acid (JA) content. This approach yielded high-quality results in the control of powdery mildew during the two growing seasons under greenhouse conditions. Additionally, relatively large statistical differences in plant growth, total yield, mineral components, and physiological traits were observed. A GC–MS analysis of *Trichoderma asperellum* (MW965676) showed hemin cation as a major component, while *Streptomyces rochei* (MN700192) contained 2,4-di-tert-butyl phenol and the hexadecenoic acid methyl ester. With respect to the morphological changes induced by powdery mildew and the treatments, plants treated with a mixture of *Trichoderma asperellum, Streptomyces rochei* and shikimic acid showed an improvement in the thickness of the midvein, increased dimensions of the main midvein bundle, a larger number of xylem rows in the main midvein bundle, greater mean diameters of vessels and of parenchyma cells in the ground tissues, as well as increased thickness of the upper and lower epidermis, lamina, palisade tissue and spongy tissue. This extensive, new study is the first step toward a more profound understanding of the use of *Trichoderma asperellum* and *Streptomyces rochei* with shikimic acid-primed seeds as a potential alternative technique for attenuating powdery mildew infection in squash.

Keywords: *Cucurbita pepo* L.; *Podosphaera xanthii*; *Trichoderma* spp.; *Streptomyces* spp.; seed priming; phenols; antioxidant enzyme

1. Introduction

Squash (*Cucurbita pepo* L.) is an important vegetable crop. It is widely cultivated and consumed as food in many countries worldwide. It is a good source of vitamins A, C, and B and is rich in antioxidants and minerals like potassium, magnesium, and manganese [1]. Powdery mildew is the most serious disease affecting squash plants, accounting for 30–50% of yield losses [2]. Resistant varieties, chemical fungicides, natural products, and biological control are used to combat plant diseases [3]. Disease control is primarily based on the use of recommended chemical fungicides that are hazardous to humans, animals, plants, and useful organisms [4].

To address these issues, appropriate control measurements and alternative safe methods, such as biological control, must be implemented. *Trichoderma* spp. and *Streptomyces* spp. are widely used as bioagents. Some *Trichoderma* species have been used to control mildew; for example, *T. viride* culture filtrates were effective in controlling *Leveillula taurica* [5], and *T. asperellum* and *Metarhizium anisopliae* were effective in controlling *L. taurica* on peppers [6]. In addition, *T. asperellum* IZR D-11 efficiently protects common oak leaves from powdery mildew caused by *Erysiphe alphitoides* [7]. The growing interest in *Trichoderma* spp results from its various complex mechanisms, *i.e.*, competition for nutrients and space, mycoparasitism, the degradation of pathogen cell walls, and the induction of plant resistance [8].

On the other hand, the genus *Streptomyces* is a promising candidate for controlling phytopathogenic microorganisms, including powdery mildew pathogens, through a range of mechanisms, including competition for nutrients, secretion of antibiotics and lytic enzymes, and stimulation of plant defenses [9]. For example [10], reported that *Streptomyces* sp. (AcH 505) reduced oak powdery mildew disease by activating plant defense responses. Additionally, wuyiencin produced by *S. albulus* CK-15 significantly reduced the disease incidence and severity of powdery mildew in cucumber plants [11].

Shikimic acid is the well-known precursor of the lignin building blocks, L-phenylalanine and L-tyrosine. Plants use phenylalanine, leucine, and tyrosine to suppress pathogen growth, promote the production of antimicrobial chemicals [12], and improve crop protection under stress conditions by altering the enzymes involved in their defense responses [13].

Several studies have shown that by activating their complex defense responses, plants can evolve appropriate defense mechanisms to recognize and resist fungal infections [14,15]. One of these reactions is the rapid formation of reactive oxygen species (ROS), such as hydrogen peroxide (H_2O_2), hydroxyl radicals (OH), and superoxide anions (O2) [15]. ROS scavenging enzymes such as superoxide dismutase (SOD), peroxidase (POD), catalase (CAT), and ascorbate peroxidase (APX) play an essential role in regulating ROS levels and the extent of oxidative damage [15]. In higher plants, another defense response is the phenylpropanoid pathway [16]. Phenylalanine ammonia-lyase is the first enzyme involved in a number of structurally diverse, defensive phenolic and lignin compounds [17]. Polyphenol oxidase (PPO) is another key enzyme involved in the formation of phenolic compounds to defend against pathogens in plants [18]. In addition, plant hormones such as jasmonic acid (JA) and salicylic acid (SA) are involved in all plant defense reactions. However, increasing JA levels in plants can improve resistance to necrotrophic infections and make plants more sensitive to biotrophic pathogens [19].

The purpose of this study was to estimate the efficacy of several potential biocontrol agents (BCAs) alone and combined with shikimic acid to control squash powdery mildew. We also examined their impact on squash plant growth, yield, and quality to determine the potential of using these two control options as an integrated disease management strategy, as well as to measure oxidative enzyme activity, plant hormones, photosynthesis, and mineral uptake to determine their role in powdery mildew control.

2. Materials and Methods

2.1. Plant Material and Tested Compounds

A squash (*Cucurbita pepo* L. cv Eskandarani) cultivar was employed as an experimental model in all experiments throughout this study. Seeds were obtained from Mecca Trade Co. (Cairo, Egypt), and shikimic acid was purchased from Sigma-Aldrich, (Humberg, Germany).

2.2. Biocontrol Agents (BCAs) Preparation

We investigated the potential benefits of *T. asperellum (MW965676) and S. rochei (MN700192)* for squash powdery mildew control. These bioagents were previously isolated from pepper [20] and sugar cane [21] plants, respectively. For improved growth of *T. asperellum*, three fresh fungal discs (5 mm) of three-day-old culture were grown in a 250 mL conical flask containing 200 mL liquid gliotoxin fermentation medium consisting of 25 g dextrose, 2 g KH_2PO_4, 2 g ammonium tartrate, and 0.01 g $FeSO_4$. The inoculated flasks were incubated in a rotary shaker at 25 °C for 11 days. On the other hand, *S. rochei* was grown in 100 mL of starch-casein medium consisting of 10 g of soluble starch, 0.3 g casein, 2 g KNO_3, 2 g NaCl, 2 g K_2HPO_4, 0.05 g $MgSO_4{\cdot}7H_2O$, 0.02 g $CaCO_3$, 0.01 g $FeSO_4{\cdot}7H_2O$, and 1000 mL distilled water in an Erlenmeyer flask (250 mL). The pH was adjusted to 7.3 and the flasks were incubated on a rotary shaker at 30 °C for 7 d.

2.3. Identification and Inoculation of *P. xanthii*

Based on its morphological features, we identified the pathogen according to the method described in [22]. *P. xanthii* conidia were collected from infected squash plants by rubbing infected leaves with a soft brush and rinsing with distilled water; this water was then collected and used as an inoculum. The conidial suspension was adjusted to 10^5 conidia mL^{-1} by a hemocytometer and sprayed onto the leaves of squash seedlings using a hand sprinkler at a rate of 5 mL per leaf.

2.4. Seed Priming Preparation

Before sowing, summer squash seeds (*Cucurbita pepo* L. cv. Eskandarani) were sterilized in 7% sodium hypochlorite for 10 min and then fully washed with distilled water. The sterilized seeds were divided into four groups. The control treatment (group 1) was primed in water, while groups 2, 3, and 4 were soaked in shikimic acid at 20, 40, and 60 ppm, respectively. For priming, summer squash seeds were immersed in a solution of shikimic acid at 20 °C for 5 h in the dark. Following the treatments, the seeds were washed four times in distilled water for 5 min each time and then dried using blotting paper. They then received a flow of dry air at 30 °C until the original moisture content was approximated [13]. Following priming, seeds were used for the pot experiments.

2.5. Pot Experiments

Two pot experiments were conducted in the greenhouse of the Plant Pathology Department, Faculty of Agriculture, Cairo University, Giza, Egypt, to evaluate the effect of shikimic acid at 20, 40, and 60 ppm and/or foliar spraying of certain bioagents (*T. asperellum*, *S. rochei*, and *T. asperellum* + *S. rochei*). The experiment comprised four main groups. The first was the control treatment, with seeds primed in water. The other three groups (2, 3, and 4) were primed with shikimic acid at 20, 40, and 60 ppm. The three groups of primed squash seeds and control were sown in pots 50 cm^2 in diameter, which were filled with a 1:1:1 mixture of peat moss, vermiculite, and perlite. Each group was subdivided into two subgroups. The first group was sprayed with water alone and served as the control. The second group was separately sprayed with *S. rochei*, *T. asperellum*, and a mixture of the two bioagents (*S. rochei* + *T. asperellum*) at a rate of 1/50 L of water (1 mL of each bioagent was adjusted to 20×10^6 CFU). The initial foliar spraying with the bioagent was performed 30 d after planting, and spraying was performed every 15 d in the early morning. All groups of plants were artificially inoculated with an aqueous conidial suspension of *P. xanthii* 24 h after the first treatment (freshly collected *P. xanthii* conidia were suspended in deionized water to a concentration of 10^5 conidia mL^{-1} water) as described in [23]. All other horticultural practices were performed as recommended for summer squash. The experiment was performed twice. A split plot design was used to analyze all of the data obtained from six replicates (total of six plants per replicate). Plant height (cm), number of leaves, and chlorophyll (SPAD) readings were assessed for all treatments in both experiments 45 d after sowing. Photosynthesis, transpiration rate, and leaf stomatal conductance analyses were performed using an infrared gas analyzer, the LICOR 6400 Portable Photosynthesis System (IRGA, Licor Inc., Lincoln, NE, USA), on the fourth leaves of 10 squash plants chosen from each treatment, with six pot replications. Measurements were taken from

10 a.m. to 2 p.m., with a light intensity of around 1300 mol m^{-2} s^{-1} and 85% RH. The leaf chamber temperature ranged from 26.2 °C to 27 °C, and the volume gas flow rate was 400 mL min^{-1}. The CO_2 content in the air was 399 μmol mol^{-1}.

2.6. Greenhouse Experiments

Two greenhouse experiments were performed in the vegetable crop greenhouse of the Department of Cairo University, Faculty of Agriculture in Giza (located at 31°1′13″ N, 30°13′5″ E), during two successive growing seasons (October to February 2020–2021 and 2021–2022) to evaluate the impact of 40 ppm shikimic acid seed priming and/or foliar spraying of certain bioagents (*T. asperellum, S. rochei*, and *T. asperellum* + *S. rochei*) on the powdery mildew, vegetative growth, and total yield of squash plants. Squash seeds primed with 40 ppm of shikimic acid were sown directly in the greenhouse on October 6 and 10 during the 2020–2021 and 2021–2022 seasons, respectively. Squash seeds were sown on ridges of 1.5 m width, 7 m length on one side of the ridge, 50 cm apart. Control squash seeds (not primed with shikimic acid at 40 ppm) were grown under the same conditions as the primed seeds. Initial foliar spraying with the bioagent was performed 30 d after planting, and spraying was performed every 15 d in the early morning. All agricultural and farming practices for squash crops in the greenhouse were performed as recommended by the Egyptian Ministry of Agriculture. A drip irrigation system was used. The farm soil type was clay loam in texture with 7.6, EC 1.2 dS m^{-1} and contained 118 ppm N, 23 ppm P, and 30 ppm K. The pH value was 7.02, and the soluble cation values were 4.5, 3.2, 2.2, and 3.3 meq L^{-1} for Ca^{++}, Mg^{++}, K^{+}, and Na^{+}, respectively, and 0.8, 4.8, and 7.1 meq L^{-1} for HCO^{3-}, Cl^{-}, and SO_4, respectively. The treatments were arranged in a randomized complete block design with three replicates.

2.7. Data Recorded

2.7.1. Disease Assessment

Disease severity, whether in pots or in the greenhouse experiments, was determined-weekly as described in [24], based on a scale ranging from 1 to 5: 1 = no detectable infection; 2 = 1–5% of leaf area infected; 3 = 6–25% of leaf area infected; 4 = 26–50% leaf area infected; and 5 = more than 50% of leaf area infected. Disease severity percentage was estimated using the following equation [25]:

$$\text{Disease severity (\%)} = [\textstyle\sum (n \times v)/5N] \times 100$$

where n refers to the number of infected leaves in each category, v denotes the numerical values of each category, and N is the total number of infected leaves.

The mean area under the disease progression curve (AUDPC) for each replicate was calculated using Equation [25] (Pandey et al., 1989).

$$\text{AUDPC} = D\,[1/2(Y1 + Yk) + (Y2 + Y3 + \ldots\ \ldots + Yk - 1)]$$

where D = time interval, Y1 = first disease severity, Yk = last disease severity, and Y2, Y3, and Yk − 1 = intermediate disease severity.

2.7.2. Vegetative Growth

Four plants were randomly picked with roots from each plot 50 days after seeding to measure the vegetative growth parameters. Plant height, the number of leaves per plant, and the fresh and dry weight of the leaves per plant were all measured. A SPAD meter was used to determine the chlorophyll content in the sample.

2.7.3. Yield and Its Components

The total number of fruits per plant, total yield per plant, average fruit weight, and total yield per hectare were calculated.

2.7.4. Mineral Content in Squash Leaves

The total nitrogen concentration of the dried samples was measured using the modified micro-Kjeldahl method [26]. Phosphorus was determined calorimetrically by using the chloro stannous molybdophosphoric blue color method in sulfuric acid, as described in [27]. The total potassium and calcium contents were determined using the flame photometer apparatus (CORNING M 410, Germany). Concentrations of Mg, Zn, B, Fe, and Mn in leaf samples were measured by atomic absorption spectrophotometry with air-acetylene and fuel (PyeUnicam, model SP-1900, USA).

2.7.5. Activity of Antioxidant Enzymes, Salicylic Acid and Jasmonic Acid

Salicylic acid and jasmonic acid were measured according to the method described in [28]. For enzyme analysis, at 45 days after sowing (24 h after the last spray), 0.5 g leaf tissues were homogenized in 3 mL of 0.05 M Tris buffer (pH 7.8), containing 0.001 M EDTA–Na2 and 7.5% Polyvinylpyrrolidone at 0–4 °C. The homogenates were centrifuged (12,000× *g* rpm, 20 min, 4 °C), and the total soluble enzyme activity in the supernatant was measured colorimetrically using a UV-160A spectrophotometer (Shimadzu, Kyoto, Japan). Catalase (CAT), superoxide dismutase (SOD), and peroxidase (POD) activities were determined according to [29]

2.7.6. GG/MS Analysis of *T. asperellum* Culture Filtrate

The chemical composition of the *T. asperellum* culture filtrate was determined using a Trace GCTSQ mass spectrometer (Thermo Scientific, Austin, TX, USA) with a TG5MS direct capillary column (30 m, 0.25 mm, 0.25 m film thickness). The column oven temperature was initially maintained at 50 °C and then increased at a rate of 5 °C/min to 250 °C and held for 2 min. It was subsequently increased at 30 °C/min to the final temperature of 30 °C and held for 2 min. The injector and MS transfer line temperatures were maintained at 270 and 260 °C, respectively. Helium was used as the carrier gas at a constant flow rate of 1 mL/min. The solvent delay was 4 min, and diluted 1 L samples were injected automatically by the AS1300 autosampler coupled with GC in split mode. EI mass spectra were recorded at 70 eV ionization voltages over the range m/z 50–650 in full-scan mode. The ion source temperature was set at 200 °C. The components were identified by comparing their mass spectra to those of the WILEY 09 and NIST 14 mass spectral databases, as described in [30].

2.7.7. Anatomical Studies

The tested materials included the blade of the fifth leaf developed on the main stem of summer squash (cv. Eskandarani) affected by powdery mildew as a control and treated with shikimic acid, bioagents (*S. rochei*+ *T. asperellum*), and mix of the two. Samples were taken throughout the second growing seasons of 2021 at 50 days after sowing. About 1.0 cm of specimens were killed and fixed in FAA solution (5 mL glacial acetic, 10 mL formalin, 35 mL water, and 50 mL ethyl alcohol 70%) for at least 48 h. The selected materials were washed in 30% ethyl alcohol, dehydrated in a normal ethanol and butyl alcohol series, embedded in paraffin wax with a melting point 56 °C, sectioned to a thickness of 15 micrometers (μm), stained with crystal violet-erythrosin, cleared in xylene, and counted in Canada balsam [31]. Transverse sections were done with a Leica Microtome RM 2125 and photographed and measured using the Leica Light Image Analysis System DM 750 at the Faculty of Agriculture, Cairo University-Research Park (CURP). The following parameters were recorded:

- Thickness of midvein (μm)
- Thickness of upper periderm (μm)
- Thickness of lower periderm (μm)
- Thickness of lamina (μm)
- Thickness of palisade tissue (μm)
- Thickness of spongy tissue (μm)
- Dimension of main midvein bundle (μm)

- No of xylem rows in main midvein bundle
- Mean diameter of vessels (μm)
- Mean diameter of parenchyma cells in the ground tissue (μm)

2.7.8. Statistical Analysis

Using the MSTAT-C computer software package prepared by [32], the data were subjected to analysis of variance. The LSD test was used to examine differences between treatment means at a 5% level of probability [33]. Furthermore, principal component analysis (PCA) was carried out using all data points of individual response variables using Origin pro-2021 version software. A heatmap was generated to visualize variations and similarities in all response variables combined with a two-way hierarchical cluster analysis using the standardized means of all results matrices for the tested treatments.

3. Results

3.1. Pot Experiments

3.1.1. Morphology of *P. xanthii*

Our microscopic examinations of the anamorphic stage of the causal pathogen showed that mycelium is ectophytic. Conidiophores were un-branched, bearing conidia in long chains varying from ellipsoid to ovoid forms of 25.30–33.20 × 14.95–22.70 μ (17.54 × 30.75 μ on the average) in size. Observations of conidia in 3% KOH indicated the presence of fibrosin bodies, which are commonly found in the cucurbit powdery mildew genus *Podosphaera*, as shown in Figure 1.

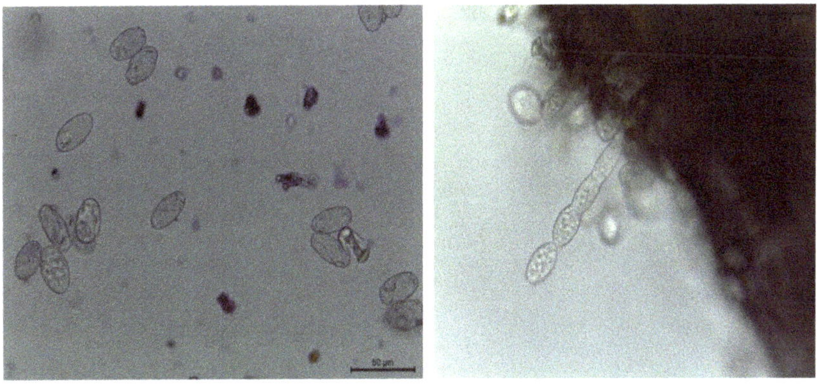

Figure 1. Morphological features of anamorphic stage of *P. xanthii*.

3.1.2. Disease Assessment

Under artificial inoculation with powdery mildew, the response of squash to shikimic acid as seed priming at 20, 40, and 60 ppm with or without foliar application of BCAs was observed. As shown in Table 1, all tested treatments had a highly significant effect on powdery mildew disease severity (DS%) in comparison to the control during the two trials. In this regard, significant disease severity reduction was obtained with 40 and 20 ppm shikimic acid, through seed priming combined with a mixture of BCAs (*S. rochei* + *T.asperellum*) as a foliar application, while the application of 60 ppm shikimic acid through seed priming, followed by the application of *S. rochei* through foliar spray as individual treatments, recorded the lowest disease severity reduction

Table 1. Effect of the interaction between seed priming of shikimic acid (0, 20, 40, 60 ppm) and foliar applications with *S. rochei* (SR), *T. asperellum* (TS), *S. rochei* + *T. asperellum* (SR + TS), and water (control) on powdery mildew severity % on squash plants during the first and second experiments.

Shikimic Acid Concentrations	Bioagents Treatments	1st Experiment		2nd Experiment	
		Disease Severity %	Reduction %	Disease Severity %	Reduction %
0 ppm	Control	37.04 ^{a}	-	36.30 ^{a}	-
	T. asperellum	30.37 ^{cd}	18.01	30.37 ^{bc}	16.34
	S. rochei	31.85 ^{bc}	14.01	32.59 ^{ab}	10.22
	T. asperellum +*S. rochei*	28.89 $^{c–e}$	22.00	25.9 $^{c–e}$	28.57
20 ppm	Control	28.15 $^{c–e}$	24.00	24.44 $^{d–f}$	32.67
	T. asperellum	23.70 ^{fg}	36.02	22.22 $^{e–g}$	38.75
	S. rochei	26.67 $^{d–f}$	28.00	22.96 $^{e–g}$	36.75
	T. asperellum +*S. rochei*	20.74 ^{g}	44.01	19.26 ^{gh}	46.94
40 ppm	Control	28.15 $^{c–e}$	24.00	24.44 $^{d–f}$	32.67
	T. asperellum	22.96 ^{fg}	38.01	19.26 ^{gh}	46.94
	S. rochei	23.70 ^{fg}	36.02	21.48 $^{e–g}$	40.83
	T. asperellum +*S. rochei*	20.00 ^{g}	46.00	16.30 ^{h}	55.10
60 ppm	Control	34.82 ^{ab}	05.99	30.37 ^{bc}	16.34
	T. asperellum	25.93 ^{ef}	29.99	22.22 $^{e–g}$	38.79
	S. rochei	29.63 $^{c–e}$	20.01	28.15 $^{b–d}$	22.45
	T. asperellum +*S. rochei*	22.96 ^{fg}	38.01	20.74 $^{f–h}$	42.87
	LSD value at 0.05:	4.17	-	4.47	-

Mean values in each column followed by a different letter are significantly different according to the LSD test ($p \leq 0.05\%$).

3.1.3. Plant Growth Parameters of Squash Plants

Under artificial inoculation with powdery mildew, the response of squash to shikimic acid as seed priming at 20, 40, and 60 ppm, with or without foliar application of BCAs, was observed. As shown in Tables 2 and 3, all tested treatments had a highly significant effect on enhancing plant growth parameters in comparison to the control in the two trials. In this regard, the maximum plant height, number of leaves, and chlorophyll content were noted following the application of 40 ppm shikimic acid through seed priming combined with exogenous foliar spray with a mixture of BCAs in both seasons.

3.1.4. Physiological Traits of Squash Plants

Transpiration rate, stomatal conductance and photosynthesis were significantly influenced by seed priming with shikimic acid and the foliar application of BCAs (Figure 2). However, no statistically significant differences were observed between 60 ppm shikimic acid and untreated squash plants in terms of transpiration rate. The values obtained for the physiological traits were higher at 40 ppm shikimic acid Overall, the mixture of BCAs as a foliar application treatment showed statistically significant ($p \leq 0.05$) transpiration rate, stomatal conductance, and photosynthesis. In this case, the untreated plants had the lowest transpiration rate, stomatal conductance and photosynthesis. The combination of seed priming with shikimic acid 40 ppm and the mixture of BCAs exhibited better transpiration rate, stomatal conductance, and photosynthesis (Figure 2A–D).

Table 2. Effect of the interaction between seed priming of shikimic acid (0, 20, 40, 60 ppm) and foliar applications with *S. rochei* (SR), *T. asperellum* (TS), *S. rochei* + *T. asperellum*(SR + TS), and water (control) on the vegetative growth characteristics of squash plants during the first experiment.

Shikimic Acid Concentrations	Bioagents Treatments	Plant Height (cm)	Leaves Numbe per Plant	Chlorophyll (SPAD) Reading
0 ppm	Control	22.73 ^{j}	9.433 ^{i}	21.73 ^{g}
	T. asperellum	47.73 $^{d–g}$	12.77 $^{f–h}$	25.10 ^{ef}
	S. rochei	38.40 ^{hi}	11.43 ^{h}	25.23 ^{ef}
	T. asperellum +*S. rochei*	51.07 ^{cde}	13.50 ^{fg}	35.60 ^{c}
20 ppm	Control	44.33 $^{f–h}$	11.20 ^{h}	27.00 ^{e}
	T. asperellum	54.33 $^{b–d}$	13.87 ^{f}	35.03 ^{c}
	S. rochei	47.00 $^{e–g}$	11.87 ^{gh}	35.57 ^{c}
	T. asperellum +*S. rochei*	61.0 ^{b}	16.87 ^{de}	46.00 ^{b}
40 ppm	Control	50.33 $^{c–f}$	16.87 ^{de}	30.73 ^{d}
	T. asperellum	59.33 ^{b}	20.20 ^{b}	46.23 ^{b}
	S. rochei	50.00 $^{c–f}$	17.87 ^{cd}	45.23 ^{b}
	T. asperellum +*S. rochei*	69.00 ^{a}	23.20 ^{a}	57.10 ^{a}
60 ppm	Control	35.60 ^{i}	13.87 ^{f}	23.73 ^{fg}
	T. asperellum	49.00 $^{d–f}$	17.87 ^{cd}	27.10 ^{e}
	S. rochei	42.00 $^{g–i}$	16.20 ^{e}	25.90 ^{ef}
	T. asperellum +*S. rochei*	56.33 ^{bc}	18.53 ^{c}	36.23 ^{c}
LSD value at 0.05:		6.7	1.646	2.6

Mean values in each column followed by a different letter are significantly different according to the LSD test ($p \leq 0.05\%$).

Table 3. Effect of the interaction between seed priming of shikimic acid (0, 20, 40, 60 ppm) and foliar applications with *S. rochei* (SR), *T. asperellum* (TS), *S. rochei* + *T. asperellum* (SR + TS), and water (control) on the vegetative growth parameters of squash plants during the second experiment.

Shikimic Acid Concentrations	Bioagents Treatments	Plant Height (cm)	Leaves Number per Plant	Chlorophyll (SPAD) Reading
0 ppm	Control	27.33 ^{h}	9.33 ^{i}	21.4 ^{i}
	T. asperellum	35.33 ^{fg}	12.67 ^{gh}	24.43 ^{gh}
	S. rochei	31.33 ^{gh}	11.33 ^{h}	26.23 ^{fg}
	T.asperellum+*S. rochei*	46.00 ^{cde}	14.00 ^{fg}	35.87 ^{d}
20 ppm	Control	43.0 ^{de}	14.67 ^{f}	27.40 ^{ef}
	T. asperellum	51.67 ^{c}	18.67 ^{cd}	34.90 ^{d}
	S. rochei	46.0 ^{cde}	17.00 ^{e}	37.23 ^{d}
	T.asperellum+*S. rochei*	60.0 ^{b}	19.33 ^{c}	47.43 ^{b}
40 ppm	Control	49.00 ^{cd}	17.67 ^{de}	29.07 ^{e}
	T. asperellum	58.0 ^{b}	21.00 ^{b}	44.23 ^{c}
	S. rochei	48.67 ^{cd}	18.67 ^{cd}	43.23 ^{c}
	T.asperellum+*S. rochei*	67.67 ^{a}	24.00 ^{a}	55.10 ^{a}
60 ppm	Control	30.67 $_{gh}$	12.00 ^{h}	22.07 ^{hi}
	T. asperellum	45.67 ^{cde}	14.67 ^{f}	25.43 ^{fg}
	S. rochei	40.00 ^{ef}	12.67 ^{gh}	27.23 ^{ef}
	T.asperellum+*S. rochei*	59.0 ^{b}	17.67 ^{de}	36.10 ^{d}
LSD value at 0.05:		6.0	1.6	2.58

Mean values in each column followed by a different letter are significantly different according to the LSD test ($p \leq 0.05\%$).

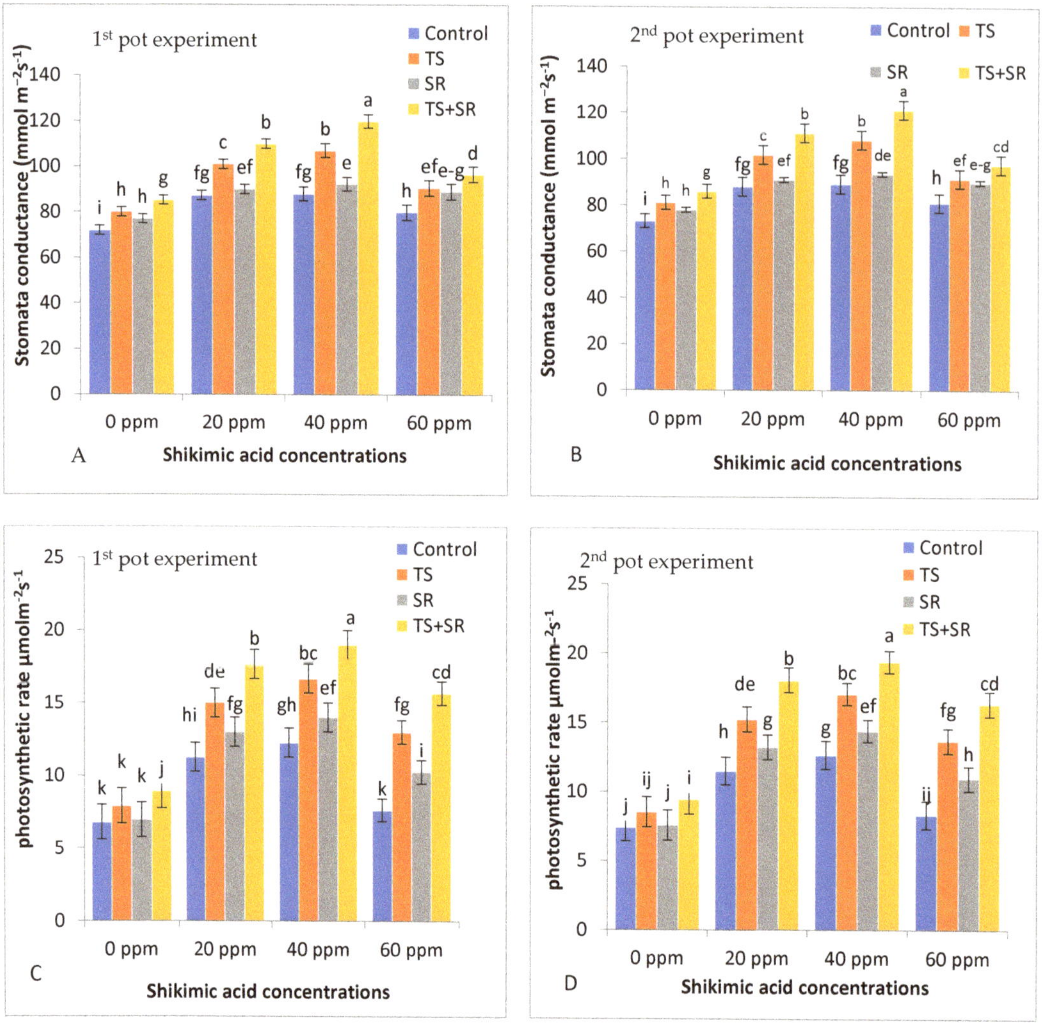

Figure 2. Effect of the interaction between seed priming of shikimic acid concentration (0, 20, 40, 60 ppm) and foliar applications with water (control), *T. asperellum* (TS), *S. rochei* (SR), *T. asperellum* + *S. rochei* + (TS + SR), and on (**A**) Stomata conductance in 1st pot experiment (**B**) Stomata conductance in 2nd pot experiment, and (**C**) photosynthetic rate in 1st pot experiment, (**D**) photosynthetic rate in 2nd pot experiment of squash plants. Standard errors of the mean are shown as vertical bars; the LSD test indicates that differences between values in each bar that are marked by different letters are significant at $p \leq 0.05$.

3.2. *Greenhouse Experiments*

3.2.1. Disease Assessment

The response of squash to 40 ppm shikimic acid seed priming with or without foliar application of BCAs was evaluated after 50 d of growth under natural infection with powdery mildew. As shown in Table 4, all treatments showed significantly ($p \leq 0.05$) decreased the final powdery mildew severity level (FDL%) and the area under the disease progress curve (AUDPC) compared to the control in both seasons. The lowest disease severity percentages were obtained with 40 ppm shikimic acid, through seed priming combined with the mixture of BCAs (*S. rochei* + *T. asperellum*) foliar spray (37.22% and 52.78%) and BCA mixture (*S. rochei* + *T. asperellum*) foliar spray alone (41.11% and 57.78%), with 37.38–30.84% and 32.14–25.71% disease reduction in the first and second season, respectively. Furthermore, the AUDPC was decreased by 51.56% and 45.72% following seed priming with 40 ppm shikimic acid treatment in combination with the BCA mixture (*T. asperellum* + *S. rochei*) compared to the control during the first and second season, respectively.

Table 4. Effect of 40 ppm shikimic acid (SA) seed priming and foliar applications of bioagents on the final powdery mildew disease severity level (FDL%) and the area under the disease progress curve (AUDPC) during the 2020–2021 and 2021–2022 growing seasons.

Treatment	FDL %	Reduction %	AUDPC	Reduction %
First season				
control	59.44 [a]	-	871.11 [a]	-
SA	49.44 [b]	16.82	672.78 [b]	22.77
SR	47.78 [bc]	19.62	614.44 [bc]	29.46
SA + SR	46.11 [b–d]	22.43	523.06 [cd]	39.95
TS	43.89 [c–e]	26.16	507.50 [de]	41.74
SA + TS	43.33 [de]	27.10	484.17 [de]	44.42
SR + TS	41.11 [ef]	30.84	455.00 [de]	47.77
SA + SR + TS	37.22 [f]	37.38	421.94 [e]	51.56
LSD.05	4.23	-	92.85	-
Second season				
control	77.78 [a]	-	1165.28 [a]	-
SA	71.67 [b]	7.86	1071.94 [ab]	8.00
SR	67.78 [bc]	12.86	974.72 [bc]	16.35
SA + SR	63.33 [cd]	18.58	869.72 [cd]	25.36
TS	61.11 [de]	21.43	801.67 [de]	31.20
SA + TS	59.44 [de]	23.58	718.06 [ef]	38.37
SR + TS	57.78 [e]	25.71	688.89 [f]	40.88
SA + SR + TS	52.78 [f]	32.14	632.50 [f]	45.72
LSD.05	4.92	-	106.58	-

Mean values in each column followed by a different letter are significantly different according to the LSD test ($p \leq 0.05\%$). SA; shikimic acid at rate 40 ppm, TS; *T. asperellum*, SR; *S. rochei*.

3.2.2. Plant Growth

The response of squash to 40 ppm shikimic acid seed priming with or without foliar application of BCAs was evaluated after 50 d of growth under natural infection with powdery mildew. As shown in Table 5, all treatments showed significantly ($p \leq 0.05$) increased plant growth compared to the control in both seasons.

Table 5. Effect of seed priming with 40 ppm shikimic acid (SA) and foliar applications of bioagents on the vegetative growth characteristics of squash plants during the 2020 and 2021 seasons.

Treatment	Plant Hight (cm)	Leaves Number/ Plant	Plant Fresh Weight (gm)	Plant Dry Weight (gm)	Chlorophyll (SPAD) Reading
2020 season					
Control	44.0 [g]	12.0 [f]	155.7 [h]	41.0 [f]	30.47 [h]
SA	64.0 [d]	21.0 [c]	310.0 [e]	85.0 [d]	38.93 [e]
TS	60.0 [e]	18.3 [d]	250.0 [f]	72. 7 [e]	36.33 [f]
SR	50.0 [f]	14.7 [e]	204.7 [g]	67.7 [e]	33.73 [g]
TS + SR	69.3 [c]	22.0 [c]	354.3 [d]	101.3 [c]	42.33 [d]
SA + TS	77.3 [b]	24.3 [b]	445.0 [b]	110.3 [b]	48.0 [b]
SA + SR	75.67 [b]	21.7 [c]	396.7 [c]	106.3 [bc]	44.7 [c]
SA + TS+ SR	86.33 [a]	30.0 [a]	543.0 [a]	146.3 [a]	52.63 [a]
LSD.05	3.34	2.2	38.24	5.69	1.20

Table 5. *Cont.*

Treatment	Plant Hight (cm)	Leaves Number/ Plant	Plant Fresh Weight (gm)	Plant Dry Weight (gm)	Chlorophyll (SPAD) Reading
2021 season					
Control	40.3 ^{h}	13.0 ^{e}	157.0 ^{g}	42.0 ^{f}	31.57 ^{h}
SA	60.3 ^{e}	22.7 ^{c}	313.3 ^{d}	85.3 ^{d}	40.0 ^{e}
TS	53.9 ^{f}	19.0 ^{d}	256.7 ^{e}	73.3 ^{e}	37.4 ^{e}
SR	45.9 ^{g}	15.3 ^{e}	207.3 ^{f}	68.3 ^{e}	34.6 ^{e}
TS + SR	71.3 ^{d}	22.7 ^{c}	352.7 ^{d}	102.7 ^{c}	43.1 ^{d}
SA + TS	86.0 ^{b}	25.3 ^{b}	446.7 ^{b}	111.7 ^{b}	48.8 ^{b}
SA + SR	79.0 ^{c}	22.3 ^{c}	399.3 ^{c}	107.3 ^{bc}	45.5 ^{c}
SA + TS + SR	95.0 ^{a}	31.7 ^{a}	546.7 ^{a}	149.0 ^{a}	53.9 ^{a}
LSD.05	1.3	2.4	39.68	6.857	1.2

Mean values in each column followed by a different letter are significantly different according to the LSD test ($p \leq 0.05\%$). SA; shikimic acid at rate 40 ppm, TS; *T. asperellum*, SR; *S. rochei*.

The greatest plant height, number of leaves, fresh weight, and dry weight were obtained with 40 ppm shikimic acid combined with the BCA mixture (*T. asperellum* + *S. rochei*) as a foliar application in both the first and second season. Furthermore, the chlorophyll content was increased by 72.7% and 70.7% following seed priming with 40 ppm shikimic acid treatment in combination with the BCAs mixture (*T. asperellum* + *S. rochei*) as a foliar spray compared to the control during the first and second season, respectively.

3.2.3. Yield and Its Components

The data in Figure 3A–D clearly show that priming of squash seeds with shikimic acid at a rate of 40 ppm and the foliar application of *T. asperellum* with or without *S. rochei* significantly ($p \leq 0.05$) influenced the total yield and its components compared to the control in both seasons. Additionally, 40 ppm of shikimic acid combined with *T. asperellum* + *S. rochei* foliar application increased the average fruit weight (32.43%, 85.8%), number of fruits per plant (62.4%, 69.7%), total fruit yield per plant (83%, 90%), and total yield per hectare, compared with the control in the first and second seasons. However, squash plants treated with 40 ppm shikimic acid seed priming or *T. asperellum* as a foliar application did not significantly differ from the conventional control treatment with respect to total fruit yield per plant in the second season.

3.2.4. Mineral Content in Squash Leaves

The results in Figure 4A–D show that all treatments significantly increased the concentrations of macronutrients, i.e., nitrogen (N), phosphorus (P), potassium (K), and magnesium (Mg), in the leaves of squash plants compared to the control. A marked increase in N (5.9% and 4.6%), P (4.5% and 4.3%), K (4% and 4.2%), and Mg (1.5% and 2.0%) was observed with 40 ppm shikimic acid seed priming combined with foliar application of the BCA mixture (*T. asperellum* + *S. rochei*) compared to the control in the first and second season, respectively. The same figures show that the highest calcium (Ca) concentrations (3.59% and 3.03%) in squash leaves were found for 40 ppm shikimic acid seed priming combined with the BCA mixture (*T. asperellum* + *S. rochei*) as a foliar spray, followed by 40 ppm of shikimic acid seed priming combined with *T. asperellum* as a foliar application in the first season. In addition, application of 40 ppm shikimic acid in combination with the foliar application of the BCA mixture (*T. asperellum* + *S. rochei*), *T. asperellum*, and *S. rochei* showed the most significant increase in Ca concentrations in the leaves (2.9%, 3.03%, and 3.33%), with no significant difference among them compared to the control treatment in the second season.

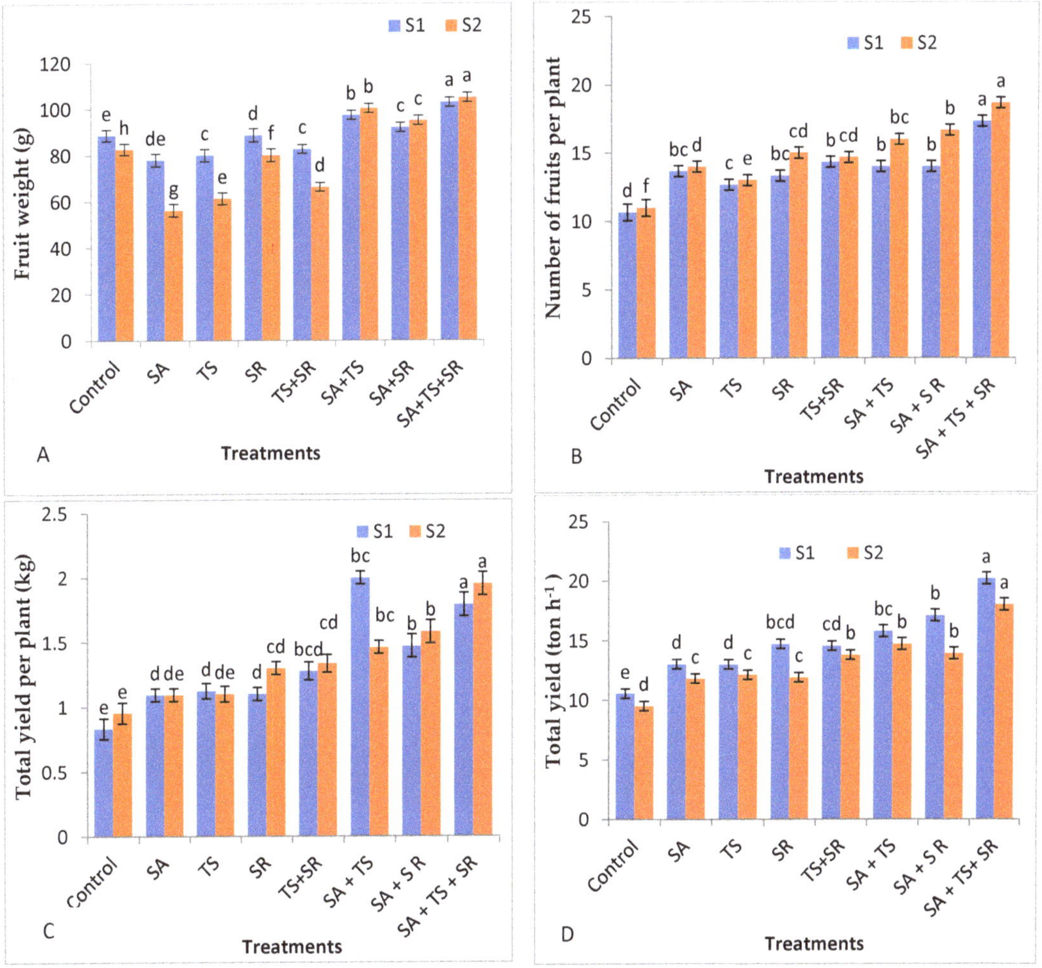

Figure 3. Effect of shikimic acid (SA) seed priming rates 40 ppm and foliar applications of water (control), *T. asperellum* (TS), *S. rochei* (SR), *T. asperellum* + *S. rochei* + (TS + SR), *shikimic acid* + *T. asperellum* (SA + TS), shikimic acid+ *S. rochei* (SA + SR), and shikimic acid + *T. asperellum* + *S. rochei* (SA + TS + SR) on (**A**) average fruit weight (g), (**B**) number of fruits per plant, (**C**) total fruit yield per plant (kg), and (**D**) total fruit yield per hectare during two seasons. Standard errors of the mean are shown as vertical bars; the LSD test indicates that differences between values in each bar that are marked by different letters are significant at $p \leq 0.05$.

In terms of the effect of treatments on the micronutrient content of squash leaves (Figure 5A–D), boron (B), iron (Fe), zinc (Zn), and manganese (Mn) were considerably affected by shikimic acid (40 ppm) seed priming, the BCA mixture (*T. asperellum*+ *S. rochei*) foliar application, and their interactions. Furthermore, the highest concentrations of B (55.47% and 57.17%), Fe (76.12% and 78.02%), Zn (56.15% and 57.18%), and Mn (66.3% and 68.0%) were obtained from plants treated with 40 ppm shikimic acid and combined BCA (*T. asperellum* + *S. rochei)* foliar application compared to the control in the first and the second season, respectively. Nevertheless, the lowest concentrations of B (14.18% and 16.08%), Fe (30.6% and 33%), Zn (13.14% and14.34%), and Mn (27.63% and 28.53%) were obtained from untreated plants during the first and second season, respectively.

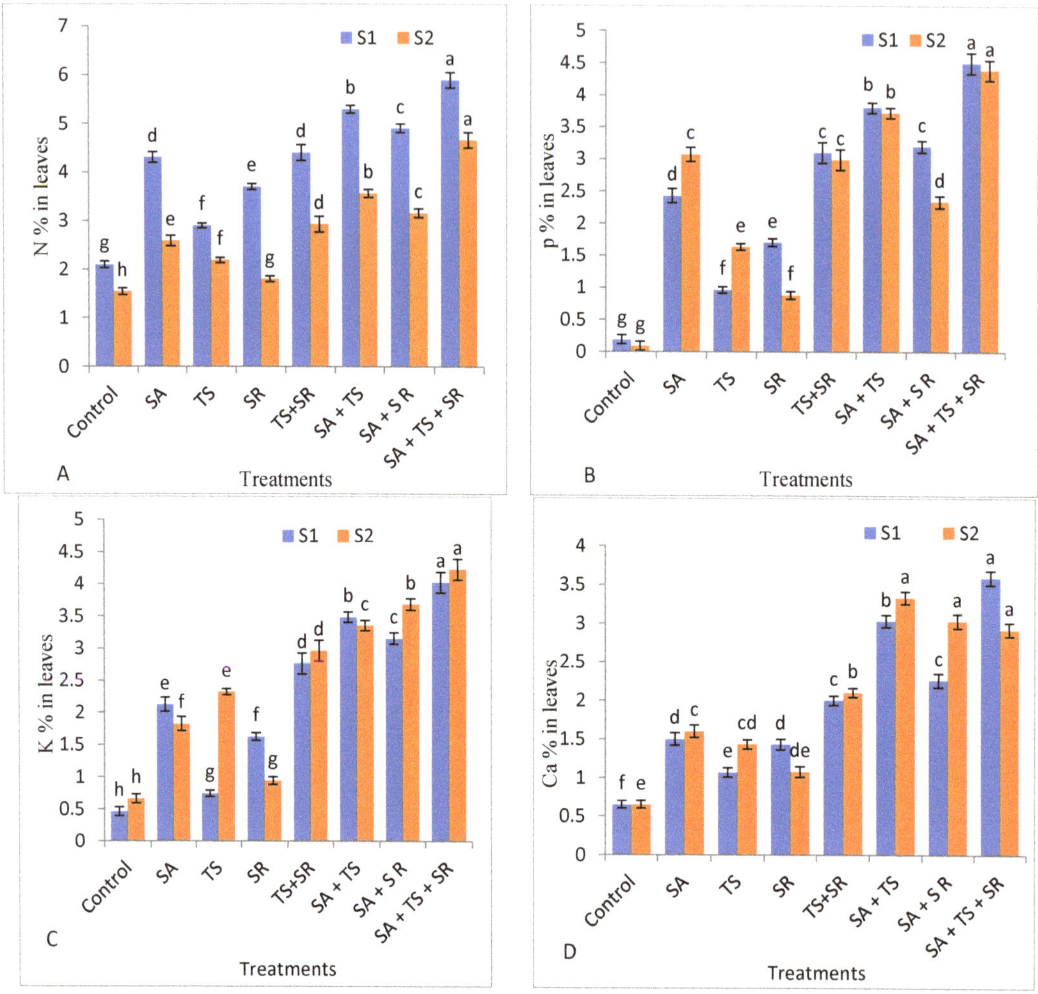

Figure 4. Effect of shikimic acid (SA) seed priming rates 40 ppm and foliar applications of water (control), *T. asperellum* (TS), *S. rochei* (SR), *T. asperellum* + *S. rochei* + (TS + SR), shikimic acid + *T. asperellum* (SA + TS), shikimic acid + *S. rochei* (SA + SR), and shikimic acid + *T. asperellum* + *S. rochei* (SA + TS + SR) on (**A**) N%, (**B**) P%, (**C**) K%, and (**D**) Ca% in leaves during two seasons. Standard errors of the mean are shown as vertical bars; the LSD test indicates that differences between values in each bar that are marked by different letters are significant at $p \leq 0.05$.

3.2.5. Antioxidant Enzymes, Total Phenols, and Plant Hormones in Squash Leaves

POD, SOD, and CAT enzyme activities were assayed in untreated and treated squash plant leaves, as shown in Figure 6A–C. Constitutive SOD, POD, and CAT activity were recorded for all treated plants and compared with the control in both seasons. Maximum SOD (346 and 353 units mg^{-1} protein), POD (686 and 675.5 units mg^{-1} protein), and CAT (189 and 183.2 units mg^{-1} protein) were recorded under a combination of seed priming with shikimic acid at 40 ppm and *T. asperellum* + *S. rochei* as a foliar application during the first and second season, respectively. The lowest activity was observed in the control plants (Figure 6A–F).

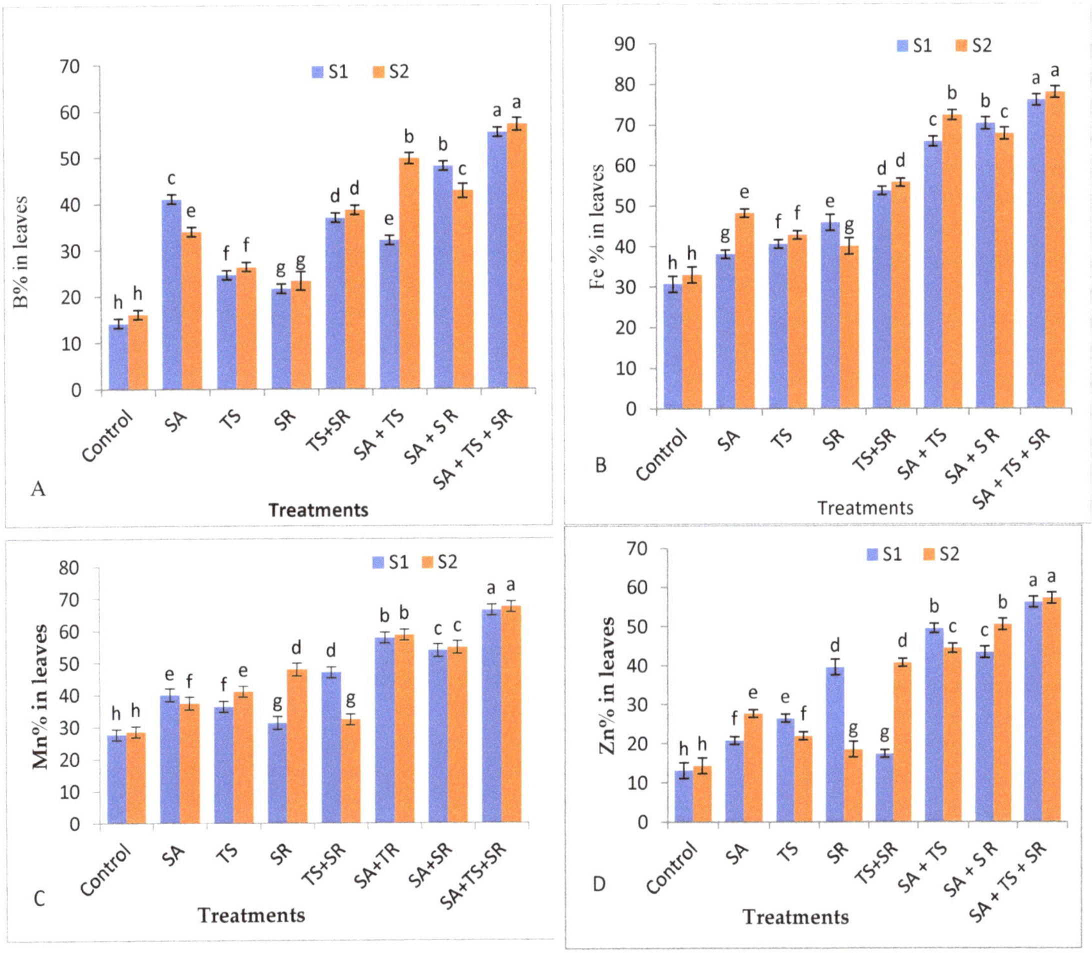

Figure 5. Effect of shikimic acid (SA) seed priming rates 40 ppm and foliar applications of water (control), *T. asperellum* (TS), *S. rochei* (SR), *T. asperellum* + *S. rochei* + (TS + SR), shikimic acid + *T. asperellum* (SA + TS), shikimic acid+ *S. rochei* (SA+SR), and shikimic acid + *T. asperellum* + *S. rochei* (SA + TS + SR) on (**A**) B%, (**B**) Fe%, (**C**) Mn%, and (**D**) Zn% in leaves during two seasons. Standard errors of the mean are shown as vertical bars; the LSD test indicates that differences between values in each bar that are marked by different letters are significant at $p \leq 0.05$.

The highest total phenol content was recorded for 40 ppm shikimic acid combined with BCA mixture foliar application (87.7 and 78.47 mg GAE eq g^{-1}), followed by 40 ppm shikimic acid combined with foliar application of *T. Asperellum* (81.08 and 71.58 mg GAE eq g^{-1}) as compared with untreated plants in the first and second season, respectively. On the other hand, the lowest content of total phenols in leaves (53.41 and 44.21 mg GAE eq g^{-1}) was recorded in untreated plants during the first and second season, respectively (Figure 6D).

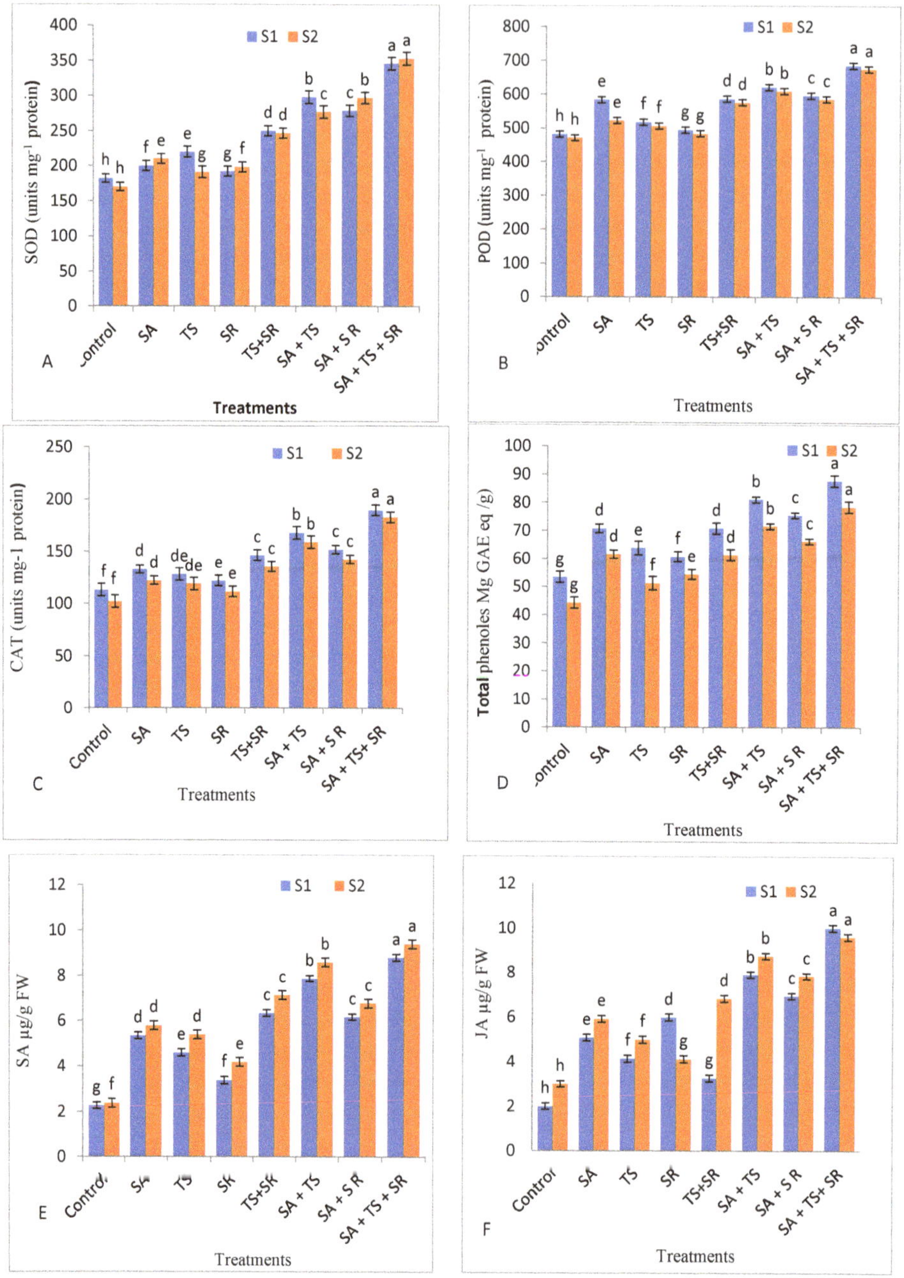

Figure 6. Effect of shikimic acid (SA) seed priming rates 40 ppm and foliar applications of water (control), *T. asperellum* (TS), *S. rochei* (SR), *T. asperellum* + *S. rochei* + (TS + SR), shikimic acid + *T. asperellum* (SA + TS), shikimic acid + *S. rochei* (SA + SR), and shikimic acid + *T. asperellum* + *S. rochei* (SA + TS + SR) on (**A**) SOD, (**B**) POD, (**C**) CAT, (**D**) phenols, (**E**) SA, and (**F**) JA in leaves during two seasons. Standard errors of the mean are shown as vertical bars; the LSD test indicates that differences between values in each bar that are marked by different letters are significant at $p \leq 0.05$.

When comparing treated and untreated plants under natural infection with powdery mildew, lower amounts of JA (2.1 and 3.0 µg g^{-1} FW) and SA (2.26, 2.38 µg g^{-1} FW) were observed in the untreated plants in the first and second season, respectively. Meanwhile, the maximum JA (10 and 9.6 µg g^{-1} FW) and SA (8.8 and 9.4 µg g^{-1} FW) values were recorded under a combination of seed priming with 40 ppm shikimic acid and *T. asperellum* + *S. rochei* foliar application during the first and the second season, respectively (Figure 6E–F). JA is often negatively correlated with SA, and we also observed that the SA level was decreased.

3.3. Clustering Analysis

Our cluster analysis included all growth traits, all yield components, and all mineral compositions of the squash plants along with the activity of antioxidant enzymes, total phenols, SA, and JA. Figure 7 presents a heatmap showing the relationships among the seed priming and foliar application of *T. asperellum*, *S. rochei*, and a mixture of the two, based on the tested parameters. A heatmap analysis clearly identified the overall variations among all treatments. The foliar application of a mixture of *T. asperellum* and *S. rochei* alone showed an increase in all studied traits compared with the control. However, seed priming with shikimic acid at a rate of 40 ppm and treatment with *T. asperellum* + *S. rochei* foliar application decreased oxidative injury and AUDPC values by enhancing plant growth, increasing antioxidant enzyme concentrations, and significantly improving plant growth and yield under natural infection with powdery mildew (Figure 7).

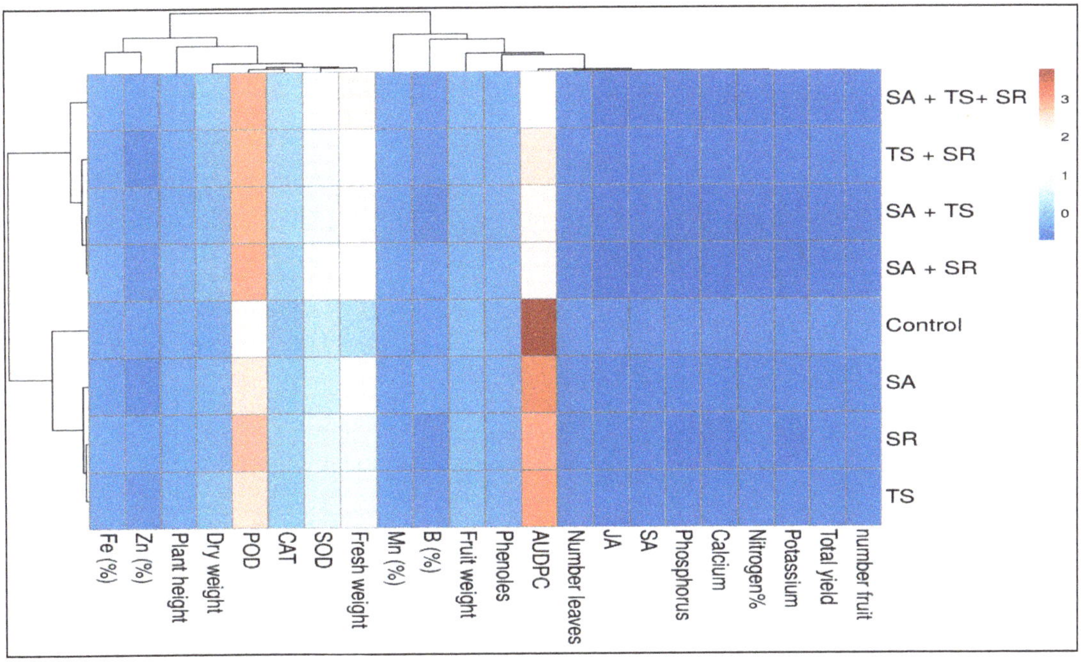

Figure 7. Heatmap of shikimic acid (SA) seed priming rates 40 ppm and foliar applications of *S. rochei* (SR), *T. asperellum* (TS), *S. rochei* + *T. asperellum* (SR + TS), and water (control) and measured parameters of squash plants. The differences in the response variables between all studied treatments are visualized in the heatmap diagram. Columns represent the individual response variables, while rows represent the treatments. Lower numerical values are colored blue, whereas higher numerical values are colored red (see the scale at the top right corner of the heatmap).

3.4. Trait Interrelationships

The association among the evaluated morphological, yield, and physiochemical traits of the squash plants were estimated based on principal component analysis (PCA; Figure 8). Data were analyzed using PCA to establish a relationship between priming with shikimic acid (40 ppm) and the application of *T. asperellum* + *S. rochei* on plant growth and yield parameters. The sum of principal components 1 and 2 (PC1 and PC2) accounted for 93.03% of the variation among plants. PC1, the first component, made up 88.93% of the total variation, and while the second component accounted for 4.1% of the total variation.

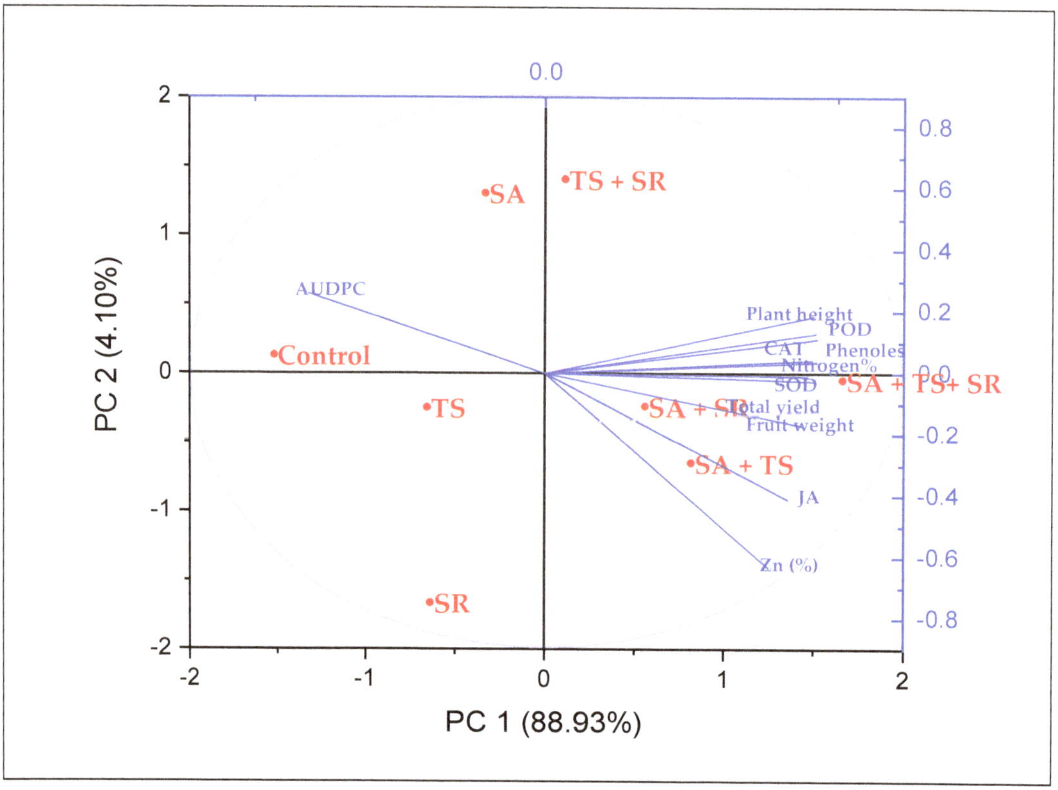

Figure 8. Biplot of the first two principal components for the morphological, yield, AUDPC, and physiochemical traits of sqaush plants. Red circles indicate the eight treatments: shikimic acid (SA) seed priming rates 40 ppm and foliar applications of *S. rochei* (SR), *T. asperellum* (TS), *S. rochei* + *T. asperellum* (SR + TS), and water (control).

3.5. GC/MS Analysis of *T. asperellum* Culture Filtrate

Our analysis of the *T. asperellum* culture filtrate led to the detection of 28 metabolites, as shown in Table 6. The most abundant component was hemin cation (Heme), at 38.95%.

3.6. Leaf Anatomy

Microscopic counts and measurements of several histological characteristics in transverse sections through the mature fifth leaf developed on the main stem of 50-day-old summer squash plants affected by powdery mildew for the control and treated with shikimic acid, bioagents *S. rochei* and *T. asperellum*, and a mixture of the two are shown in Table 7 and Figures 9–11.

Table 6. Composition of the *T. asperellum* (MW965676) culture filtrate.

No.	RT	Compounds	Area %
1	3.44	1-methoxy-2-acetoxypropa NE	4.97
2	4.27	Cyclohexane, (ethoxymethoxy)-	1.40
3	4.35	Tetracarbonyl) [1-(t-buty l)-2,4-bis{[2′,4′,6′-tris(t-butyl) phenyl] imino}-1-aza-2,4-dip hospha-3-molybdacyclobutane	0.96
4	4.59	Z)-6-deoxy-1,2-o-isopropyliDene-3,4-di-o-methyl-5-à-d-xYlo-hepteno-furannurononItrile	0.96
5	5.49	2H-Pyran, 3,6-dihydro-4-methyl-2-(2-methyl-1-p ropenyl)-	0.86
6	7.13	1-pentene, 3,4-dimethyl-	1.28
7	8.14	1,5-pentanediamine	1.18
8	9.91	6-benzyloxy-7-methoxy-1,2,3,4-tetrahydroisoquinoline	4.91
9	13.08	1,5-Hexadien-3-yne, 2-methyl-	2.11
10	13.13	Benzeneacetic acid	1.02
11	13.20	Cycloheptatrienylium, iodide	2.50
12	13.46	1,5-Hexadien-3-yne, 2-methyl-	0.47
13	14.21	1,6-Heptadien-3-yne	0.78
14	15.29	3-Benzylsulfanyl-3-fluoro-2-trifluoro Methyl-acrylonitrile	1.11
15	15.48	3-pyridinemethanol	0.68
16	15.92	Nà-Z-D-2,3-Diaminopropionic acid	1.23
17	16.18	Progesterone 3-biotin	2.33
18	16.45	Tetrahydro-1,3-oxazine-2-thione	0.60
19	17.52	[5-(4′-Formyl phenyl) -10,15,20-tris(4″-tolyl) porphyrinat o] zinc(ii)	1.46
20	18.67	2)-n-cyclohexyl carbamoyl)-4-methoxy-4-methyl-6,6-diphe nyl-2,9,10-triazatricyclo [5. 2.2.0(1,5)] undeca-4,8,10-trien-3-one	0.64
21	19.45	Gln-pro-arg	1.75
22	22.40	7-Oxabicyclo [4.1.0] heptan-2-one	2.93
23	24.74	1-butanol, 2-nitro	2.77
24	25.80	Oxalic acid, cyclobutyl nonyl ester	1.72
25	28.22	4-hydroxyhexenal	3.48
26	35.84	Norcollatone	2.31
27	36.26	2,4,5-tris)Isopropyl)-1,1,3,3-tetrachlOro-2,4,5-triphospha-1,3-dig Ermolane	4.06
28	36.58	Hemin cation	38.95

Table 7. Counts and measurements in micrometers (μm) of certain histological characters in transverse sections through the blade of the fifth leaf developed on the main stem of summer squash (cv. Eskandarani), aged 40 days, affected by powdery mildew as a control and treated with shikimic acid, bioagent (*S. rochei* + *T. Asperullum*), and a mix between them (Means of three sections from three specimens).

Histological Aspects	Treatments			
	Control	40 ppm (SA)	(SR + TS)	40 ppm SA and (SR + TS)
Thickness of midvein	2020.766	2404.648	2501.365	2812.411
Thickness of upper epidermis	12.132	12.418	13.459	15.422
Thickness of lower epidermis	6.987	7.196	8.633	8.951
Thickness of lamina	120.173	120.688	124.944	145.919
Thickness of palisade tissue	40.700	44.202	42.102	50.911
Thickness of spongy tissue	53.877	57.404	62.465	69.911
No of vascular bundle	4.000	4.000	5.000	6.000
Dimentions of main midvein bundle	454.521	469.419	490.742	559.504
No of xylem rows in main midvein bundle	3.2	3.2	3.3	3.5
Mean diameter of vessels	30.614	30.699	30.888	30.921
Mean diameter of paranchyma cells in ground tissues	104.666	111.945	108.887	120.111

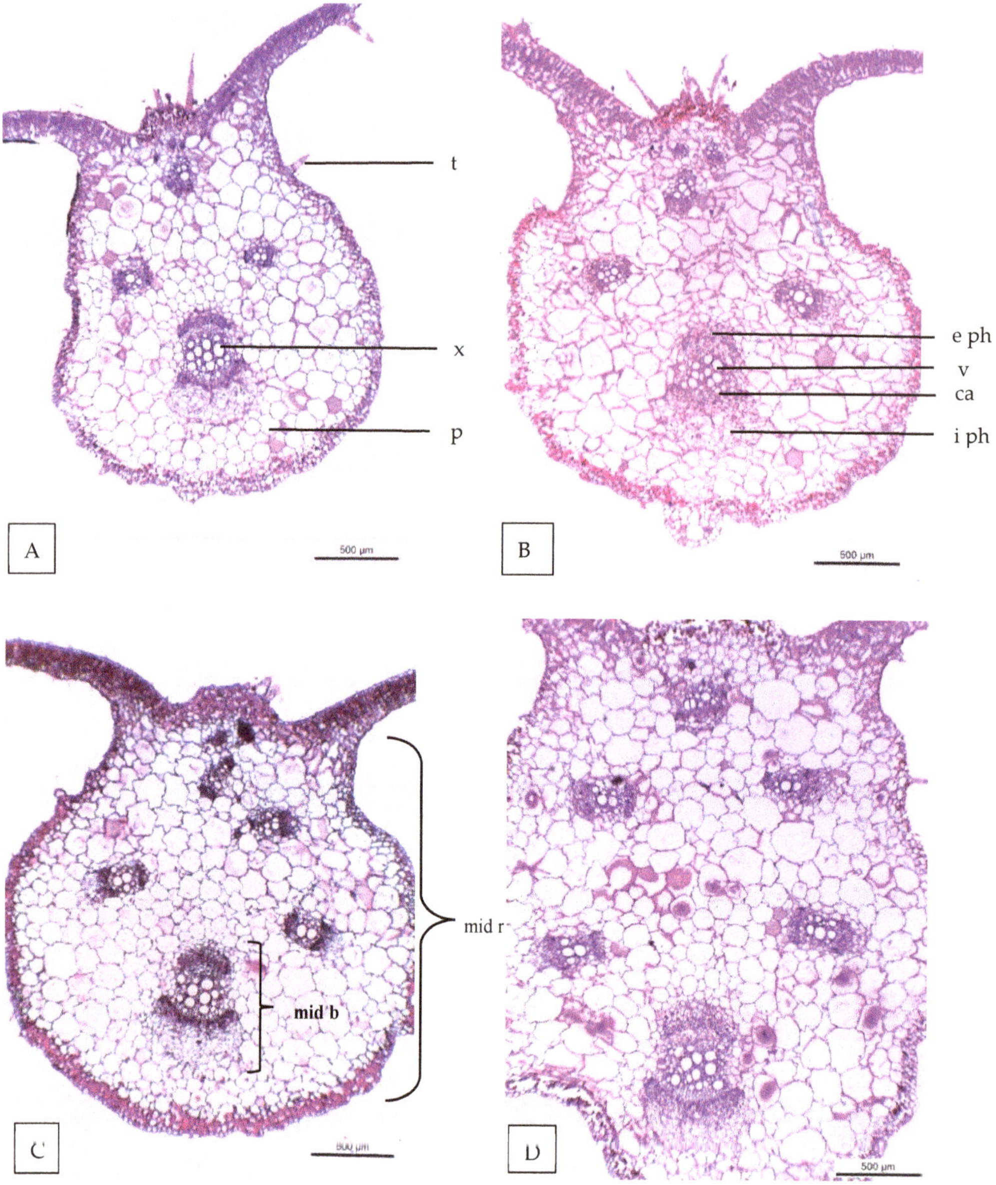

Figure 9. Microphotographs of cross sections through the blade of the fifth leaf developed on the main stem of summer squash (cv. Eskandarani), aged 50 days. Scale bars = 500 µm. (**A**) Plants affected by powdery mildew (control). (**B**) Plants treated by 40 ppm shikimic acid (SA). (**C**) Plants treated by *S. rochei* + *T. Asperullum* (SR + TS). (**D**) Plants treated by mix between 40 ppm SA and (SR + TS). Abbreviations: mid b—midvein bundle; mid r—midvein region; t—trichome; p—paranchyma cells; ca—cambium zone; e ph—external phloem; i ph—internal phloem; v—vessels; and x—xylem.

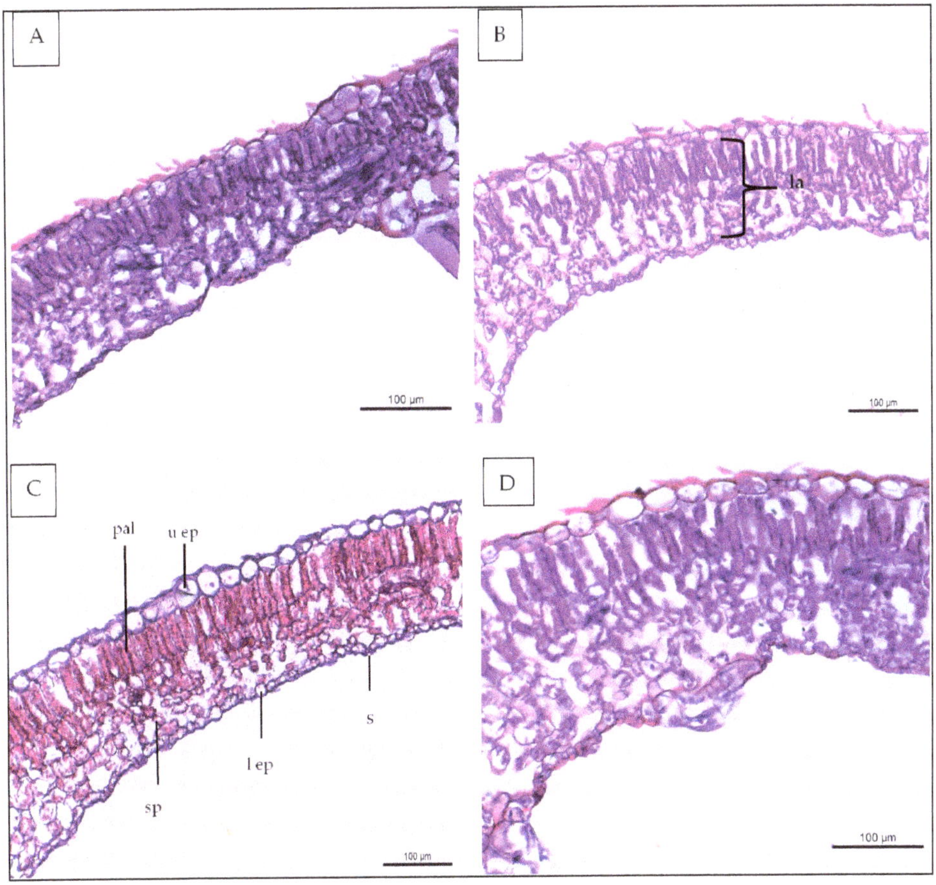

Figure 10. Magnified portions of the leaf blade lamina of summer squash (cv. Eskandarani), aged 50 days. Scale bars = 100 μm. (**A**) Plants affected by powdery mildew (control). (**B**) Plants treated by 40 ppm shikimic acid (SA). (**C**) Plants treated by *S. rochei* + *T. Asperullum* (SR + TS). (**D**) Plants treated by mix between 40 ppm SA and (SR + TS). Abbreviations: la—lamina; l ep—lower epidermis; spo—spongy tissue; pal—palisade tissue; u ep—upper epidermis; and s—stomata.

The data in Table 7 and Figures 9 and 10 reveal that treatment with shikimic acid (40 ppm), bioagents *S. rochei* and *T. asperellum*, and a mixture of the two improved all histological aspects under study. The values for thickness of midvein, dimension of main midvein bundles, number of xylem rows in the main midvein bundle, mean diameter of vessels, and mean diameter of parenchyma cells in the ground tissue of plants treated with a combination of shikimic acid (40 ppm) and *S. rochei* + *T. asperellum* were increased by 39.17%, 23.09%, 9.37%, 1.00%, and 14.75% compared to the control, respectively (Figure 9A,D). A combination of shikimic acid (40 ppm) and *S. rochei* + *T. asperellum* yielded an increase in thickness of the upper epidermis, lower epidermis, lamina, palisade tissue, and spongy tissue of 27.11%, 28.10%, 21.42%, 25.08%, and 29.76% compared to the control, respectively (Figure 10A,D). Plants that were treated with *S. rochei* + *T. asperellum* showed an increase in the thickness of the midvein, upper epidermis, lower epidermis, lamina, spongy tissue, number of vascular bundles, dimension of the main midvein bundle, and mean diameter of vessels of 4.02%, 8.38%, 19.96%, 3.52%, 8.81%, 25%, 4.54%, and 0.61% compared to those treated with shikimic acid (40 ppm) alone. In contrast, treatment with shikimic acid (40 ppm) showed increases of 4.75% and 2.7% in the thickness of palisade tissue and mean diameter of parenchyma cells in the ground tissues, respectively, compared with plants treated with *S. rochei* + *T. asperellum* (Figures 9 and 10B,C). This treatment caused recovery of the reduction in all examined tissues of the leaf blade that were affected by powdery mildew. It is clear from Figure 11A and B that plants affected by powdery mildew had an accumulation of conidia on the epidermal cells, while those treated with the bioagents (*S. rochei* + *T. asperellum*) and shikimic acid (40 ppm) appeared to be largely free of conidia on the epidermal cells. In addition, the cutical layer appeared thicker in the plants treated with the bioagents (*S. rochei* + *T. asperellum*) with shikimic acid 40 ppm compared to the control. The cells of the palisade tissue appeared closely packed in two rows in the plants treated with the bioagents (*S. rochei* + *T. asperellum*), while the cells of the palisade tissue of the control plants were dilapidated and undefined.

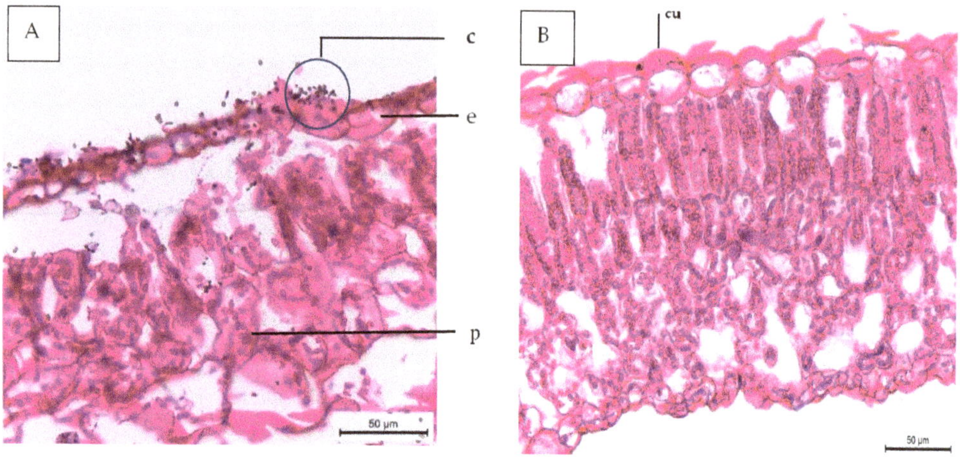

Figure 11. Magnified portions of the leaf blade lamina of summer squash (cv. Eskandarani), aged 50 days. Scale bars = 50 µm. (**A**) Plants affected by powdery mildew (control). (**B**) Plants treated by mix between 40 ppm SA and (SR + TS). Abbreviations: cu—cuticle; e—epidermis; c—conidia; and p—plastids.

4. Discussion

In crop protection against powdery mildew, restrictions on fungicides have encouraged research into biological management as an alternative. Recent discoveries support the idea that some BCAs may have beneficial impacts on plants, including disease control, plant growth stimulation, increased yield, and improved nutrient uptake and crop quality [34]. In particular, plant growth-promoting fungi such as *Trichoderma* spp. are capable of producing a variety of bioactive secondary metabolites that stimulate plant growth and protect plants from numerous phytopathogens [35]. The stimulatory effect of *Trichoderma* spp. on plants is probably related to their participation in the crosstalk between the growth hormones synthesized by these fungi and the defense hormones they induce in the plant [36]

In this study, *T. asperellum* alone reduced powdery mildew severity under controlled conditions by 18.01% and 16.34% in the first and second experiments, respectively. Moreover, *S. rochei* alone was less effective than *T. asperellum*, as it reduced the percentage of powdery mildew by 14.01% and 10.22% in the first and second experiments, respectively. Surprisingly, when applied together, they activated one another; this positive effect was reflected in an increase in resistance to the disease, with a decrease in powdery mildew severity of 22.00% and 28.57% in the first and second pot experiments, respectively. Accordingly, we can infer that a combination of *S. rochei* and *T. asperellum* as a foliar spray may have a synergistic effect against powdery mildew. This observation led us to quantify the components of *T. asperellum* and *S. rochei* culture filtrates by GC-MS analysis. An analysis of *T. asperellum* culture filtrate revealed 28 compounds. Hemin cation, a naturally occurring metalloporphyrin, was identified during GC-MS analysis of the *T. asperellum* culture filtrate (area percentages). This compound is primarily indicated as a basis for the new fungicides owing to its recognized pro-oxidant capabilities, which also likely contribute to its antibacterial activity [37]. The antifungal effect of myeloperoxidase is due to the synthesis of ROS [38]. Therefore, the presence of hemin may be responsible for the antifungal activity of *T. asperellum* culture filtrate against the powdery mildew pathogen observed in this study.

On the other hand, our GC-MS analysis of the *S. rochei* culture filtrate (area percentages) revealed the presence of 54 compounds. In peak 5, the volatile phenolic compound 2,4-di-tert-butylphenol (1.04%), often a major component of violates that exhibits potent antifungal, antibacterial, antiviral, antimicrobial, antioxidant, and phytotoxic activities. The second most abundant compound, in peak 16 (0.96%), was the fatty acid hexadecanoic acid methyl ester which has a strong odor and which acts as an enzyme inhibitor [39]. Fatty acids such as 9-heptadecenoic acid and 6-methyl-9-heptadecenoic acid have anti-powdery mildew activity [40]. Antifungal fatty acids enter powdery mildew cells and disrupt the membrane, causing the release of intracellular components, cytoplasmic disorder, and, eventually, cell disintegration [40]. Therefore, a possible explanation for the synergistic effect between *S. rochei* and *T. asperellum* found in this study may be the interaction between hemin cations from *T. asperellum* and the antifungal compounds, hexadecanoic acid methyl ester and 2,4-di-tert-butylphenol from *S. rochei*. This finding coincides with that of [41], who found that a combination of *Streptomyces* spp. and *Trichoderma* spp. could suppress the intensity of leaf spot disease by up to 16.2% in chili fruits and enhance plant growth compared to either one as a single treatment.

We further hypothesized that seed priming with shikimic acid, a natural organic acid, in combination with the tested BCAs as a foliar spray would increase plant resistance to powdery mildew infection and promote plant growth. When applied as a seed primer under controlled conditions, shikimic acid alone at 20, 40, and 60 ppm reduced powdery mildew severity by 24.00%, 24.00%, and 5.99% in the first experiment and by 32.67%, 32.67%, and 16.34%, in the second experiment, respectively. It was noted that by increasing shikimic acid concentrations to 60 ppm, the disease severity increased while the plant growth parameters (plant height, number of leaves, chlorophyll content), photosynthesis, transpiration rate, and stomatal conductance decreased. The increased severity of the infection may be attributed to the decrease in chlorophyll content and the efficiency of photosynthesis, which leads to a decrease in disease resistance

In most cases, the improvement in powdery mildew control, plant height, number of leaves, and chlorophyll reading (SPAD) was due to a combination of seed priming with shikimic acid at 40 ppm and a BCA mixture (*T. asperellum* + *S. rochei*) as a foliar application treatment. Throughout the two trials, this effect was consistent. The additive effects were most likely due to a combination of mechanisms involved in disease control by the combined treatment, as opposed to the modes of action provided by shikimic acid or individual antagonists alone [7,11]. According to [42], shikimic acid is effective as a seed soak to cure chocolate spot disease in faba beans. Similarly, various authors have reported that exogenous shikimic acid application can improve plant growth, improve antioxidant defense, and reduce oxidative damage [13].

Another finding was that untreated squash leaves showed a significant reduction in plant growth and productivity with powdery mildew infection. This finding is consistent with that of [43], who found that the powdery mildew fungal infection resulted in a significant reduction in photosynthetic rate, leaf pigments, transpiration rate, and stomatal conductance. A possible explanation is that mildew colonies on the leaf surface limit the amount of light reaching the mesophyll cells, resulting in a considerable drop in leaf chlorophyll and carotenoids [44]. Photosynthetic pigment reduction is proportional to the leaf surface's photosynthetic efficiency [43]. As a result, the photosynthesis rate is significantly decreased in untreated infected plants. Furthermore, the photosynthesis rate is inversely proportional to the transpiration rate and stomatal conductance [45]. Fungal colonization results in partial coverage of the stomata cavity or hole by the hyphae and/or spores of the fungus. As a result, transpiration rates in fungus-infected plants are consistently low [44]. In addition, an increase in chlorophyll content, which is a reliable indicator of plant health, improves the efficiency of the photosynthetic system and increases disease resistance [46].

This study examined, for the first time, the effect of seed priming with shikimic acid and the foliar application of BCAs to improve the growth and productivity of squash. Improved growth of squash plants in response to shikimic acid might be due to improvement in leaf survival by maintaining photosynthesis and increasing mineral content, which may increase plant growth [13]. Plant growth parameters are a perfect indicator for evaluating abiotic and biotic stresses. *Trichoderma* spp. root colonization has been linked to increased plant nutrient uptake as a result of more efficient macro- and micronutrient solubilization [47]. These results are in agreement with those obtained by [48] in squash. Plant growth-promoting streptomycetes promote a variety of direct and indirect biosynthetic processes in plants, including inorganic phosphate solubilization, chemical biosynthesis chelation, phytohormone production, pathogen suppression, and abiotic stress reduction [49]. In this study, priming seeds of squash plants with 40 ppm of shikimic acid combined with the foliar application of *T. asperellum* and *S. rochei* yielded significantly higher fruit numbers, fruit weight, and total yield per plant.

We further studied the responses of squash plants to the treatments by determining the level of plant hormones in the leaves. SA is a plant hormone that plays a pivotal role in plant defense against biotrophic and semi-biotrophic pathogens [23]. When comparing treated and non-treated plants under natural infection with powdery mildew, lower amounts of JA and SA were observed in the water-treated control. Surprisingly, in plants grown after seed priming with 40 ppm of shikimic acid combined with foliar spray of *S. rochei* + *T. asperellum*, the levels of JA and SA were much higher than in the water-treated control. We also observed that because JA is often negatively correlated with SA, the SA level was decreased. Elevated levels of JA and SA in treated plants compared to the control were to be expected, because both hormones are usually a part of resistance mechanisms. The JA and SA contents increase in infected tissues, and the application of shikimic acid through seed priming combined with foliar spray of BCAs reduces disease symptoms. The work conducted by [50] studied some of the physiological mechanisms resulting from powdery mildew inoculation of the sensitive Ingrid barley cultivar and near-isogenic lines carrying various resistant genes (Mla, Mlg, and mlo). They found that in the leaves of the cultivar Ingrid, the JA content was significantly decreased after inoculation compared to the non-inoculated control. Meanwhile, powdery mildew inoculation of Mla and Mlg barley that showed hypersensitive reactions led to notably elevated levels of JA and SA compared to the Ingrid cultivar.

Phenolic compounds are formed by plants primarily for growth, protection, and development. These aromatic benzene ring components are very important during the interactions between plants and abiotic stresses. They represent an essential component of phytochemicals and play an important role in various mechanical and physiological activities [51]. In the present study, the total phenol content in squash leaves was increased in all treatments compared with control plants, whereas plants primed with 40 ppm of shikimic acid combined with foliar spray of *S. rochei* + *T. asperellum* showed a greater accumulation of phenolics, which enhance resistance against invasion of P. xanthii. This finding is in line with that obtained by [52], who found an increase in total phenols, yield, and powdery mildew resistance in squash plants treated with certain BCAs. Furthermore, disease resistance induction is a powerful strategy to enhance the defense mechanisms of plants. This mechanism could be linked to the activities of defense-associated enzymes such as PPO and POD, which are key proteins related to pathogenesis in plant tissues owing to their ability to disrupt the cell wall structure of pathogens and contribute to resistance to pathogen invasion [53]. PPO and POD are terminal enzymes; the former can oxidize phenolic substances and produce toxic quinines that can restrict and kill invading pathogens, while the latter is linked to the synthesis of lignin [54]. SOD is essential to plant stress resistance; it provides the first line of defense against the harmful effects of elevated levels of ROS [55]. In addition, catalase is an oxygen-binding enzyme that protects cells from the harmful effects of H_2O_2 and acts as a signaling molecule to increase plant defense genes, which helps protect plants from infection [56].

Our results are in line with earlier studies that demonstrated enhanced activities of POD, SOD, and CAT under different treatments. Shikimic acid, when combined with the foliar application of a BCA mixture (*T. asperellum* + *S. rochei*), increased the activity of antioxidant enzymes such as POD, SOD, and CAT, thereby protecting against powdery mildew [20,42]. Antioxidant enzymes are proteins involved in the catalytic conversion of ROS and their byproducts into stable, non-toxic molecules. Therefore, they represent the main defense mechanism against cellular damage caused by oxidative stress. The authors of [57] found that bioagents *Bacillus* spp. and *Trichoderma* spp. could significantly reduce the severity of squash powdery mildew (*P. axanthii*). However, the role of these bioagents was attributed to the upregulation of defense-related enzymes such as CAT, POD, and PPO, which stimulate growth and yield characteristics.

The interrelationship among the evaluated parameters (Figures 7 and 8) indicates that the yield parameters were positively associated with plant height and the mineral content of leaves, including N%, P%, and K%. We speculate that the high levels of antioxidant activity, including SOD, POD, and CAT, were associated with the greater total yield and its related traits, especially after seed priming with 40 ppm of shikimic acid combined with foliar spray of *S. rochei* + *T. asperellum*. In accordance with these results, it is important to note that specific biochemical and physiological traits are closely associated with yield-related traits under natural infection with powdery mildew.

With regard to the anatomy of squash leaves, this study demonstrated that plants infected with *P. xanthii* that received no treatment showed a reduction in the thickness of the upper epidermis, lower epidermis, and midvein, dimension of the main midvein bundle, number of xylem rows in main midvein bundle, mean diameter of vessels, and mean diameter of parenchyma cells in the ground tissues. A possible explanation for this might be that the powdery mildew mycelium grows on the surface of the plant, with nutrients being obtained via the haustoria in the plant epidermal cells, leading to nutrient uptake. In addition, powdery mildew colonies on the leaf surface limit the amount of light reaching the mesophyll cells (palisade and spongy), resulting in a considerable drop in leaf chlorophyll and carotenoids, reduced photosynthesis, increased respiration and transpiration, and impaired growth [44]. In general, the alteration of anatomical traits of squash leaves with the applied treatments is of great interest, because these alterations included increases in the thickness of upper epidermis, lower epidermis, and midvein, dimension of the main midvein bundle, number of xylem rows in main midvein bundle, mean diameter of vessels, and mean diameter of parenchyma cells in the ground tissues. The treatments improved the process of photosynthesis by increasing the area of epidermis and mesophyll tissues exposed to the sunlight. They also reflect those of [42], who found that seed priming with shikimic acid improved the growth parameters of cowpea plants by simulating effects on leaf expansion and photosynthetic pigments as well as on the transpiration rate of cowpea plants during growth periods.

On the other hand, shikimic acid might exert effects on the photosynthetic machinery at the mesophyll and chloroplast level by increasing plastid biogenesis via an increase in the biosynthesis of indole acetic acid from tryptophan. Furthermore, some reports have indicated that a combination of antagonistic bacteria and antagonistic fungi, especially *Trichoderma* spp., provides greater plant protection than their use individually [58]. Additionally, the authors of [59] reported that application of a mixture of *T. harzianum* and *S. rochei* is more effective at controlling Phytophthora root rot in pepper. In the present study, a combination of shikimic acid (40 ppm) seed priming and *S. rochei* + *T. asperellum* as a foliar application yielded the best results in terms of the anatomical features of leaves. Additionally, the upper surface of the plant showed a reduction of powdery mildew symptoms. This improved the plants' ability to receive sunlight and improved the mechanism of chloroplast function, thereby enhancing the process of photosynthesis. To our knowledge, this study is the first to report the effects of shikimic acid (40 ppm) and *S. rochei* + *T. asperellum* as a foliar application on the anatomical structure of squash leaves.

5. Conclusions

This study showed the effects of *T. asperellum*, *S. rochei*, and shikimic acid-primed seeds, and their combination treatments on squash productivity under artificial and natural infection with powdery mildew. The use of *T. asperellum* + *S. rochei* with shikimic acid primed seeds treatment was more effective than using each treatment separately, while *T. asperellum* outperformed *S. rochei*. The use of *T. asperellum* + *S. rochei* with shikimic acid primed seeds treatment showed an improvement in terms of plant growth, total yield, mineral components, physiological traits, and antioxidant activity, as well as in all histological aspects of the studied squash plants.Moreever, maximum disease severity reduction and a reduction in AUDPC values were observed following treatment of *T. asperellum* + *S. rochei* with 40 ppm shikimic acid primed seeds. Therefore, this treatment could be used as an alternative method to lessen powdery mildew infection in squash plants.

Author Contributions: Conceptualization, E.G.S., S.N.A., N.E.-H.A.R. and A.M.A.; methodology N.E.-H.A.R. and E.G.S.; software, E.G.S., S.N.A. and N.E.-H.A.R. validation, S.N.A., E.G.S., N.E.-H.A.R. and A.M.A.; formal analysis, E.G.S.; investigation, E.G.S.; resources, E.G.S., S.N.A., N.E.-H.A.R. and A.M.A.; data curation, E.G.S., N.E.-H.A.R.; writing—original draft preparation E.G.S., N.E.-H.A.R. and A.M.A.; writing E.G.S., N.E.-H.A.R., S.N.A. and A.M.A. review and editing, N.E.-H.A.R.; visualization E.G.S.; supervision E.G.S., N.E.-H.A.R. and S.N.A.; project administration E.G.S., S.N.A. and N.E.-H.A.R.; funding acquisition, E.G.S., S.N.A., A.M.A. and N.E.-H.A.R. All authors have read and agreed to the published version of the manuscript.

Funding: This research received no external funding.

Institutional Review Board Statement: Not applicable.

Informed Consent Statement: Not applicable.

Data Availability Statement: Not applicable.

Acknowledgments: The authors are appreciative to the Department of Vegetable Crops, and the Department of plant pathology, the Faculty of Agriculture, and Cairo University for supplying some of the equipment and facilities to accomplish this research.

Conflicts of Interest: The authors declare no conflict of interest.

References

1. Abd El-All, H.M.; Ali, S.M.; Shahin, S.M. Improvement growth, yield and quality of squash (*Cucurbita pepo* L.) plant under salinity conditions by magnetized water, amino acids and selenium. *J. Appl. Sci. Res.* **2013**, *9*, 937–944.
2. El-Naggar, M.; El-Deeb, H.; Ragab, S. Applied approach for controlling powdery mildew disease of cucumber under plastic houses. *Pak. J. Agric.* **2012**, *28*, 54–64.
3. Hafez, Y.M.; Bayoumi, Y.A.; Pap, Z.; Kappel, N. Role of hydrogen peroxide and Pharmaplant-turbo against cucumber powdery mildew fungus under organic and inorganic production. *Int. J. Hortic. Sci.* **2008**, *14*, 39–44. [CrossRef]
4. Abdel-Monaim, M.; Abdel-Gaid, M.; Armanious, H. Effect of chemical inducers on root rot and wilt diseases, yield and quality of tomato. *Int. J. Agric. Sci.* **2012**, *7*, 211–220.
5. Deore, P.B.; Sawant, D.M. Management of guar powdery mildew by *Trichoderma* spp. culture filtrates. *J. Maharashtra Agric. Univ.* **2001**, *25*, 253–254.
6. Guigón López, C.; Muñoz Castellanos, L.N.; Flores Ortiz, N.A.; Judith Adriana, G.G. Control of powdery mildew (*Leveillula taurica*) using *Trichoderma asperellum* and *Metarhizium anisopliae* in different pepper types. *BioControl* **2019**, *64*, 77–89. [CrossRef]
7. Oszako, T.; Voitka, D.; Stocki, M.; Stocka, N.; Nowakowska, J.; Linkiewicz, A.; Hsiang, T.; Belbahri, L.; Daria Berezovska, D.; Malewski, T. *Trichoderma asperellum* efficiently protects *Quercus robur* leaves against *Erysiphe alphitoides*. *Eur. J. Plant Pathol.* **2021**, *159*, 295–308. [CrossRef]
8. Tyskiewicz, R.; Nowak, A.; Ozimek, E.; Jaroszuk-Ściseł, J. *Trichoderma*: The Current Status of Its Application in Agriculture for the Biocontrol of Fungal Phytopathogens and Stimulation of Plant Growth. *Int. J. Mol. Sci.* **2022**, *23*, 2329. [CrossRef]

9. Pacios-Michelena, S.; Aguilar González, C.N.; Alvarez-Perez, O.B.; Rodriguez-Herrera, R.; Chávez-González, M.; Arredondo Valdés, R.; Ascacio Valdés, J.A.; Govea Salas, M.; Ilyina, A. Application of Streptomyces Antimicrobial Compounds for the Control of Phytopathogens. Front. Sustain. *Food Syst.* **2021**, *5*, 696518. [CrossRef]
10. Kurth, F.; Mailänder, S.; Bönn, M.; Feldhahn, L.; Herrmann, S.; Große, I.; Buscot, F.; Schrey, S.D.; Tarkka, M.T. Streptomyces-induced resistance against oak powdery mildew involves host plant responses in defense, photosynthesis, and secondary metabolism pathways. *Mol. Plant-Microbe Interact.* **2014**, *27*, 891–900. [CrossRef] [PubMed]
11. Yang, M.; Wei, Q.; Shi, L.; Wei, Z.; Lv, Z.; Asim, N.; Zhang, K.; Ge, B. Wuyiencin produced by *Streptomyces albulus* CK-15 displays biocontrol activities against cucumber powdery mildew. *J. Appl. Microbiol.* **2021**, *131*, 2957–2970. [CrossRef] [PubMed]
12. Hildebrandt, T.M.; Nunes Nesi, A.; Araujo, W.L.; Braun, H.P. Amino Acid Catabolism in Plants. *Mol. Plant* **2015**, *8*, 1563–1579. [CrossRef]
13. Sayed, E.G.; Mahmoud, A.W.M.; Abdel-Wahab, A.; Elbahbohy, R.M. Rootstock Priming with Shikimic Acid and *Streptomyces griseus* for Growth, Productivity, Physio-Biochemical, and Anatomical Characterisation of Tomato Grown under Cold Stress. *Plants* **2022**, *11*, 2822. [CrossRef]
14. Dangl, J.; Jones, J. Plant pathogens and integrated defense responses to infection. *Nature* **2001**, *411*, 826–833. [CrossRef]
15. Zhang, S.; Liu, J.; Xu, B.; Zhou, J. Differential Responses of *Cucurbita pepo* to *Podosphaera xanthii* Reveal the Mechanism of Powdery Mildew Disease Resistance in Pumpkin. *Front. Plant Sci.* **2021**, *12*, 633221. [CrossRef] [PubMed]
16. Hou, L.; Wang, L.N.; Wu, X.L.; Gao, W.; Zhang, J.X.; Huang, C.Y. Expression patterns of two Pal genes of *Pleurotus ostreatus* across developmental stages and under heat stress. *BMC Microbiol.* **2019**, *19*, 231. [CrossRef]
17. Han, C.; Li, J.; Jin, P.; Li, X.; Wang, L.; Zheng, Y. The effect of temperature on phenolic content in wounded carrots. *Food Chem.* **2017**, *215*, 116–123. [CrossRef]
18. Niu, J.H.; Cao, Y.; Lin, X.G.; Leng, Q.Y.; Chen, Y.M.; Yin, J.M. Field and laboratory screening of anthurium cultivars for resistance to foliar bacterial blight and the induced activities of defense-related enzymes. *Folia Hortic.* **2018**, *30*, 129–137. [CrossRef]
19. Pieterse, C.M.J.; Van der Does, D.; Zamioudis, C.; Leon-Reyes, A.; Van Wees, S.C.M. Hormonal modulation of plant immunity. *Annu. Rev. Cell Dev. Biol.* **2012**, *28*, 489–495. [CrossRef]
20. Ali, A.A.; Abd EL-Kader, A.E.S.; Ghoneemi, K.H.M. Two *Trichoderma* species and *Bacillus subtilis* as biocontrol agents against Rhizoctonia disease and their influenceon potato productivity. *Egypt. J. Agric. Res.* **2017**, *95*, 2017.
21. Gebily, D.A.S.; Ghanem, G.A.M.; Ragab, M.M.; Ali, A.M.; Soliman, N.E.K.; Abd El-Moity, T.H. Characterization and potential antifungal activities of three *Streptomyces* spp. As biocontrol agents against *Sclerotinia sclerotium* (Lib.) de Bary infecting green bean. *Egypt. J. Biol. Pest Control.* **2021**, *31*, 33. [CrossRef]
22. Shin, H.D. *Erysiphaceae of Korea*; National Institute of Agricultural Science and Technology: Suwon, Korea, 2000; pp. 227–235.
23. Margaritopoulou, T.; Toufexi, E.; Kizis, D. *Reynoutria sachalinensis* extract elicits SA-dependent defense responses in courgette genotypes against powdery mildew caused by *Podosphaera xanthii. Sci. Rep.* **2020**, *10*, 3354. [CrossRef]
24. Morishita, M.; Sugiyama, K.; Saito, T.; Sakata, Y. Powdery mildew resistance in cucumber. *Jpn. Agric. Res. Q.* **2003**, *37*, 7–14. [CrossRef]
25. Pandey, H.N.; Menon, T.C.M.; Rao, M.V. Asimple formula for calculating area under disease progress curve. *Rachis* **1989**, *8*, 38–39.
26. Helrich, K. *Official Methods of Analysis*, 15th ed.; Association of Official Agricultural Chemist: Arlington, VA, USA, 1990; Volume 1, p. 673.
27. Jackson, M.L. *Soil Chemical Analysis*; Text Book; Printice-Hall of India: New Delhi, India; Privat Limited: New Delhi, India, 1973; pp. 144–197, 381.
28. Mur, L.; Kenton, P.; Atzorn, R.; Miersch, O.; Wasternack, C. The Outcomes of Concentration-Specific Interactions between Salicylate and Jasmonate Signaling Include Synergy, Antagonism, and Oxidative Stress Leading to Cell Death. *Plant Physiol.* **2006**, *140*, 249–262. [CrossRef] [PubMed]
29. Bergmeyer, H.U. *Methods of Enzymatic Analysis*, 2nd ed.; Academic Press: New York, NY, USA, 1974.
30. Abd El-Kareem, M.S.M.; Rabbih, M.A.; Selim, E.T.M.; Elsherbiny, A.E.; El-Khateeb, A.Y. Application of GC/EIMS in combination with semi-empirical calculations for identification and investigation of some volatile components in basil essential oil. *Int. J. Anal. Mass Spectrom. Chromatogr.* **2016**, *4*, 14–25. [CrossRef]
31. Mohammed, I.A.; Guma, A.N. Anatomical diversity among certain genera of Family Cucurbitaceae. *Int. J. Res. Stud. Biosci.* **2015**, *3*, 85–91.
32. Bricker, B. *MSTATC: A Micro Computer Program from the Design Management and Analysis of Agronomic Research Experiments*; Michigan State University: East Lansing, MI, USA, 1991.
33. Snedecor, G.W.; Cochran, W.G. *Statistical Methods*, 7th ed.; Iowa State University Press: Ames, IA, USA, 1980; p. 507.
34. Marra, R.; Lombardi, N.; d'Errico, G.; Troisi, J.; Scala, G.; Vinale, F. Application of *Trichoderma* strains and metabolites enhances soybean productivity and nutrient content. *J. Agric. Food Chem.* **2019**, *67*, 1814–1822. [CrossRef] [PubMed]
35. Vinale, F.; Sivasithamparam, K.; Ghisalberti, E.L.; Woo, S.L.; Nigro, M.; Marra, R.; Lombardi, N.; Pascale, A.; Ruocco, M.; Lanzuise, S.; et al. *Trichoderma* secondary metabolites active on plants and fungal pathogens. *Open Mycol. J.* **2014**, *8* (Suppl. 1), 127–139. [CrossRef]
36. Carná, M.; Repka, V.; Skůpa, P.; Šturdík, E. Auxins in defense strategies. *Biologia* **2014**, *69*, 1255–1263. [CrossRef]
37. Everse, J.; Hsia, N. The toxicities of native and modified hemoglobins. *Free. Radic. Biol. Med.* **1997**, *22*, 1075–1099. [CrossRef]
38. Rogovin, V.V.; Piruzyan, L.A.; Murav'ev, R.A. *Peroksidazosomy (Peroxysomes)*; Nauka: Moscow, Russia, 1977.

39. Ghanem, G.A.M.; Gebily, D.A.S.; Ragab, M.M.; Ali, A.M.; Soliman, N.E.K.; Abd El-Moity, T.H. Efficacy of antifungal substances of three *Streptomyces* spp. against different plant pathogenic fungi. Journal of Biological Pest Control. *Egypt. J. Biol. Pest Control.* **2022**, *32*, 112. [CrossRef]
40. Avis, T.J.; Bélanger, R.R. Mechanisms and means of detection of biocontrol activity of Pseudozyma yeasts against plant-pathogenic fungi. *FEMS Yeast Res.* **2002**, *2*, 5–8. [CrossRef] [PubMed]
41. Astiko, W.; Muthahanas, I. Biological Control Techniques for Chili Plant Diseaseby Using *Streptomyces* sp. and *Trichoderma* sp. *Int. J. Innov. Sci. Res. Technol.* **2019**, *4*, 155–160.
42. Aldesuquy, H.S.; Ibrahim, H.A. The role of shikimic acid in regulation of growth, transpiration, pigmentation, photosynthetic activity and productivity of *Vigna sinensis* plants. *Phyton* **2000**, *40*, 277–292.
43. Khan, M.R.; Rizvi, T.F. Effect of elevated levels of CO2 on powdery mildew development in five cucurbit species. *Sci. Rep.* **2020**, *10*, 4986. [CrossRef]
44. Percival, G.C.; Fraser, G.A. The influence of powdery mildew infection on photosynthesis, chlorophyll fluorescence, leaf chlorophyll and carotenoid content of three woody plant species. *Arboric. J.* **2002**, *26*, 333–346. [CrossRef]
45. Taub, D. Effects of Rising Atmospheric Concentrations of Carbon Dioxide on Plants. *Nat. Educ. Knowl.* **2010**, *3*, 21.
46. Amaresh, C.; Bhatt, R.K. Biochemical and physiological response to salicylic acid in reaction to systemic acquired resistance. *Photosynthetica* **1998**, *35*, 255–258.
47. Yedidia, I.; Srivastva, A.K.; Kapulnik, Y.; Chet, I. Effect of *Trichoderma harzianum* on microelement concentrations and increased growth of cucumber plants. *Plant Soil* **2001**, *235*, 235–242. [CrossRef]
48. Shehata, M.N.; Abdelgawad, K.F. Potassium Silicate and Amino Acids Improve Growth, Flowering and Productivity of Summer Squash under High Temperature Condition. *Am. Eurasian J. Agric. Environ. Sci.* **2019**, *19*, 74–86.
49. Sousa, J.A.J.; Olivares, F.L. Plant growth promotion by streptomycetes: Ecophysiology, mechanisms and applications. *Chem. Biol. Technol. Agric.* **2016**, *3*, 24. [CrossRef]
50. Saja, D.; Janeczko, A.; Barna, B.; Skoczowski, A.; Dziurka, M.; Kornaś, A.; Gullner, G. Powdery Mildew-Induced Hormonal and Photosynthetic Changes in Barley Near Isogenic Lines Carrying Various Resistant Genes. *Int. J. Mol. Sci.* **2020**, *21*, 4536. [CrossRef]
51. Pratyusha, S. Phenolic Compounds in the Plant Development and Defense: An Overview. In *Plant Stress Physiology—Perspectives in Agriculture*; IntechOpen: London, UK, 2022. [CrossRef]
52. Elsisi, A.A. Evaluation of biological control agents for managing squash powdery mildew under greenhouse conditions. *Egypt. J. Biol. Pest Control.* **2019**, *29*, 89. [CrossRef]
53. Wu, Y.; Lin, H.; Lin, Y.; Shi, J.; Xue, S.; Hung, Y.; Chen, Y.; Wang, H. Effects of biocontrol bacteria *Bacillus amyloliquefaciens* LY-1 culture broth on qualityattributes and storability of harvested litchi fruit. *Postharvest Biol. Technol.* **2017**, *132*, 81–87. [CrossRef]
54. Wang, F.; Xiao, J.; Zhang, Y.; Li, R.; Liu, L.; Deng, J. Biocontrol ability and action mechanism of *Bacillus halotolerans* against *Botrytis cinerea* causing grey mould in postharvest strawberry fruit. *Postharvest Biol. Technol.* **2021**, *174*, 111456. [CrossRef]
55. Kaur, H.; Salh, P.K.; Singh, B. Role of defense enzymes and phenolics in resistance of wheat crop (*Triticum aestivum* L.) towards aphid complex. *J. Plant Interact.* **2017**, *12*, 304–311. [CrossRef]
56. Nandi, A.; Yan, L.-J.; Jana, C.K.; Das, N. Role of Catalase in Oxidative Stress- and Age-Associated Degenerative Diseases. *Oxidative Med. Cell. Longev.* **2019**, *2019*, 9613090. [CrossRef]
57. Hafez, Y.M.; El-Nagar, A.S.; Elzaawely, A.A.; Kamel, S.; Maswada, H.F. Biological control of *Podosphaeraxanthii* the causal agent of squash powdery mildew disease by upregulation of defense-related enzymes. *Egypt. J. Biol. Pest Control.* **2018**, *28*, 57. [CrossRef]
58. Brewer, M.T.; Larkin, R.P. Efficacy of several potential biocontrol organisms against *Rhizoctonia solani* on potato. *Crop Prot.* **2005**, *24*, 939–950. [CrossRef]
59. Ezziyyani, M.; Requena, M.E.; Gilabert, C.E.; Candela, M.E. Biological control of phytophthora root rot of pepper using *Trichoderma harzianum* and *Streptomyces rochei* in combination. *J. Phytopathol.* **2007**, *155*, 342–349. [CrossRef]

horticulturae

Article

Identification and Comparative Analysis of the Rosaceae RCI2 Gene Family and Characterization of the Cold Stress Response in *Prunus mume*

Lichen Yang [1], Ping Li [2], Like Qiu [1], Sagheer Ahmad [3], Jia Wang [1] and Tangchun Zheng [1,*]

1 Beijing Key Laboratory of Ornamental Plants Germplasm Innovation and Molecular Breeding, National Engineering Research Center for Floriculture, Beijing Laboratory of Urban and Rural Ecological Environment, Engineering Research Center of Landscape Environment of Ministry of Education, Key Laboratory of Genetics and Breeding in Forest Trees and Ornamental Plants of Ministry of Education, School of Landscape Architecture, Beijing Forestry University, Beijing 100083, China
2 College of Landscape and Travel, Agricultural University of Hebei, Baoding 071000, China
3 Key Laboratory of National Forestry and Grassland Administration for Orchid Conservation and Utilization, College of Landscape Architecture, Fujian Agriculture and Forestry University, Fuzhou 035002, China
* Correspondence: zhengtangchun@bjfu.edu.cn; Tel.: +86-10-62336321

Abstract: Rare cold inducible 2 (RCI2) proteins are a group of low molecular weight proteins that widely exist in various tissues of plants and play crucial roles in plant growth and development and abiotic stress responses. Genome-wide identification and analysis of *RCI2* have not been documented in Rosaceae plants. Therefore, we identified 23 *RCI2* genes from seven Rosaceae plants, which were classified into three subfamilies. The RoRCI2 protein encodes a highly conserved domain of Pmp3. Three homologous *PmRCI2*s genes from *Prunus mume* were cloned and named *PmRCI2-1*, *PmRCI2-2*, and *PmRCI2-3*. The results of subcellular localization prediction showed that three PmRCI2s localized to membrane structures, and the abscisic acid response element were found to have the largest number in the promoter sequences of *PmRCI2*s. The results of quantitative real-time PCR (qRT-PCR) showed that *PmRCI2-3* was significantly induced by low temperature and highly expressed in stems and buds during the endodormancy stage. Our study improves the understanding of the RCI2 family of Rosaceae plants regarding the cold responses and provides a theoretical basis for the cold-resistant breeding of *P. mume*.

Keywords: *Prunus mume*; RCI2 gene family; phylogenetic analysis; cold stress

1. Introduction

Temperature is an important limiting factor for plant distribution and growth and development [1]. Low temperatures (cold), especially freezing, can disturb ionic homeostasis by impairing cell membrane permeability, thereby inducing electrolyte leakage [2]. Rare cold inducible 2 (RCI2) proteins are important regulators of temperature-mediated signaling pathways, which are active in maintaining cellular ion homeostasis and plasma membrane potential [3]. The genome-wide identification and analysis of *RCI2* have been accomplished in many model plants [3,4] and horticultural herbaceous crops [5]. However, there are few studies on perennial woody plants, especially in Rosaceae.

RCI2 proteins, also named low-temperature induced proteins 6 (Lti6) or plasma membrane proteins 3 (Pmp3), are present as multigene family encoding proteins in plants [6]. RCI2 family gene transcripts are accumulated in response to cold stress differently in different organs, for example, the *AtRC12B* mRNA accumulates in stems, flowers, and siliques of plants exposed to 4 °C, but not in roots. The transcription of *AtRC12A* showed the highest levels in stems, while the lowest in roots. The *RCI2* genes respond to various abiotic stresses and may affect seed germination and root growth in response to low-temperature stress [7,8]. In both *Arabidopsis thaliana* and *Aeluropus littoralis*, *RCI2* homologs respond to membrane dehydration, caused by freezing injury, by stabilizing membrane proteins and reducing electrolyte leakage [9]. In *A. littoralis*, *AlTMP1/2* overexpressed tobacco improved tolerance to freezing stress at the seedling stage [10,11], and *AlTMP1* transgenic tobacco plants mostly recovered to normal growth after 2 h at −20 °C [11]. In addition to cold stress, these genes can also enhance the tolerance of transgenic plants to other abiotic stresses, such as heat, salt, and drought. Recently, more and more studies show that transgenic plants overexpressing *RCI2* genes showed high tolerance to abiotic stresses [7,8,12]. C-repeat binding factor/dehydration-responsive element binding (*CBF/DREB1*) transcription factors can regulate some cold-induced genes of *Arabidopsis*. The *OsLti6b* is highly expressed in *CBF1/DREB1b* transgenic rice, suggesting that it may play a role in the signaling pathway downstream of the rice *CBF1/DREB1b* ortholog [13]. Therefore, the *RCI2* gene plays a crucial role in regulating plant growth and development and abiotic stress responses [10].

Prunus mume is an important woody ornamental plant in early spring with an attractive aroma, colorful petals, and diverse petal shapes. However, traditional *P. mume* varieties cannot naturally overwinter on land in the north temperate zone, which seriously limits their economic and ornamental value [14]. Breeding super cold-resistant *P. mume* varieties is an important topic at present; however, the molecular mechanism of *P. mume* responding to low temperatures is still unclear. Here, we identified 23 members of the *RCI2*s family in 7 Rosaceae plants and analyzed their gene sequences, phylogenetic trees, evolutionary relationships, and expression patterns under different low-temperature conditions. The findings of this study provide a basis for a comprehensive understanding of the phylogenetic relationship of the *RCI2*s family in Rosaceae and their functions in response to low-temperature stress in *P. mume*.

2. Materials and Methods

2.1. Plant Materials

The plant materials used for qRT-PCR were collected from the greenhouse of Beijing Forestry University. Healthy plants of *P. mume* "Zao Lve" with the age of two years were selected. Plant material was treated at different temperatures (4, 0, −4, −8 °C) for 4 h and at 4 °C for different durations (0, 1, 3, 5, 7, 9, 11 d), respectively. Annual shoots from five plants with different treatments were collected and mixed for RNA-seq with three biological repeats. All samples were immediately frozen in liquid nitrogen and stored at −80 °C for further usage.

2.2. Plants Genome Resources

The high-quality genome assembly and annotation files of *A. thaliana* (TAIR10.41), *Prunus dulcis* (v4.0), *Prunus salicina* (v2.0), *Prunus armeniaca* (v1.0), *Prunus persica* (v2.0), *Prunus avium* (v1.0.a1), *Rosa chinensis* (v1.0), and *P. mume* (v1.0) were downloaded from Ensembl Plants (https://plants.ensembl.org/index.html, accessed on 22 July 2022) [15] and the Genome Database for Rosaceae (GDR, https://www.rosaceae.org, accessed on 22 July 2022) [16].

2.3. Identification of RCI2s Gene Family

According to the *A. thaliana* RCI2 family protein sequence, the BLAST GUI Wrapper-Two Sequences Files (E = 10^{-5}) in TB tools (v1.0987663) software [17] were used to search for RCI2 family members in 7 Rosaceae plants. We used phylogenetic trees, sequence size, and conserved domain calibration to check the genetic members and finally screened out *RCI2*s members of 7 Rosaceae plants. With the aid of a phylogenetic tree, sequence size, and conserved domains, we deleted the false gene members. Molecular weights (MW) and isoelectric points (pI) of the RCI2s family members were calculated using the online tool ExPASy (https://www.expasy.org/, accessed on 23 July 2022), The subcellular localization of 7 Rosaceae plants was predicted using the subcellular localization tool WoLF PSORT (https://wolfpsort.hgc.jp/, accessed on 25 July 2022).

2.4. Gene Structure and Protein Conserved Motif Analysis

For investigating the RCI2 domains of the closely related *P. persica*, *P. mume*, *P. armeniaca*, and *P. salicina* in Prunoideae, the conserved domains of the RCI2s gene family were obtained by NCBI CD-Search (https://www.ncbi.nlm.nih.gov/cdd/, accessed on 23 July 2022) and Pfam (https://pfam.xfam.org, accessed on 23 July 2022). The RCI2 proteins were submitted to MEME (v4.12.0) [18] with settings (-nmotifs 3 -minw 8 -maxw 50, accessed on 23 July 2022) to search for conversed motifs. At the same time, Muscl was used to perform multiple sequence alignments of the full-length RCI2 protein sequences of seven Rosaceae plants. The trimAI (v2.1.3) was used to trim the alignment results, and IQ-tree (v2.1.3) was used to automatically screen amino acid replacement models [19]. Finally, TB tools [17] was used to conduct the tree-structure-motif map.

2.5. Chromosome Location and Synteny Analysis

Gene tandems were analyzed using the Multiple Collinearity Scan Toolkit (MCScanX) in TB tools with the default parameters. McscanX analyzed the synteny of the RCI2s across *A. thaliana*, *P. salicina*, *P. armeniaca*, *P. persica*, and *P. mume*, which were visualized in TB tools. The chromosomal length and location information of *RCI2*s were extracted from the gff and fasta files downloaded from the *P. mume* genome project. TB tools [17] were used to map chromosome distribution of RCI2s family members of *P. mume*.

2.6. Cis-Acting Element Analysis of RCI2 Gene Promoters

A cis-element analysis of the 2000 bp upstream genomic sequences of *PmRCI2*s genes, retrieved from the *P. mume* genome database. was performed by PlantCARE (https://bioinformatics.psb.ugent.be/webtools/plantcare/html/, accessed on 20 August 2022) [20], and these sequences were used as putative promoter sequences (Table S1). Then, we screened the results and visualized them by constructing the element distribution with TB tools [17].

2.7. Phylogenetic Analysis of RCI2s

The protein sequences in eight species (*A. thaliana*, *P. dulcis*, *P. salicina*, *P. armeniaca*, *P. persica*, *P. avium*, *R. chinensis*, and *P. mume*) were aligned by ClustalX [21]. At the same time, we constructed the maximum likelihood tree using IQ-tree (v2.1.3) [22]. Finally, the Chiplot (https://www.chiplot.online/, accessed on 15 July 2022) was used to decorate this phylogenetic tree.

2.8. Expression Pattern of PmRCI2s

Aiming to investigate the expression pattern of *PmRCI2*s involved in tissues development and low-temperature response, raw data (Tables S2–S4) from the RNA-seq were collected from different tissues, different dormancy periods, and natural overwintering conditions of *P. mume*. The different tissues of *P. mume* included flower buds, fruits, leaves, roots, and stems [23]. The different dormancy periods of *P. mume* "Zao Lve" buds overwintering in open fields in Beijing area included EDI (Endodormancy I, November), EDII (Endodormancy II, December), EDIII (Endodormancy III, January), and NF stage (Natural Flush, February) [24]. Using the stem of "Songchun" cultivar as material, the transcriptome data of *PmRCI2*s was acquired in three different seasons (autumn, October; winter, January; spring, March) under natural cold at three different places: Beijing (BJ, 39°54′ N, 116°28′ E), Chifeng (CF, 42°17′ N, 118°58′ E), and Gongzhuling (GZL, 43°42′ N, 124°47′ E). We used TB tools [17] to analyze these data and drew gene expression heatmaps.

In order to investigate the expression pattern of *PmRCI2*s under different cold treatment conditions in *P. mume* "Zao Lve", the grafted annual stems of "Zao Lve" were used. Additionally, whole plants were treated at 4 °C for 0, 1, 3, 5, 7, 9, 11 d, and 4, 0, −4, and −8 °C for 6 h. The culture material at 24 °C was used as the inner group control. RNA extraction was performed on the treated material using RNAprep Pure Plant Plus Kit (TIANGEN, Beijing, China). First-strand cDNA synthesis was performed using a TIANScript

First Strand cDNA Synthesis Kit (Tiangen, Beijing, China) according to the manufacturer's instructions. The Integrated DNA Technologies (IDT, https://sg.idtdna.com, accessed on 23 July 2022) was used to design qRT-PCR primers (Table S5), and TB Green chimeric fluorescence method was used for Real-Time PCR. qRT-PCR was carried out using a PikoReal real-time PCR system (Thermo Fisher Scientific, Waltham, MA, USA). *PmPP2A* gene of *P. mume* was selected as the internal reference gene [25]. The experiment was carried out with three biological replicates and three technical replicates, and the expression level of target genes was calculated by the $2^{-\Delta\Delta Ct}$ method [26].

3. Results

3.1. Identification of Rosaceae RCI2s Gene Family

Three members of the RCI2 family were obtained by BLSATp in *P. mume* genome and named *PmRCI2-1* (*Pm027750*), *PmRCI2-2* (*Pm003262*), and *PmRCI2-3* (*Pm003263*). We also identified members of the *RCI2* family in the genomes of almond (*Prunus dulcis*), plum (*Prunus salicina*), peach (*Prunus persica*), apricot (*Prunus armeniaca*), cherry (*Prunus avium*) and rose (*Rosa chinensis*). The amino acid length of *RCI2* in seven Rosaceae plants was between 55 aa and 149 aa, and the molecular weight (MW) was between 5.85 kDa and 15.74 kDa. *Pav_sc0000749.1_g020.1.mk:mrna* in *P. armeniaca* encoded the RCI2 protein with the largest MW among the seven plants, and the predicted isoelectric point (pI) was 4.09~9.95 (Table 1). Subcellular localization prediction results showed that the *RCI2* genes of seven plants were all located in the membrane structures, of which 39% were located in the plasma membrane, and 57% were located in the vacuole.

Table 1. Basic characteristics of the RCI2 gene family in seven Rosaceae plants.

Species	Gene ID	Basic Features			
		pI	MW/kDa	Length/aa	Prediction of Subcellular Localization
Prunus dulcis	*VVA22294*	7.87	12.37	118	Plasma membrane
Prunus persica	*ONH98753*	5.09	8.53	77	Vacuole
	ONI04115	5.43	6.33	58	Plasma membrane
	ONI04114	4.68	6.01	55	Vacuole
	ONI03135	9.69	7.33	69	Plasma membrane
	ONI24010	7.81	13.94	140	Plasma membrane
Prunus armeniaca	*PARG28267m01*	6.15	14.13	123	Plasma membrane
	PARG02720m01	9.95	12.73	115	Mitochondrion
	PARG18782m01	4.25	5.93	55	Vacuole
Prunus avium	*Pav_sc0001938.1_g430.1.mk:mrna*	4.68	6.01	55	Vacuole
	Pav_sc0001938.1_g440.1.mk:mrna	6.5	6.03	55	Plasma membrane
	Pav_sc0000749.1_g020.1.mk:mrna	8.79	15.74	149	Plasma membrane
	Pav_sc0000582.1_g860.1.mk:mrna	8.86	15.12	136	Plasma membrane
Rosa chinensis	*PRQ17753*	4.75	6.6	64	Vacuole
	PRQ42527	5.82	6.02	55	Vacuole
	PRQ42528	5.93	6.36	58	Vacuole
	PRQ58884	4.25	5.85	55	Vacuole
	PRQ43472	4.09	7.53	71	Vacuole
	PRQ46105	5.01	8.59	79	Vacuole
Prunus mume	*Pm027750*	4.79	8.65	77	Vacuole
	Pm003262	5.43	6.35	58	Plasma membrane
	Pm003263	9.58	15.27	135	Vacuole
Prunus salicina	*evm.model.LG07.71*	4.79	8.65	77	Vacuole

3.2. Gene Structure Analysis and Phylogenetic of RCI2 Genes

Chromosome position analysis showed that three *PmRCI2*s were located in two different chromosomes in the *P. mume* genome, two on chromosome 1 and one on chromosome 8 (Figure 1a). In order to better analyze the gene structure of *RCI2*s in *P. mume*, we analyzed the RCI2s domains of *P. armeniaca*, *P. salicina*, *P. persica*, and *P. mume* (Figure 1b), which were closely related in the Rosaceae Prunoideae. A high degree of similarity in the structure of the RCI2s was found among these four plants. The motif was analyzed by MEME software, and three different motifs were identified (Figure 1b). Except for *ONI03135* in *P. persica*, other *RCI2*s all contained two motifs (Motif 1 and Motif 2), indicating that these two motifs were conserved in the RCI2s family in *P. armeniaca*, *P. salicina*, and *P. mume*. In the four species, motif 3 was generally at the end of the peptide segment and it was only present in clade A of the RCI2s.

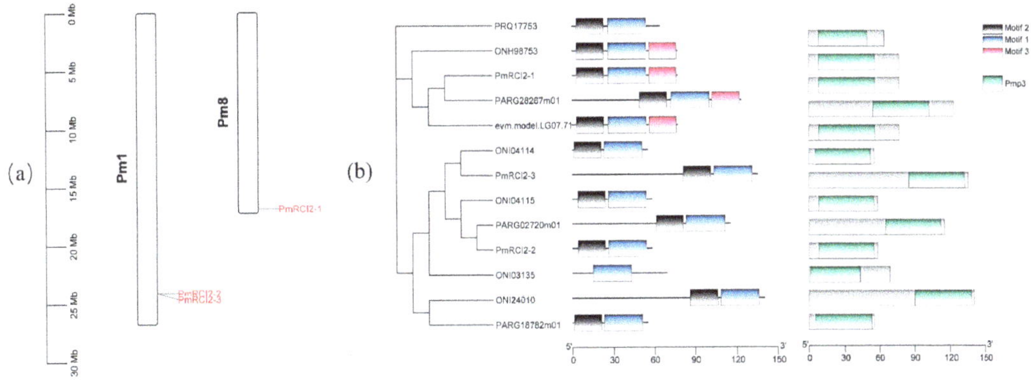

Figure 1. Gene locations and structures of *RCI2*s in Rosaceae. (**a**) Gene locations of *PmRCI2*s on the *P. mume* chromosome. (**b**) The motif composition and conserved domains were identified in RCI2s of *P. mume*, *P. armeniaca*, *P. salicina*, and *P. persica*.

In order to explore the phylogenetic relationship of RCI2s, we used a maximum likelihood method to create a phylogenetic tree of eight plants (Figure 2). The 23 identified RoRCI2 proteins were divided into three distinct clades. Clade A was the largest group and contained the most members (twelve) as compared to clade B containing only eight members (Figure 2). The PmRCI2s were only found in clades A and C, where PmRCI2-1 was homologous to PARG28267 in the *P. armeniaca* and evm.model. LG07.71 in *P. salicina*, PmRCI2-2 was homologous to PARG02720 in *P. armeniaca*, and PmRCI2-3 was homologous to ONI04114 in *P. persica*.

3.3. Analysis of Cis-Elements in the Promoters of PmRCI2s

Identification of cis-elements in the promoter region can help to explore the possible regulatory mechanisms of *PmRCI2*s in *P. mume*. Predicted results of PlantCARE (Table S6) showed that stress response elements were the most abundant (57.65%) among the other elements. multiple number of elements were involved in the light response, such as G-box, GT1-motif, and I-box. They were widely distributed in *PmRCI2*s promoters (Figure 3a,b). Anaerobic induced element (ARE) and MBS, involved in drought inducibility, were found in *PmRCI2-1* and *PmRCI2-2*. Low-temperature response element (LTR) was found only in *PmRCI2-2*. Notably, *PmRCI2-2* only contained some light response elements in stress responses (Figure 3c). In addition, hormone-related elements such as CGTCA-motif, TGACG-motif, P-box, TGA-box, and TCA-element participating in different hormone responses including MeJA, auxin and salicylic acid were found (Figure 3c). All *PmRCI2*s genes had abscisic acid response element (ABRE), and the number of this element was the largest (Figure 3b), *PmRCI2-2* contained seven ABREs, and *PmRCI2-3* contained six ABREs.

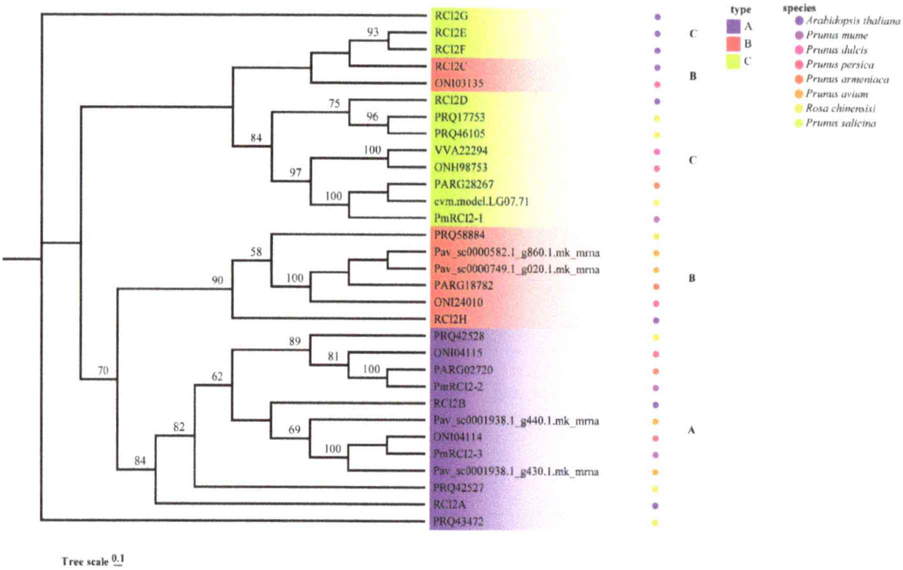

Figure 2. Phylogenetic tree of RCI2s in Rosaceae. The ML phylogenetic tree of RCI2s family in *P. dulcis*, *P. salicina*, *P. persica*, *P. armeniaca*, *R. chinensis*, *P. mume*, and *A. thaliana*.

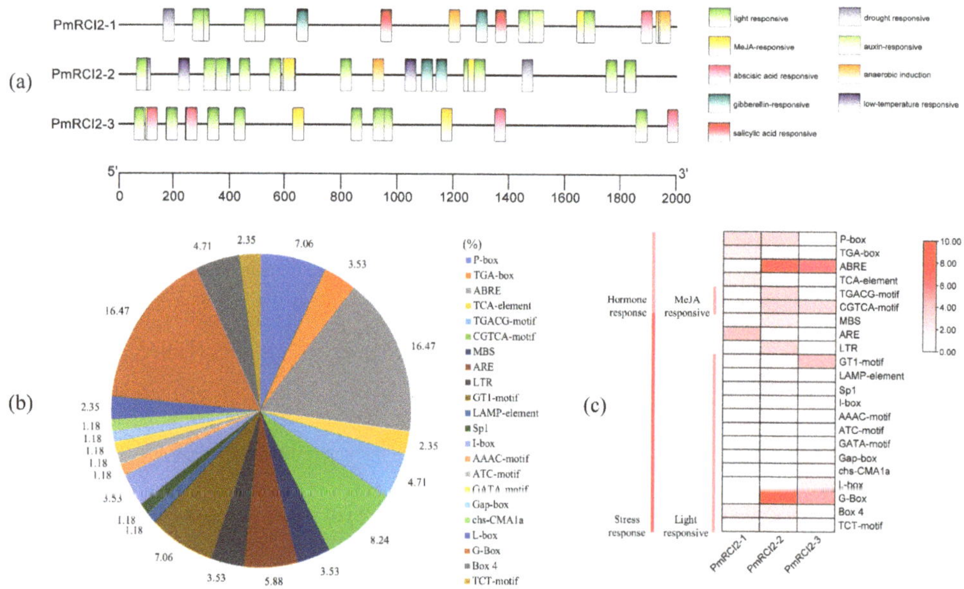

Figure 3. Cis-elements analysis of *PmRCI2s* promoters. (**a**) The distribution of the main cis-elements on promoter of identified *PmRCI2s* gene. (**b**) The proportion (%) of cis-elements in the *PmRCI2s* promoters involved in stress and hormone responses. (**c**) Heatmap of the cis-elements involved in stress responsiveness, and hormone responsiveness.

3.4. Collinear Analysis of RCI2s in Rosaceae and A. thaliana

In order to further analyze the evolution of *RCI2s* in Rosaceae plants, we performed an inter-genome comparison and collinearity analysis of *A. thaliana*, *P. mume*, *P. armeniaca*,

P. salicina and *P. persica* (Figure 4). There were 4 *RCI2*s homologous gene pairs between *A. thaliana* and *P. mume*. *PmRCI2*s (*PmRCI2-1*, *PmRCI2-2*) had two collinearities in *A. thaliana* (*AT3G05880.1*/*AT1G57550.1*, *AT2G24040.1*/*AT4G30650.1*), one collinearity in *P. persica* (*ONH98753*, *ONI04114*) and in *P. armeniaca* (*PARG28267*, *PARG02720*). In addition, there was only one pair of RCI2 family homologous genes between *P. mume* and *P. persica*, suggesting that RCI2s family was contracted in some Rosacea genomes. *PmRCI2-3* homologous gene could not be found in these species, suggesting that *PmRCI2-3* was a unique gene in *P. mume*, which might be mutated.

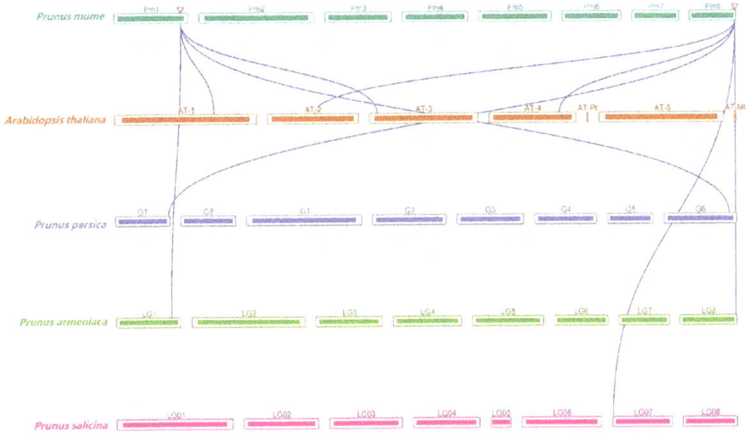

Figure 4. Syntenic analyses of *RCI2*s in *P. mume* and other plants. Blue lines represent syntenic *RCI2* gene pairs.

3.5. Expression Pattern of PmRCI2s under Cold Stress

*PmRCI2*s was expressed differently in different tissues of *P. mume*, such as *PmRCI2-3* had a low expression level in roots, but relatively high expression in stems and buds, and *PmRCI2-2* had a relatively high expression level in fruits and leaves (Figure 5a). The expression level of *PmRCI2-1* and *PmRCI2-2* gradually accumulated and reached the maximum at the NF stage when the bud dormancy was released, and the expression level of *PmRCI2-2* was higher than that of *PmRCI2-1*. On the contrary, the expression level of *PmRCI2-3* gradually upregulated from EDI to NF stage (Figure 5b). Moreover, we examined the expression pattern of "Songchun" cultivar in stems in different seasons at three places in BJ, CF, and GZL. In the same season, the expression levels of the three *PmRCI2*s increased with the decrease in the average temperature in different locations. The variation trend of gene expression of the same gene in different places was different. For example, in Beijing and Chifeng, the expression level of *PmRCI2-1* decreased from autumn to spring, but in Gongzhuling, the expression level of *PmRCI2-1* increased during the cold acclimation period and decreased after the cold acclimation period (from winter to spring). Geographical factors and the cold resistance of the plants may explain differences in gene expression. In addition, we found that the overall expression level of *PmRCI2-3* was higher than that of the other two genes during the cold acclimation period (from autumn to winter) at the same place (Figure 5c).

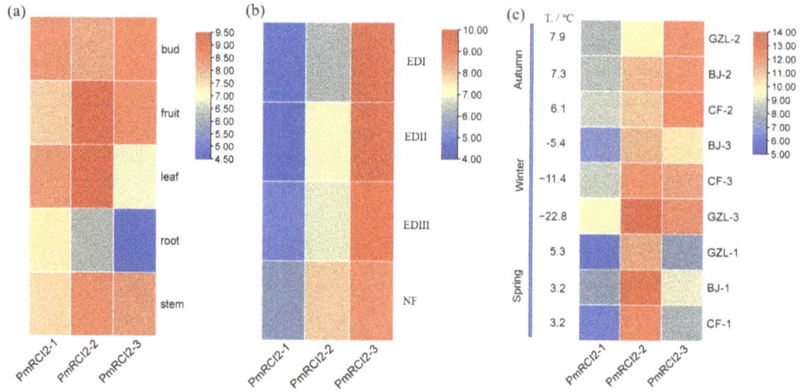

Figure 5. The expression pattern of *PmRCI2*s. (**a**) The expression level of *PmRCI2*s in different tissues of *P. mume*. (**b**) An examination of the expression levels of *PmRCI2*s in flowers of *P. mume* "Zao Lve" during three stages of endodormancy at low temperatures. (**c**) Three regional test sites (Beijing, BJ; Chifeng, CF; Gongzhuling, GZL) were used to collect data on the expression accumulation of *P. mume* "Songchun" stems during three different periods (autumn, winter, and spring). In the same season, the average temperature of different locations is ranked from high to low.

For further understanding the role of *PmRCI2* genes in response to cold stress, the expression level of *PmRCI2* genes under different treatments was detected by qRT-PCR (Figure 6). Under the condition of cold treatment at 4 °C, the expression level of *PmRCI2-3* was significantly higher than that of without cold treatment. The expression level of *PmRCI2-3* increased gradually with the prolonged cold treatment time and reached the maximum value at 9 d. The increase in the expression of *PmRCI2-3* was also significantly higher than the other two genes. Under different temperature treatments, the expression levels of *PmRCI2-1* and *PmRCI2-2* fluctuated to a certain extent but did not show a significant upward or downward trend. However, the expression level of *PmRCI2-3* gradually increased with the severity of cold stress and reached a peak at −8 °C. These results demonstrate that *PmRCI2-3* is significantly induced by cold stress and plays an important role in resisting constant low-temperature conditions.

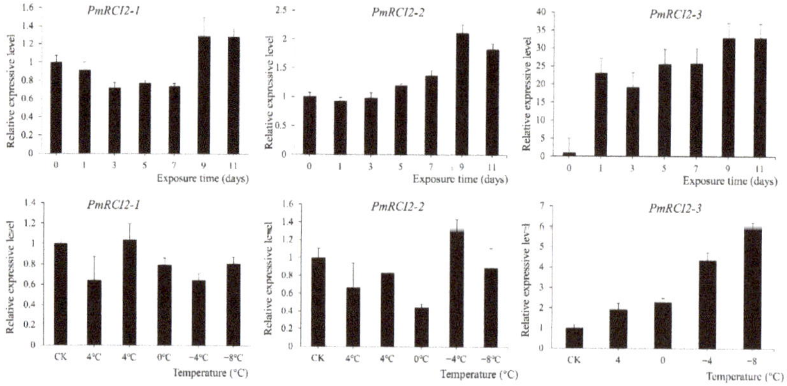

Figure 6. qRT-PCR analysis of *PmRCI2*s in an annual branch of the cultivar "Zao Lve" under low-temperature treatment. The grafted annual stems of "Zao Lve" were treated at 4 °C for 0, 1, 3, 5, 7, 9, 11 d, and 4, 0, −4, and −8 °C for 6 h, the culture material at 24 °C was used as the inner group control.

4. Discussion

Low-temperature stress severely restricts grain production, and China loses 300–500 million tons of rice each year due to cold damage [27]. Low temperatures in early spring and in late spring are the main disastrous weather for agricultural production in the northern temperate zone. Meanwhile, low-temperature stress in early spring is also one of the limiting factors for the growth and distribution of *Prunus* genus. At present, RCI2 family genes have been identified in *A. thaliana* [9], *Oryza sativa* [13], *Cucumis sativus* [5], and other plants, and related studies have shown that *RCI2*s are involved in the biological activities of plants in response to cold stress. In this study, we identified 23 *RCI2* genes from seven Rosaceae plants, of which 3 *PmRCI2*s genes (*PmRCI2-1, PmRCI2-2, PmRCI2-3*) belonged to *P. mume* (Table 1). Among the four species of *P. armeniaca, P. salicina, P. persica,* and *P. mume, P. salicina* had only one member of the RCI2 gene family, while *P. persica* had the most members, containing five RCI2s. An equal number of RCI2 gene family members were found in *P. armeniaca* and *P. mume*. This is in line with the evolutionary relationship of these four plants, indicating the expansion of the family during the evolution of the *Prunus* genus. In addition, the collinearity analysis showed that *P. salicina, P. armeniaca,* and *P. mume* all contained part of the RCI2 gene family members homologous to *AT3G05880.1/AT1G57550.1, AT2G24040.1/AT4G30650.1* of *A. thaliana*, while the *evm.model.LG07.71* of *P. salicina* was homologous to *AT3G05880.1*, suggesting that the genes homologous to *AT3G05880.1/AT1G57550.1* in these four plants are all generated by genome amplification (Figure 4).

The *RCI2*s play a role in different membrane structures in the cell [4]. Abiotic stress can affect the normal physiological activities of plants by changing the membrane structure in cells. For instance, heat stress alters thylakoid structure and reduces photosystem activity [28]. In this study, we predicted that *PmRCI2-1* and *PmRCI2-3* were located in the vacuole, whereas *PmRCI2-2* was located on the plasma membrane (Table 1). Plant cell vacuole contains various substances such as inorganic salts, organic acids, and sugars, which play an important role in maintaining intracellular ion homeostasis [29]. Therefore, *PmRCI2-1* and *PmRCI2-3* may have special functions in response to abiotic stresses and maintain normal plant growth and development. Like *OsLti6b* in *O. sativa* [13,30] and *AlTMP2* in *A. littoralis* [10], these genes are also localized in the vacuole and exhibit resistance to cold stress.

Plenty of evidence indicates that salt stress and cold stress are major abiotic stressors that induce *RCI2*s [30,31]. The plant materials used are perennial plants with robust growth, which can respond to cold stress normally; however, in our study, only *PmRCI2-3* was found to have a strong response to low-temperature stress (Figure 6). The expression level of *PmRCI2-3* always maintains a high expression level under long-term low-temperature treatment. The expression level of *PmRCI2-2* increased with the prolongation of the cold treatment time, reached a peak expression in 9 d, and then began to decline. However, the expression trend of *PmRCI2-1* was opposite to that of *PmRCI2-2*, and it increased after 9 d. Under different temperature treatments, the expression level of *PmRCI2-3* gradually increased with the decrease in temperature, while the expression level of *PmRCI2-1* and *PmRCI2-2* showed a trend of first decreasing and then increasing. All three *PmRCI2*s had Pmp3 conserved domains unique to the RCI2s gene family and contained motif 1 and motif 2. Through the analysis of cis-acting elements, no cold-responsive elements were found for *PmRCI2-3* with the highest expression under low-temperature stress (Figure 3). Similar results have been found in cucumber [5], which may be due to the presence of an unknown cold-responsive cis-element in the promoter region.

All three *PmRCI2*s promoter regions contain many abscisic acid response elements (ABRE), which is consistent with the report of a large number of ABRE found in the *RCI2* gene promoters of *Cucumis sativus* [5] and *Brassica napus* [32]. ABA induces second messengers to activate defense responses by generating ROS, and the expression of antioxidant enzyme genes and non-enzymatic defense system genes is also activated by the ABA signaling-inducible mechanism [33]. This shows that *P. mume* may employ ABA-induced *RCI2*s expression in response to cold stress.

5. Conclusions

Twenty-three *RCI2* genes were identified from seven Rosaceae plants by genome-wide screening. Three *PmRCI2* genes (*PmRCI2-1, PmRCI2-2, PmRCI2-3*) from *P. mume* were cloned and systematically analyzed. The results of qRT-PCR showed that *PmRCI2-3* is significantly induced by low temperature and highly expressed in stems and buds during endodormancy stage. Our findings provide key candidate genes for cold resistance breeding of *P. mume* and other *Prunus* species.

Supplementary Materials: The following supporting information can be downloaded at: https://www.mdpi.com/article/10.3390/horticulturae8110997/s1. Table S1: The promoter sequence of *PmRCI2*s. Table S2: Expression profiles of *PmRCI2* genes in different tissues. Table S3: Expression profiles of *PmRCI2* genes during the process of flower bud dormancy release. Table S4: Expression profiles of *PmRCI2* genes in different regions and seasons. Table S5: The Primer of qRT-PCR. Table S6: Cis-acting element on the promoter of *PmRCI2*s.

Author Contributions: T.Z. conceived the manuscript. L.Y. and T.Z. drafted the manuscript. L.Y., L.Q., S.A. and P.L. analyzed the data. T.Z. and J.W. finalized the manuscript. All authors have read and agreed to the published version of the manuscript.

Funding: The research was supported by the Opening Project of State Key Laboratory of Tree Genetics and Breeding (No. K2021101) and the National Natural Science Foundation of China (No. 31800595), and the Special Fund for Beijing Common Construction Project.

Institutional Review Board Statement: Not applicable.

Informed Consent Statement: Not applicable.

Data Availability Statement: Not applicable.

Conflicts of Interest: The authors declare no conflict of interest.

References

1. Weiser, C. Cold resistance and injury in woody plants: Knowledge of hardy plant adaptations to freezing stress may help us to reduce winter damage. *Science* **1970**, *169*, 1269–1278. [CrossRef]
2. Yeshvekar, R.K.; Nitnavare, R.B.; Chakradhar, T.; Bhatnagar-Mathur, P.; Reddy, M.K.; Reddy, P.S. Molecular characterization and expression analysis of pearl millet *plasma membrane proteolipid 3* (*Pmp3*) genes in response to abiotic stress conditions. *Plant Gene* **2017**, *10*, 37–44. [CrossRef]
3. Navarre, C.; Goffeau, A. Membrane hyperpolarization and salt sensitivity induced by deletion of PMP3, a highly conserved small protein of yeast plasma membrane. *EMBO J.* **2000**, *19*, 2515–2524. [CrossRef] [PubMed]
4. Medina, J.; Ballesteros, M.L.; Salinas, J. Phylogenetic and functional analysis of *Arabidopsis RCI2* genes. *J. Exp. Bot.* **2007**, *58*, 4333–4346. [CrossRef] [PubMed]
5. Zhou, Y.; Ge, L.; Li, G.; He, P.; Yang, Y.; Liu, S. In silico identification and expression analysis of Rare Cold Inducible 2 (RCI2) gene family in cucumber. *J. Plant Biochem. Biotechnol.* **2020**, *29*, 56–66. [CrossRef]
6. Won, S.Y.; Jung, J.A.; Kim, J.S. Genome-wide analysis of the MADS Box gene family in *Chrysanthemum*. *Comput. Biol. Chem.* **2021**, *90*, 107424. [CrossRef]
7. Mitsuya, S.; Taniguchi, M.; Miyake, H.; Takabe, T. Disruption of RCI2A leads to over-accumulation of Na+ and increased salt sensitivity in *Arabidopsis thaliana* plants. *Planta* **2005**, *222*, 1001–1009. [CrossRef]
8. Fu, J.; Zhang, D.; Liu, Y.; Ying, S.; Shi, Y.; Song, Y.; Li, Y.; Wang, T. Isolation and characterization of maize *PMP3* genes involved in salt stress tolerance. *PLoS ONE* **2012**, *7*, e31101. [CrossRef]
9. Capel, J.; Jarillo, J.A.; Salinas, J.; Martinez-Zapater, J.M. Two homologous low-temperature-inducible genes from *Arabidopsis* encode highly hydrophobic proteins. *Plant Physiol.* **1997**, *115*, 569–576. [CrossRef]
10. Ben-Romdhane, W.; Ben-Saad, R.; Meynard, D.; Zouari, N.; Mahjoub, A.; Fki, L.; Guiderdoni, E.; Al-Doss, A.; Hassairi, A. Overexpression of *AlTMP2* gene from the halophyte grass *Aeluropus littoralis* in transgenic tobacco enhances tolerance to different abiotic stresses by improving membrane stability and deregulating some stress-related genes. *Protoplasma* **2018**, *255*, 1161–1177. [CrossRef]

11. Ben Romdhane, W.; Ben-Saad, R.; Meynard, D.; Verdeil, J.L.; Azaza, J.; Zouari, N.; Fki, L.; Guiderdoni, E.; Al-Doss, A.; Hassairi, A. Ectopic expression of *Aeluropus littoralis* plasma membrane protein gene *AlTMP1* confers abiotic stress tolerance in transgenic tobacco by improving water status and cation homeostasis. *Int. J. Mol. Sci.* **2017**, *18*, 692. [CrossRef] [PubMed]
12. Kim, Y.O.; Kim, H.S.; Lim, H.G.; Jang, H.Y.; Kim, E.; Ahn, S.J. Functional characterization of salt-stress induced rare cold inducible gene from *Camelina sativa (CsRCI2D)*. *J. Plant Biol.* **2021**, *65*, 279–289. [CrossRef]
13. Kim, S.H.; Kim, J.Y.; Kim, S.J.; An, K.S.; An, G.; Kim, S.R. Isolation of cold stress-responsive genes in the reproductive organs, and characterization of the *OsLti6b* gene from rice (*Oryza sativa* L.). *Plant Cell Rep.* **2007**, *26*, 1097–1110. [CrossRef] [PubMed]
14. Zhang, Q.; Zhang, H.; Sun, L.; Fan, G.; Ye, M.; Jiang, L.; Liu, X.; Ma, K.; Shi, C.; Bao, F. The genetic architecture of floral traits in the woody plant *Prunus mume*. *Nat. Commun.* **2018**, *9*, 1702. [CrossRef]
15. Bolser, D.; Staines, D.M.; Pritchard, E.; Kersey, P. Ensembl plants: Integrating tools for visualizing, mining, and analyzing plant genomics data. In *Plant Bioinformatics*; Springer: Berlin/Heidelberg, Germany, 2016; pp. 115–140.
16. Jung, S.; Lee, T.; Cheng, C.-H.; Buble, K.; Zheng, P.; Yu, J.; Humann, J.; Ficklin, S.P.; Gasic, K.; Scott, K. 15 years of GDR: New data and functionality in the Genome Database for Rosaceae. *Nucleic Acids Res.* **2019**, *47*, D1137–D1145. [CrossRef] [PubMed]
17. Chen, C.; Chen, H.; Zhang, Y.; Thomas, H.R.; Frank, M.H.; He, Y.; Xia, R. TBtools: An integrative toolkit developed for interactive analyses of big biological data. *Mol. Plant* **2020**, *13*, 1194–1202. [CrossRef]
18. Bailey, T.L.; Williams, N.; Misleh, C.; Li, W.W. MEME: Discovering and analyzing DNA and protein sequence motifs. *Nucleic Acids Res.* **2006**, *34*, W369–W373. [CrossRef] [PubMed]
19. Nguyen, L.T.; Schmidt, H.A.; von Haeseler, A.; Minh, B.Q. IQ-TREE: A fast and effective stochastic algorithm for estimating maximum-likelihood phylogenies. *Mol. Biol. Evol.* **2015**, *32*, 268–274. [CrossRef]
20. Walsh, B. Population-genetic models of the fates of duplicate genes. In *Origin and Evolution of New Gene Functions*; Springer: Berlin/Heidelberg, Germany, 2003; pp. 279–294.
21. Thompson, J.D.; Gibson, T.J.; Higgins, D.G. Multiple sequence alignment using ClustalW and ClustalX. *Curr. Protoc. Bioinform.* **2003**. [CrossRef]
22. Trifinopoulos, J.; Nguyen, L.T.; von Haeseler, A.; Minh, B.Q. W-IQ-TREE: A fast online phylogenetic tool for maximum likelihood analysis. *Nucleic Acids Res.* **2016**, *44*, W232–W235. [CrossRef]
23. Zhang, Q.; Chen, W.; Sun, L.; Zhao, F.; Huang, B.; Yang, W.; Tao, Y.; Wang, J.; Yuan, Z.; Fan, G. The genome of *Prunus mume*. *Nat. Commun.* **2012**, *3*, 1318. [CrossRef] [PubMed]
24. Zhang, Z.; Zhuo, X.; Zhao, K.; Zheng, T.; Han, Y.; Yuan, C.; Zhang, Q. Transcriptome profiles reveal the crucial roles of hormone and sugar in the bud dormancy of *Prunus mume*. *Sci. Rep.* **2018**, *8*, 5090. [CrossRef] [PubMed]
25. Wang, T.; Lu, J.; Xu, Z.; Yang, W.; Wang, J.; Cheng, T.; Zhang, Q. Selection of suitable reference genes for miRNA expression normalization by qRT-PCR during flower development and different genotypes of *Prunus mume*. *Sci. Hortic.* **2014**, *169*, 130–137. [CrossRef]
26. Livak, K.J.; Schmittgen, T.D. Analysis of relative gene expression data using real-time quantitative PCR and the 2(-Delta Delta C(T)) Method. *Methods* **2001**, *25*, 402–408. [CrossRef] [PubMed]
27. Zhang, Z.; Li, J.; Pan, Y.; Li, J.; Shi, H.; Zeng, Y.; Guo, H.; Yang, S.; Zheng, W.; Yu, J. Natural variation in *CTB4a* enhances rice adaptation to cold habitats. *Nat. Commun.* **2017**, *8*, 14788. [CrossRef]
28. Hasanuzzaman, M.; Hakim, K.; Nahar, K.; Alharby, H.F. *Plant Abiotic Stress Tolerance*; Springer: Berlin/Heidelberg, Germany, 2019.
29. Blumwald, E. Sodium transport and salt tolerance in plants. *Curr. Opin. Cell Biol.* **2000**, *12*, 431–434. [CrossRef]
30. Kim, H.S.; Lee, J.E.; Jang, H.Y.; Kwak, K.J.; Ahn, S.J. *CsRCI2A* and *CsRCI2E* genes show opposite salt sensitivity reaction due to membrane potential control. *Acta Physiol. Plant.* **2016**, *38*, 50. [CrossRef]
31. Rocha, P.S. Plant abiotic stress-related RCI2/PMP3s: Multigenes for multiple roles. *Planta* **2016**, *243*, 1–12. [CrossRef]
32. Sun, W.; Li, M.; Wang, J. Genome-wide identification and characterization of the RCI2 gene family in allotetraploid *Brassica napus* compared with its diploid progenitors. *Int. J. Mol. Sci.* **2022**, *23*, 614. [CrossRef]
33. Vishwakarma, K.; Upadhyay, N.; Kumar, N.; Yadav, G.; Singh, J.; Mishra, R.K.; Kumar, V.; Verma, R.; Upadhyay, R.; Pandey, M. Abscisic acid signaling and abiotic stress tolerance in plants: A review on current knowledge and future prospects. *Front. Plant Sci.* **2017**, *8*, 161. [CrossRef]

horticulturae

MDPI

Article

Response of Warm Season Turf Grasses to Combined Cold and Salinity Stress under Foliar Applying Organic and Inorganic Amendments

Dina Taher [1], Emam Nofal [1], Mahmoud Hegazi [1], Mohamed Abd El-Gaied [2], Hassan El-Ramady [3,*] and Svein Ø. Solberg [4,*]

1 Horticulture Department, Faculty of Agriculture, University of Kafrelsheikh, Kafr El-Sheikh 33516, Egypt
2 Sakha Horticulture Research Station, Horticulture Research Institute, Agriculture Research Center, Giza 12619, Egypt
3 Soil and Water Department, Faculty of Agriculture, University of Kafrelsheikh, Kafr El-Sheikh 33516, Egypt
4 Faculty of Applied Ecology and Agricultural Sciences, Inland Norway University of Applied Sciences, P.O. Box 400, 2418 Elverum, Norway
* Correspondence: hassan.elramady@agr.kfs.edu.eg (H.E.-R.); svein.solberg@inn.no (S.Ø.S.)

Abstract: Turfgrasses are considered an important part of the landscape and ecological system of golf courses, sports fields, parks, and home lawns. Turfgrass species are affected by many abiotic stresses (e.g., drought, salinity, cold, heat, waterlogging, and heavy metals) and biotic stresses (mainly diseases and pests). In the current study, seashore paspalum (*Paspalum vaginatum* Sw.) and Tifway bermudagrass (*Cynodon transvaalensis* Burtt Davy × C. Dactylon) were selected because they are popular turfgrasses frequently used for outdoor lawns and sport fields. The effect of the combined stress from both soil salinity and cold on these warm season grasses was investigated. Some selected organic and inorganic amendments (i.e., humic acid, ferrous sulphate, and silicon) were applied as foliar sprays five times during the winter season from late October to March. This was repeated over two years in field trials involving salt-affected soils. The physiological and chemical parameters of the plants, including plant height; fresh and dry weight per plot; total chlorophyll content; and nitrogen, phosphorus, iron, and potassium content, were measured. The results showed that all the studied amendments improved the growth of seashore paspalum and Tifway bermudagrass during this period compared to the control, with a greater improvement observed when using ferrous sulphate and humic acid compared to silicon. For seashore paspalum, the highest chlorophyll content in April was recorded after the application of ferrous sulphate at a level of 1000 ppm. The current research indicates that when grown on salt-affected soils, these amendments can be used in warm-season grasses to maintain turf quality during cold periods of the year. Further research is needed to examine any negative long-term effects of these amendments and to explain their mechanisms.

Keywords: *Cynodon transvaalensis*; diatomite; *Paspalum vaginatum*; seashore paspalum; Tifway bermudagrass; salt-affected soil

1. Introduction

Turfgrasses have been utilized by humans for centuries in outdoor sports fields such as football fields and golf courts due to their low cost and safe recreational surface [1]. These grasses also have other important functions; they help to reduce surface water runoff and work as an eco-barrier that protects urban wildlife [2]. Among the 40 most important turfgrasses, two—Tifway bermudagrass (*Cynodon transvaalensis* Burtt-Davy × C. Dactylon) and seashore paspalum (*Paspalum vaginatum* Sw.)—were considered warm-season species. Tifway bermudagrass is a hybrid developed by Glenn Burton and released by the University of Georgia in 1960 [3]. It has a dark green color with medium-sized, fine-textured leaves and a high shoot density. Tifway bermudagrass is commonly grown in the tropics and subtropics as a warm-season grass and provides an excellent surface for golf course fairways and athletic fields [4]. Seashore paspalum is native to tropical coastal environments. Despite its many desirable characteristics, including salinity tolerance [5] and turf quality [6], the species is less widely used on golf courses and sports fields. Turfgrass quality is particularly important for managers of recreation and sports facilities [7]. Turfgrass quality includes functional characteristics such as uniformity, density, leaf texture, leaf color, and ground cover [8]. Turfgrass species may vary in terms of appearance, use, requirements, and stress tolerance [9].

Abiotic stresses (e.g., drought, salinity, and cold stress) inhibit a plant's water uptake by creating osmotic imbalance or osmotic stress at the root–soil interface, and are considered to be the main factors that limit the growth and development of turfgrasses [10]. There are few studies published about the effect of combined abiotic stresses on turfgrasses (e.g., Torun et al. [11]), whereas the effect of individual stresses such as salinity stress is more widely investigated [10,12]. Cold temperature is a major abiotic stress affecting warm-season turfgrasses that ultimately leads to reduced growth and quality due to wilting and leaf firing [13]. Cold stress may also induce dormancy and affects CO_2 absorption during photosynthesis, thus promoting a decline in the photosynthetic production [14]. Warm-season grasses grow most actively within the temperature range 25–35 °C [1], whereas the optimal range for growth of cool-season grasses is 16–24 °C [15]. Turfgrasses of tropical and subtropical origin are susceptible to injuries at temperatures below 12 °C [16]. Plant nutrients (i.e., exogenous and indigenous) have beneficial effects on grass metabolism, including stimulating vital processes that result in greater turfgrass tolerance to abiotic and biotic stresses such as applied calcium [13] or foliar application of amino acids [17]. Thus, the application of amendments could benefit warm-season grasses during the winter in arid/semi-arid regions.

Agricultural production involves many abiotic and biotic stresses. These stresses can be ameliorated using many different organic materials (e.g., humic substances, organic matter, compost, etc.) and inorganic amendments (e.g., silicon sources, sulfuric compounds, selenium, etc.) [18]. These amendments have various roles in supporting crop production under stressful conditions [19]. Concerning silicon, it is the second-most abundant element in the earth's crust, and can be found in biological systems in the form of amorphous silica [20]. Silicon is considered a vital soil amendment which can improve resistance/tolerance to biotic/abiotic stresses in many plant species [20]. Silicon can improve the production of many crops under different stresses, such as water deficit [21], heavy metals toxicity [22], drought stress [23], and biotic stress [24]. Humic substances (mainly humic, fulvic, and acids) significantly impact plant cultivation by stimulating plant nutrient uptake, phytohormone signaling, enzyme antioxidants, and photosynthetic efficiency and regulating reactive oxygen species [25]. Humic substances can promote plant growth and development under different stresses such as salinity [25], soil remediation [26], drought and salinity [27], and biotic stress [28]. Sulfur compounds, which function like soil amendments or conditioners of salt-affected soils when oxidized into sulfuric acid, also have a vital role [29]. Sulfur is also effective at remediating alkaline/sodic soil by decreasing soil pH and increasing the concentration of sulfate anions [30].

Therefore, this study sought to answer the following specific research questions:

(1) Which turfgrass is more tolerant to combined cold and salinity stress?
(2) Which applied amendment is more effective in mitigating these previous stresses on the studied turfgrasses?
(3) Which source of silicon and its applied dose is the best to ameliorate the growth and quality of the studied turfgrasses?
(4) Which dose of the applied amendment can be used under salt-affected soil conditions?

2. Materials and Methods

2.1. Experimental Design and Growth Conditions

Two turfgrasses (i.e., seashore paspalum and Tifway bermudagrass; hereby referred to in the text as SP and TB, respectively) were examined in field trials during winter months at the experimental farm of the Faculty of Agriculture at Kafrelsheikh University (31°05′54″ N and 30°57′00″ E). The daily minimum and maximum temperatures (at 2 m from the ground) during the experimental period ranged from 7 to 23 °C and 11 to 32 °C, respectively, in the 2018/2019 season and from 7 to 25 °C and 11 to 35 °C, respectively, in the 2019/2020 season. The soil used was salt-affected soil. The soil texture (0–20 cm) was classified as clay, with the particle size distribution being 19.7% sand, 25.0% silt, and 55.3% clay. The soil salinity (EC), pH, and sodium adsorption ratio (SAR) were 4.49 dS m^{-1}, 8.65, and 19.0, respectively. The soil organic matter and soil cation exchange capacity (CEC) were 14.5 g kg^{-1} and 40.5 $cmol_c$ kg^{-1} soil, respectively, and the water table was at 90 cm from the soil surface. The available N, P, and K values were 30, 12, and 185 mg kg^{-1}, respectively. The main soil moisture parameters, which were field capacity, wilting point, and available water, had values of 43.25, 23.11, and 20.14%, respectively. The previous soil parameters were determined according to the methods described by Sparks et al. [31] and Campbell [32]. In the current study, selected amendments, including both organic (humic acid) and inorganic amendments (sources of sulfate and silica), were tested. Compost (2.38 kg m^{-2}) was applied to the research area on November 1st of each season. To both reduce the surface tension of the leaves and to avoid surface runoff, 0.1% Tween 20 (Polysorbate 20) was added into the sprinkler system.

The irrigation water used was a non-saline fresh water with pH and EC values of 7.71 and 225 ppm, respectively. The contents of available N and P in the irrigation water were 0.09 and 28 mg L^{-1}, respectively. Irrigation was performed according to the recommendations of the Ministry of Agriculture and Land Reclamation of Egypt. Throughout the experiment, the turfgrasses were mowed at a height of 3 cm using the appropriate mowing machine 2–3 times a week during the warm months of the year and from 0 times to 1 time a week during the cold months. The plants were sprinkler-irrigated for 15 min three times a week and fertilized with NPK (20:20:20) once a week at a rate of 1.5 g L^{-1} through the sprinkler irrigation system. The temperature during the study was recorded and the relevant data were provided as the average maximum and minimum per month (Figure 1).

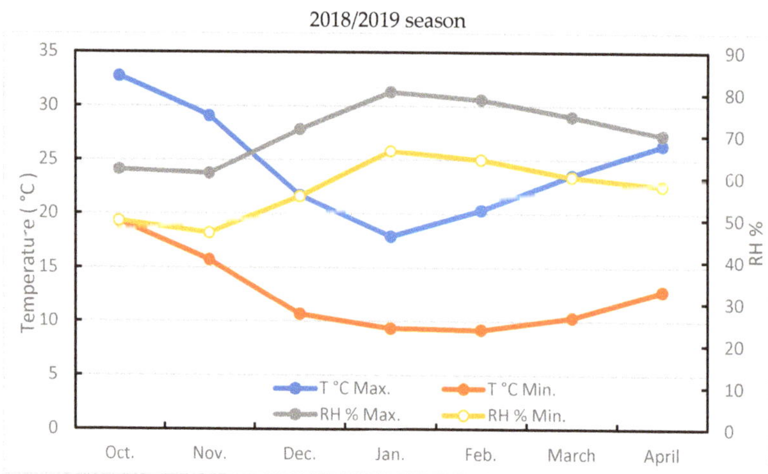

Figure 1. *Cont.*

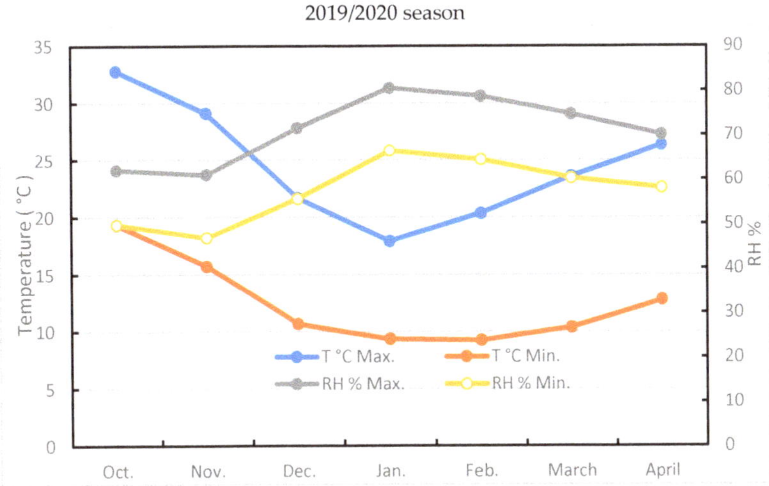

Figure 1. Average maximum and minimum temperature and relative humidity per month during the study in both seasons.

2.2. *Treatments and Their Sources*

The plant materials were obtained from the Horticulture Department at Kafrelsheikh University. A field study was run from October to April over two separate cold seasons. Some of the amendments that were investigated in the current study included humic acid (organic) and sources of iron and silicon (inorganic). Both types of grasses were treated with humic acid, ferrous sulphate ($FeSO_4$), and two silica (S) sources (low level or Citrok Plus silica as 3% SiO_2 and high level or diatomite silica as 86–89% SiO_2) in two doses (Table 1 and Figure 2). Each experimental plot consisted of one grass unit with an area of 10 m^2. Each unit received five sprays every two weeks starting from 9 October until 9 December. After 9 December, the units were sprayed until the point of run-off at 20 days intervals. Treatments in both seasons were arranged according to a randomized complete block design (RCBD) with three replicates.

Table 1. Overview of the different treatments and their applied doses.

Code	Treatments (Applied Dose)	Details of the Amendments Used	Active Ingredient in the Applied Products
C	Control	Tap water	---------
HA1	Humic acid (1000 ppm)	Humic (20%) from GrowTech for Agricultural Development, Cairo, Egypt	Humic acid (0.02%)
HA2	Humic acid (2000 ppm)	Humic (20%) from GrowTech for Agricultural Development, Cairo, Egypt	Humic acid (0.04%)
FS1	Ferrous sulphate (250 ppm)	X-xtra Iron (10%) from Growth Products Ltd., White Plains, NY, USA	$FeSO_4$ (0.0025%)
FS2	Ferrous sulphate (1000 ppm)	X-xtra Iron (10%), Growth Products Ltd., White Plains, NY, USA, 1 cm L^{-1}	$FeSO_4$ (0.01%)
S1	Silica (3000 ppm)	Citrok plus (3% silica) from Novac Bio Science, El Mansurá, Egypt, 3%, SiO_2	SiO_2 (0.009%)
S2	Silica (6000 ppm)	Citrok plus (3% silica) from Novac Bio Science, El Mansurá, Egypt, 3%, SiO_2	SiO_2 (0.018%)
D1	Diatomite (1000 ppm)	Diatomite (86–89% SiO_2) from Shengmai Diatomite Functional Material Co. Ltd., Linjiang, China, 86–89% SiO_2	SiO_2 (0.875 g L^{-1})
D2	Diatomite (2000 ppm)	Diatomite (86–89% SiO_2) from Shengmai Diatomite Functional Material Co. Ltd., Linjiang, China, 86–89% SiO_2	SiO_2 (1.75 g L^{-1})

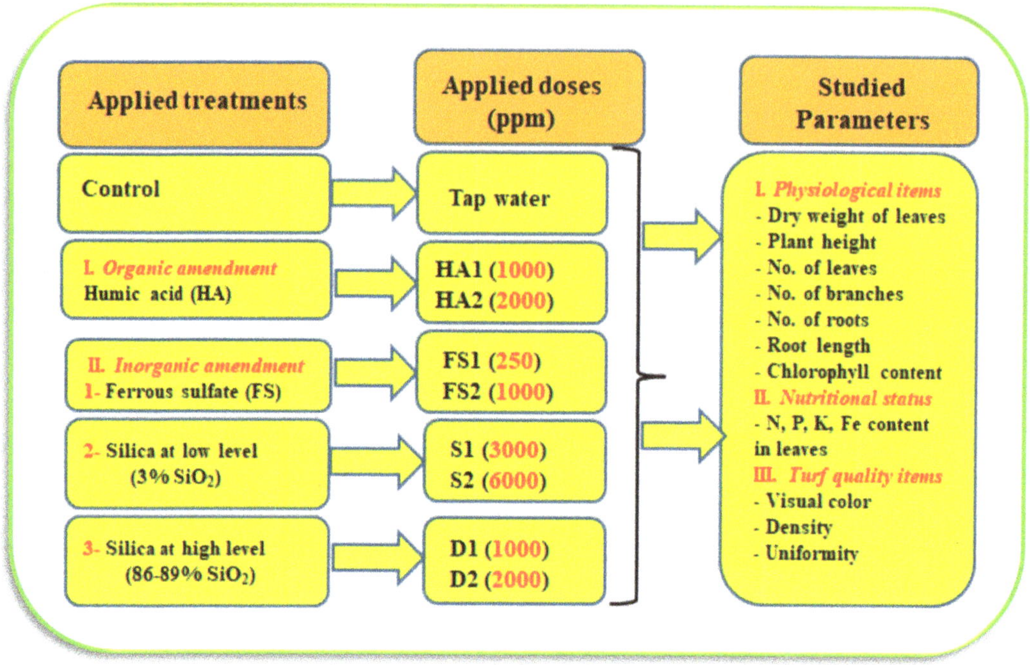

Figure 2. Overview of the experimental design including the main treatments, different applied doses of each treatment, and studied measurements.

2.3. Plant Physiological and Chemical Parameters

Different vegetative characteristics, including the dry weight of leaves; plant height; number of leaves, branches, and roots; and finally, the length of the longest root, were measured. The area of each plot used to measure these parameters was 100 cm^2. For chemical characteristics, total chlorophyll was measured in fresh leaf samples using an SPAD instrument (SPAD 501 leaf chlorophyll meter, Minolta, Co., Ltd., Tokyo, Japan) according to the method described by Netto et al. [33]. Nitrogen (N) content was determined using the micro-Kjeldahl instrument (UDK 159, Velp Scientifica, Usmate, Italy), phosphorus (P) content was determined using a spectrophotometer (Libra S80PC, Biochrom, Cambridge, UK), and iron (Fe) and potassium (K) contents were measured using an inductively coupled plasma-optical emission spectrometry (ICP-OES) apparatus (Prodigy 7, Leeman Labs., Hudson, NH, USA) according to the method described by Sparks et al. [31]. Only plant height and total chlorophyll were recorded twice in each season during January and April; the remaining vegetative characteristics as well as the chemical composition were both measured only once in January.

2.4. Plant Quality Parameters

The main parameters of turfgrass quality were measured with the naked eye and included the visual turf color, turf density, and turf uniformity. The scorings were done according to the National Turfgrass Evaluation Program or NTEP guidelines [34]. The rating scale for most visual parameters (mainly turf color and density) ranged from 1 to 9, where 1 is the poorest or lowest value and 9 is the highest or best rating. In the winter, turf color is around a 1 on the visual rating scale if it is straw-brown or has not retained any color and is considered a 9 if it is dark green. All quality parameters were recorded only once in January.

2.5. Statistical Analyses

Statistical analyses were performed using the statistical software SAS (version 9.1; SAS Institute, Cary, NC, USA). One-way analysis of variance (ANOVA) and species-wise Duncan's multiple range tests were used to compare the mean values between the two seasons [35].

3. Results

3.1. Applied Amendments and Vegetative Growth

For each of the two studied turfgrass species, six vegetative parameters, including the dry weight of leaves (g); plant height (cm); number of leaves, branches, and roots; and finally, the length of the longest root (cm), were measured. The mean values of plant height and the number of branches and roots were tabulated in Tables 2 and 3, respectively, while the mean values for the dry weight of leaves, number of leaves, and the length of the longest root were presented in Figure 3. From the data in Table 2, it is apparent that the height of both turfgrasses was significantly influenced by the applied amendments in January and April. In both species and across both seasons, the measured mean values of plant height in January were lower than those in April. For each season, the highest respective values of plant height in January were 2.7 and 3.6 cm for SP and 2.7 and 3.5 cm for TB, whereas the highest respective mean values in April were 3.7 and 6.7 cm for SP and 3.6 and 5.6 cm for TB. For both plant species, it was noted that across both seasons, the previous highest values were recorded when ferrous sulphate was applied at a concentration of 250 or 1000 ppm. It was also noted that different sources of silicon were less effective compared to other amendments (humic and iron).

Table 2. Measured mean values of plant height (cm) of seashore paspalum and Tifway bermudagrass treated with different amendments during January and April in two different seasons.

Treatments	Seashore Paspalum (SP)		Tifway Bermudagrass (TB)	
	Plant Height (cm) in the First Season			
	January	April	January	April
Control (water)	2.5 c	3.1 c	2.6 b	2.9 c
Humic acid (1000 ppm)	2.6 abc	3.4 b	2.6 b	3.2 bc
Humic acid (2000 ppm)	2.7 a	3.5 b	2.6 ab	3.5 ab
Ferrous sulphate (250 ppm)	2.7 abc	3.5 ab	2.7 a	3.6 a
Ferrous sulphate (1000 ppm)	2.7 ab	3.7 a	2.7 a	3.4 ab
Silicon (3000 ppm)	2.6 abc	3.3 b	2.6 ab	3.1 bc
Silicon (6000 ppm)	2.6 abc	3.5 b	2.6 ab	3.0 bc
Diatomite (1000 ppm)	2.6 bc	3.3 b	2.6 ab	3.3 ab
Diatomite (2000 ppm)	2.6 abc	3.4 b	2.6 ab	3.4 ab
	Plant height (cm) in the second season			
Control (water)	3.0 b	4.6 d	3.0 c	4.1 d
Humic acid (1000 ppm)	3.3 ab	4.7 d	3.2 bc	4.8 bc
Humic acid (2000 ppm)	3.2 ab	5.1 cd	3.3 ab	4.9 bc
Ferrous sulphate (250 ppm)	3.4 ab	6.0 b	3.5 a	5.6 a
Ferrous sulphate (1000 ppm)	3.6 a	6.7 a	3.5 a	5.4 ab
Silicon (3000 ppm)	3.3 ab	5.1 cd	3.4 abc	5.0 b
Silicon (6000 ppm)	3.5 ab	5.7 bc	3.4 bc	5.1 ab
Diatomite (1000 ppm)	3.3 ab	5.5 bc	3.3 abc	4.3 cd
Diatomite (2000 ppm)	3.4 ab	5.6 bc	3.3 abc	5.0 ab

Means within each column that have the same letters are not significantly different from one another according to the Duncan's Multiple Range Test (at $p < 0.05$).

Table 3. Number of roots and branches of both species treated with different amendments during the studied seasons (area used for measuring was 100 cm^2).

Treatments	Seashore Paspalum (SP) No. of Roots	Seashore Paspalum (SP) No. of Branches	Tifway Bermudagrass (TB) No. of Roots	Tifway Bermudagrass (TB) No. of Branches
First season				
Control (water)	53 i	28 i	17 g	20 h
Humic acid (1000 ppm)	97 c	65 g	35 d	40 d
Humic acid (2000 ppm)	106 b	115 a	38 c	43 c
Ferrous sulphate (250 ppm)	119 a	103 b	41b	46 b
Ferrous sulphate (1000 ppm)	87 e	80 e	40 a	49 a
Silicon (3000 ppm)	61 h	49 h	29f	31 g
Silicon (6000 ppm)	93 d	88 d	33 e	36 e
Diatomite (1000 ppm)	80 f	72 f	30 f	35 e
Diatomite (2000 ppm)	78 g	95 c	29 f	33 f
Second season				
Control (water)	60 i	30 h	21 h	23 i
Humic acid (1000 ppm)	106 e	72 f	37 d	45 d
Humic acid (2000 ppm)	125 c	117 a	42 c	49 c
Ferrous sulphate (250 ppm)	174 a	111 b	47 b	55 b
Ferrous sulphate (1000 ppm)	149 b	103 d	58 a	64 a
Silicon (3000 ppm)	140 h	54 g	28 g	33 h
Silicon (6000 ppm)	113 d	107 c	35 e	41 e
Diatomite (1000 ppm)	91 g	80 e	34 e	37 f
Diatomite (2000 ppm)	95 f	110 b	32 f	35 g

Means within each column that have the same letters are not significantly different from one another according to the Duncan's Multiple Range Test (at $p < 0.05$).

The most striking result to emerge from the data in Figure 3 (which presents values as an arithmetic mean of the two seasons) is that the mean values of the studied parameters of SP, including the number of leaves, root length, and dry weight of leaves, were higher than those of TB. Concerning the number of leaves, there were significant differences between the studied species for all the amendments that were studied. Compared to other amendments, the highest mean values of leaves were obtained after applying iron at 250 ppm, with the values obtained (near to 800 leaves per 100 cm^2) being about four-fold greater than those of the control treatment. All amendments increased the root length compared to the control treatment. Although ferrous sulphate (both doses) produced the longest roots (about 22 cm) in both species, SP had the highest values, especially after humic or iron treatments were applied. Compared to the control treatment, the dry weight of leaves increased after applying each of the studied organic and inorganic amendments; the mean values of SP were higher compared to the other species. The inorganic ferrous sulphate still gave us the highest values (more than 5.2 g) of dry weight of leaves per 100 cm^2, which represents more than a 2.5-fold increase compared to the control treatment.

In both species, the highest numbers of leaves were retrieved with the application of ferrous sulphate. For SP, the highest numbers of branches (115 and 117 per 100 cm^2 in each season, respectively) were found for humic acid at 2000 ppm, followed by ferrous sulphate at 250 ppm. For TB, the highest numbers of branches (49 and 64 branches per 10 cm^2 in each season, respectively) were recorded after applying ferrous sulphate at 1000 ppm, followed by the same treatment at a dose of 250 ppm. In general, the mean values of number of roots of SP were some folds higher than the values of other species (TB), with a similar trend being observed for the number of branches (Table 3).

3.2. *Applied Amendments and Turf Quality Parameters*

Three different parameters of turf quality were evaluated: turf color, turf density, and turf uniformity. In terms of leaf color, all amendments resulted in darker green leaves than the control treatment, with the darkest colors being produced by ferrous sulphate, followed by humic acid. Visual scoring of the turf density showed the same pattern as described above, with the iron amendment resulting in the highest values. In general, the mean values of turf color and turf density of SP were higher than those of the other species, whereas the mean values of turf uniformity in both seasons were similar for both species

(Table 4). In both grasses and for both seasons, all treatments resulted in significantly higher turf quality (i.e., color, density, and uniformity) scores compared to the control treatment.

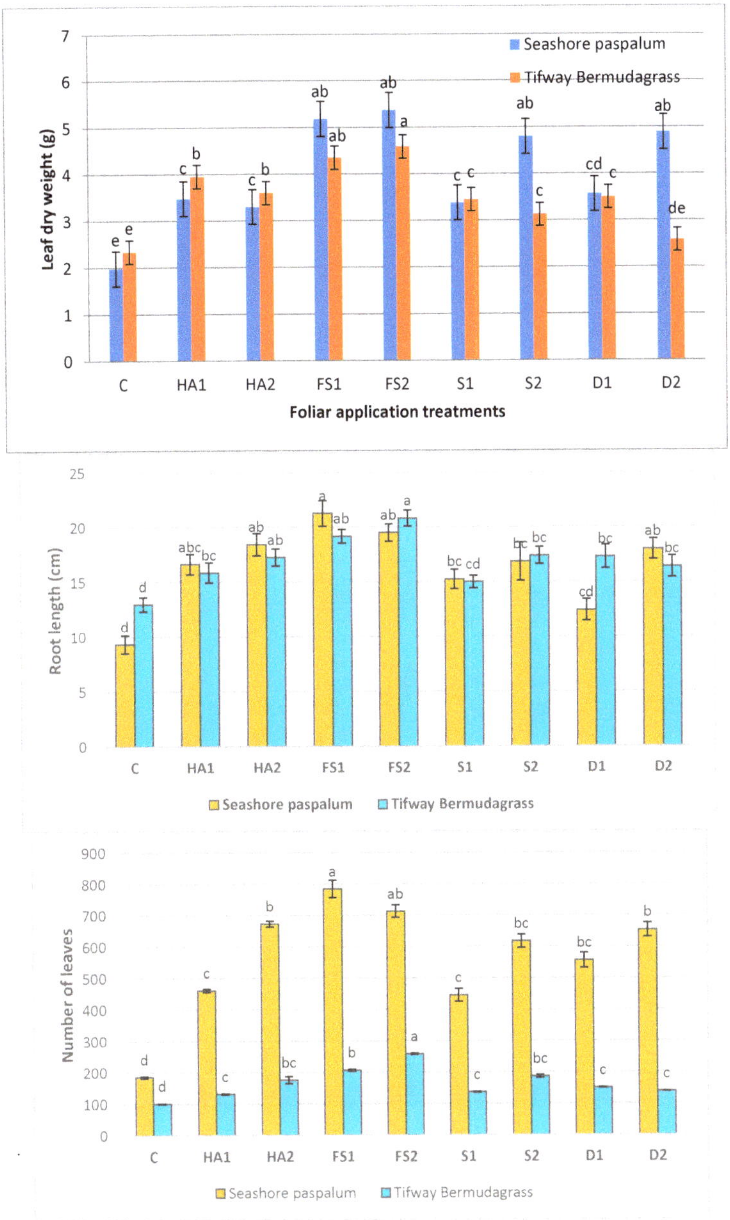

Figure 3. Mean values of leaf dry weight (g), root length, and number of leaves measured in January for the two species treated with different amendments obtained from combining the data from two seasons. C = control, HA = humic acid, FS = ferrous sulphate, S = silica, D = diatomite. For more details about the treatments used, please refer to Table 1. Values are presented as mean ± standard error of the mean, with the letter(s) from the species-wise Duncan's multiple range tests indicated at the top of each bar. The area used for measuring the parameters was 100 cm^2.

Table 4. Visual turf color score, turf density, and turf uniformity in seashore paspalum and Tifway bermudagrass treated with different amendments (area used for measuring was 100 cm^2).

Treatments	Seashore Paspalum (SP)			Tifway Bermudagrass (TB)		
	Turf Color	Turf Density	Turf Uniformity	Turf Color	Turf Density	Turf Uniformity
			First season			
Control (water)	1.0 d	1.0 c	1.0 b	1.0 c	1.0 b	1.0 b
Humic acid (1000 ppm)	5.3 ab	3.0 a	2.0 a	5.0 ab	2.7 a	2.0 a
Humic acid (2000 ppm)	6.0 a	3.0 a	2.0 a	5.3 a	2.7 a	2.0 a
Ferrous sulphate (250 ppm)	6.0 a	3.0 a	2.0 a	5.0 ab	2.7 a	2.0 a
Ferrous sulphate (1000 ppm)	5.0 ab	2.7 ab	2.0 a	5.0 ab	2.7 a	2.0 a
Silicon (3000 ppm)	4.0 bc	2.3 ab	2.0 a	4.0 ab	2.3 a	2.0 a
Silicon (6000 ppm)	4.3 bc	2.3 ab	2.0 a	4.3 ab	2.3 a	2.0 a
Diatomite (1000 ppm)	3.3 c	2.0 b	2.0 a	4.0 ab	2.3 a	2.0 a
Diatomite (2000 ppm)	4.3 bc	2.3 ab	2.0 a	3.0 b	1.7 ab	2.0 a
			Second season			
Control (water)	1.0 d	1.0 c	1.0 b	1.0 b	1.0 b	1.0 b
Humic acid (1000 ppm)	5.7 a	3.0 a	2.0 a	5.0 a	2.7 a	2.0 a
Humic acid (2000 ppm)	5.7 a	3.0 a	2.0 a	5.3 a	2.7 a	2.0 a
Ferrous sulphate (250 ppm)	5.7 a	3.0 a	2.0 a	5.3 a	2.7 a	2.0 a
Ferrous sulphate (1000 ppm)	5.3 ab	3.0 a	2.0 a	5.0 a	2.7 a	2.0 a
Silicon (3000 ppm)	4.3 abc	2.3 b	2.0 a	4.3 a	2.7 a	2.0 a
Silicon (6000 ppm)	4.3 abc	2.3 b	2.0 a	4.3 a	2.3 a	2.0 a
Diatomite (1000 ppm)	3.7 c	2.0 b	2.0 a	4.3 a	2.3 a	2.0 a
Diatomite (2000 ppm)	4.0 bc	2.3 b	2.0 a	4.7 a	2.3 a	2.0 a

Means within each column that have the same letters are not significantly different from one another according to the Duncan's Multiple Range Test (at $p < 0.05$).

3.3. Applied Amendments and Chlorophyll Content

Foliar application of the amendments significantly increased the leaf total chlorophyll content (measured as SPAD values) in the two turfgrass species compared to the control treatment (Figure 4; values given as a mean of the two seasons) during January and April. In the case of the January measurements, both applied doses of ferrous sulphate gave the highest values for SP (27 and 23 for each dose, respectively). For TB, ferrous sulphate at 1000 ppm (mean of both seasons of 15) produced the highest values, followed by 250 ppm ferrous sulphate. In the case of April, all measured values were higher compared to the values in January for all the treatments (including the control). All the measured values of SP were higher in both January and April compared to the other species (TB). Dolomite treatments resulted in the lowest mean values of SPAD in both species, but these values were still higher than those of the control treatment. For both species, applying ferrous sulphate at 1000 ppm produced the highest values of SPAD for April, followed by ferrous sulphate at a dose of 250 ppm.

3.4. Applied Amendments and Chemical Composition of Leaves

The data illustrated in Figure 5 show that the applied amendments significantly increased the content of nitrogen and other mineral nutrients in the two turf grasses. The mean values of all the studied nutrients (i.e., N, P, K, and Fe) were higher for SP compared to the other plant species for all treatments including the control. In general, for all the studied nutrients, the application of iron sources and humic acid led to the highest leaf nutrient contents in both species. For SP, the highest mean values of nutrient content of N, P, K, and Fe were 4, 0.24, 2.4%, and 7.6 mg kg^{-1}, respectively. All these values resulted from the application of humic acid or iron sources, and the same trend was also noticed for TB. After the control treatment, application of silicon sources produced the lowest mean values of nutrients contents.

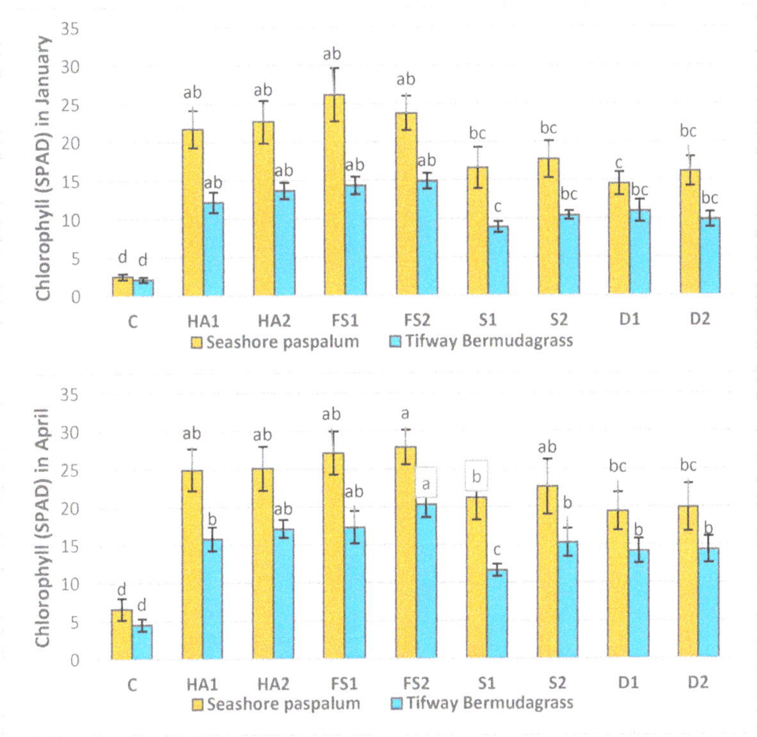

Figure 4. Mean values of chlorophyll content in leaves in both January and April (SPAD value ± SE) for seashore paspalum and Tifway bermudagrass treated with different amendments; values represent a mean of the two seasons. C = control, HA = humic acid, FS = ferrous sulphate, S = silica, D = diatomite. For more details about the treatments, please refer to Table 1. Data are presented as mean ± standard error of the mean, with the letter(s) from the species-wise Duncan's multiple range tests indicated at the top of each bar.

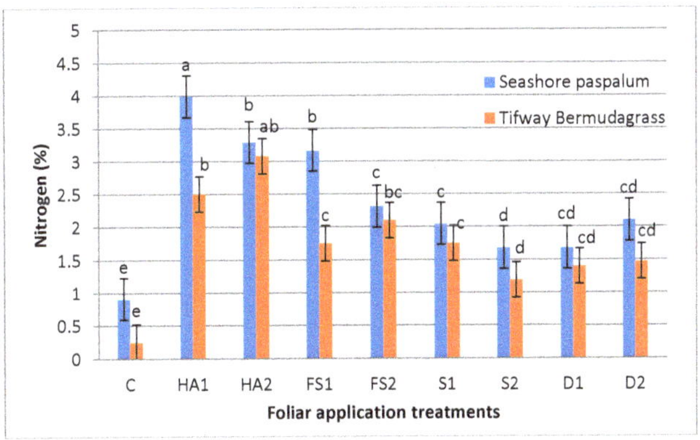

Figure 5. *Cont.*

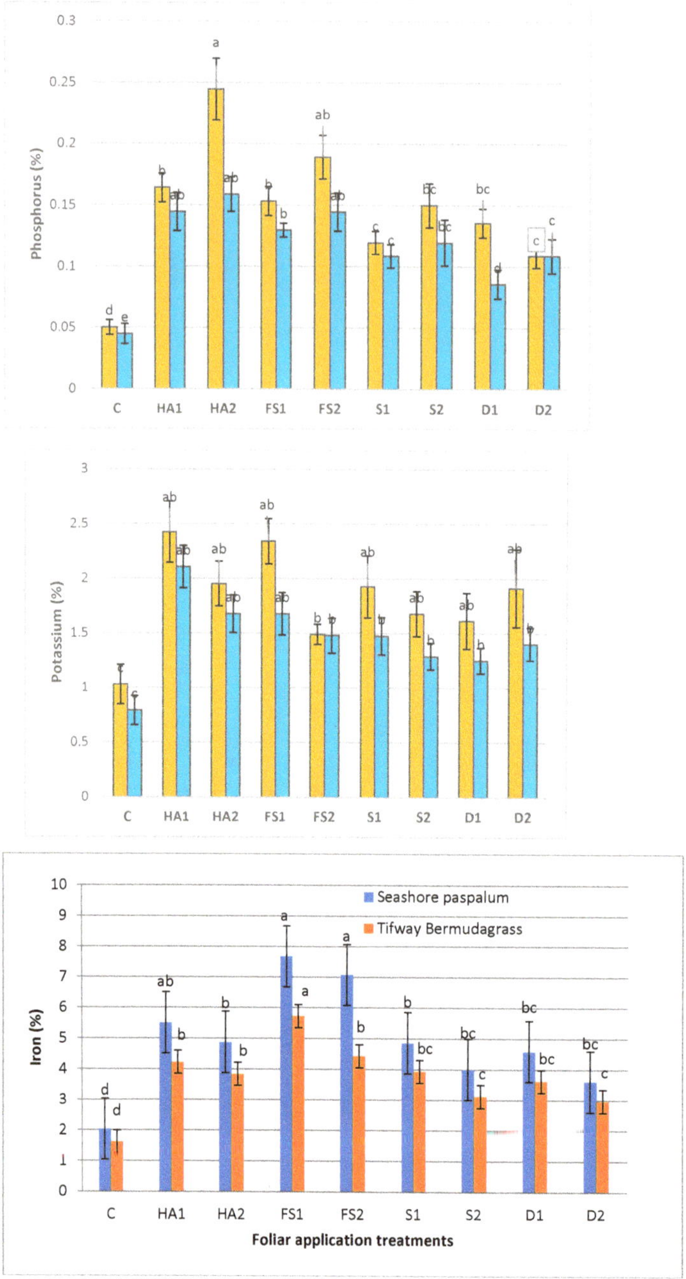

Figure 5. Mean leaf content of nitrogen (%), phosphorus (%), potassium (%) and iron (mg kg^{-1}) $\pm$ SE in seashore paspalum (orange bars) and Tifway bermudagrass (blue bars) treated with different amendments; values represent a mean of the two seasons. C = control, HA = humic acid, FS = ferrous sulphate, S = silica, D = diatomite. For more details about the treatments, please refer to Table 1. Data are presented as mean value $\pm$ standard error of the mean, with the letter(s) from the species-wise Duncan's multiple range tests indicated above each bar.

4. Discussion

In this section, it is important to answer our previous questions, which we added as the main objectives of the current study:

Which turfgrass is more tolerant to combined cold and salinity stress? In the present study, the main foliar treatments or amendments applied on two turfgrass species (i.e., SP and TB) under combined cold and soil salinity stress included humic acid, ferrous sulphate, and two sources of silicon. All amendments were applied during the period from October to December, in two successive seasons, after which the plant samples were taken for physiological and chemical analyses. The turfgrass of SP was more tolerant to both cold and salinity stress compared to TB, as shown in all the studied vegetative parameters, namely, the leaf chlorophyll content, chemical composition of plant leaves, and quality characteristics (Tables 1–4, and Figures 1–3). This might be linked to the ability of this tolerant grass to grow under many environmental stresses, especially salinity stress [36], which may cause problems such as winterkill for bermudagrasses [37,38].

Which applied amendment is more effective in mitigating these previous stresses on the studied turfgrasses? In general, one could order the effectiveness of the applied amendments as follows: iron sources (ferrous sulphate in 250 and 1000 ppm), then humic acid, and finally, sources of silicon. For example, ferrous sulphate produced the highest mean values of N and Fe content in leaves, chlorophyll content, quality parameters and almost all the vegetative characteristics. What is the role of exogenous iron under turfgrass stress? Iron has distinct functions in different plant physiological processes like photosynthesis because it is essential for many enzymes involved in chlorophyll biosynthesis [39]. On the other hand, as a source of sulfate, ferrous sulphate reaches the cultivated soil via sprinkler irrigation. Once there, the sulfate can react with the Na^+ and Cl^- ions present in saline/alkaline soil to sustain optimal $Ca^{2+}/Na+$ and K^+/Na^+ ratios, leading to a reduction in soil pH and saving of nutrients for plant uptake [40]. This supporting role in nutrient uptake can indirectly enhance plant photosynthesis and other physiological attributes [40]. Under cold temperatures, iron fertilization has been shown to enhance recovery from winter desiccation injury in creeping bentgrass (*Agrostis palustris* L.). However, the amount of iron needed within plants can vary according to the local soil and plant conditions, and the response to iron fertilization can differ among species and varieties [41]. To the best of our knowledge, currently only one study on the growth of turfgrasses under cold and salinity stress has been published [42], and there are no publications on the effect of different amendments under these combined stresses. What is the main role of humic acid under plant stress? Humic acid can enhance phosphorus bioavailability and increase the availability of micronutrients and may also act as a plant growth stimulator by forming organo-iron complexes [43]. Similar results have been obtained in tall fescue (*Festuca arundinaceae* Schreb.) and creeping bentgrass [44].

Which source and dose of silicon is the best to ameliorate the growth and quality of the studied turfgrasses? Why did silicon treatments produce lower values compared to other treatments? In general, treatment with sources of silicon had a positive impact on the turfgrasses compared to the control treatment, but produced lower mean values compared to ferrous sulphate and humic acid. Some studies have shown beneficial effects of silicon (or silica) when it is used as a plant promotor against abiotic/biotic stresses such as salinity and cold stress, but the current results showed lower mean values with silicon compared to the other treatments. This could be because the higher applied doses of silicon used in this study might have had toxic effects on the turfgrasses. Concerning the applied doses of silicon for other crops, plant growth was promoted by applying 200 ppm under conditions of combined soil salinity and heat stress for cucumbers [45], 50–200 kg ha^{-1} under salinity and drought stress for wheat [46], and 2000 to 4000 ppm under water deficit for squash [47]. The suggested mechanism for enhancing cucumber salt tolerance using silicon may involve increasing the accumulation of polyamine, enhancing antioxidants, and managing osmotic and oxidative stress to decrease the level of oxidative damage [48–50]. In our study, there was no clear trend regarding the use of different sources of silicon for all the measured parameters. Thus, both the source of silicon and the applied dose that can best ameliorate the growth and quality of the studied turfgrasses are not clear. The applied Si-sources had different effects on the studied parameters, resulting in significantly higher values for many of the vegetative growth attributes, but no significant trend was observed in the case of turf quality. Silicon is considered to be a beneficial or quasi-essential element that can impact the plant–soil system by both protecting plants against biotic/abiotic stresses and optimizing soil fertility by improving the soil water status and maintaining the availability of nutrients to plants [20]. For the nutrient content, there was no clear difference between the silicon and diatomite treatments. Although many studies on the impact of different sources of silicon on some cultivated plants under stress, such as canola under water deficit [51], wheat under water deficit [52], and feverfew under drought [53], have been published, studies on turfgrass are rare (Figure 6).

Thus far, only a few studies have attempted to explain the production of warm-season turfgrasses under combined stresses. In one study, Liu et al confirmed that exposing turfgrasses to combined cold and salinity stress may cause more severe damage compared to growing turfgrasses under other conditions [42]. The results obtained in this study are in agreement with those of Chavarria et al. [49], who confirmed that turfgrass species and their cultivars differ in their tolerance to salinity stress. Concerning the role of humic acid under conditions of combined stress, its positive effects may be related to the enhancement of various antioxidants and subsequent scavenging of superoxide anions in the leaves, which protects the plant cell from damage caused by reactive oxygen and free radicals that are formed under stress. Our results are in agreement with the results of Abdel Fatah et al. [50], who concluded that the application of humic acid increased the content of nitrogen, phosphorus, and potassium in the leaves of bermudagrass. In the current study, it seems that treatment with iron sources not only increased the iron content in the leaves, but also increased the content of other nutrients. The stimulatory effects of humic acid on the nitrogen, phosphorus, and potassium contents of leaves have been observed in other studies. In addition to enhancing the bioavailability of phosphorus, humic acid has also been shown to increase both the availability of micronutrients and the phosphorus content of leaves [54,55] and improve nutrient uptake by plants.

After presenting the suggested mechanisms in Figure 6, it is clear that the selected organic and inorganic amendments have different modes of action. The specific mode of action for each amendment depends on how the amendment can contribute to tolerance against combined cold and salinity stress in cultivated plants. These differences between the studied amendments may lead to several questions including the following one: which applied dose of each studied amendment can be used? Currently, more studies under salt-affected soils are needed before the recommended dose of each applied amendment can be determined, this is especially true for treatments that use sources of silicon. The interaction between these different amendments also requires further investigation, including additional research using different soil application methods and under salt-affected conditions.

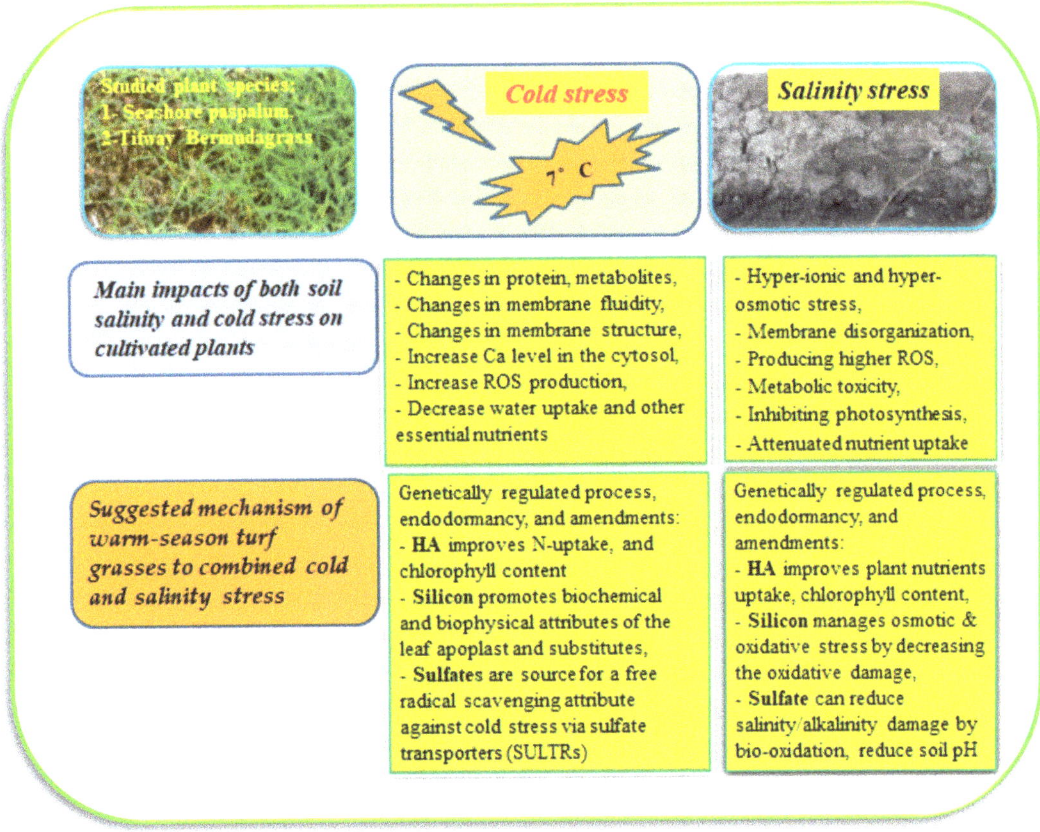

Figure 6. An overview of the suggested main damage to studied plants resulting from both cold and soil salinity and the mechanisms suggested to contribute to tolerance of these stresses (sources: Pompeiano et al. [56], Prokopiuk et al. [57], Sharma et al. [58] Bello et al. [40], Hajiboland et al. [59], Nguyen et al. [60]).

5. Conclusions

This research has shown that the foliar application of humic acid, ferrous sulphate, and to some extent, silicon, on warm season turfgrasses (i.e., SP and TB) is beneficial for the turf growth and quality during the cold season in salt-affected soils. Ferrous sulphate and humic acid were the best options and worked at both applied doses. From an economic point of view, lower doses might be sufficient, but further research is needed to fine-tune the application regimes. Under the studied conditions, seashore paspalum had the highest values of all parameters of growth and quality compared to Tifway bermudagrass, which was due to its tolerance to salinity stress. A lot of well-defined evidence, which is clearly reflected in all the tables and figures, was collected in this study. Such evidence verified that seashore paspalum is superior to Tifway bermudagrass in many studied measurement (e.g., chlorophyll content, turf quality, nutrient contents, and vegetative parameters). The most striking result is that in both turfgrasses, the effect of silicon was lower than that of humic acid and ferro sulphate; this was due to the use of high applied doses, which may have negatively impacted plant growth. This study offers some important insights into the cultivation of warm-season turfgrasses under combined cold and soil salinity stress, which we have shown can be regulated by increasing nutrient uptake, chlorophyll content in leaves, and turf quality. More studies using lower applied doses of silicon sources and different combinations of applied amendments are needed.

Author Contributions: Conceptualization, E.N., M.H., M.A.E.-G. and H.E.-R.; methodology D.T., E.N., M.H., M.A.E.-G. and H.E.-R.; investigation, D.T.; data curation, D.T.; writing—original draft preparation, D.T.; writing—review and editing, S.Ø.S.; visualization, D.T. and S.Ø.S.; supervision, E.N., M.H., M.A.E.-G. and H.E.-R. All authors have read and agreed to the published version of the manuscript.

Funding: This research received no external funding.

Data Availability Statement: Not applicable.

Acknowledgments: We would like to thank the institutions involved for providing administrative and technical support needed to conduct the field trials and analysis.

Conflicts of Interest: The authors declare no conflict of interest.

References

1. Jiuxin, L.; Liebao, H. Progress and Challenges in China Turfgrass Abiotic Stress Resistance Research. *Front. Plant Sci.* **2022**, *13*, 922175. [CrossRef] [PubMed]
2. Clark, J.; Kenna, M. *Lawn and Turf Hayes' Handbook of Pesticide Toxicology*; Academic Press: London, UK, 2010; pp. 1047–1076.
3. Beard, J.B.; DiPaola, J.M.; Karnok, K.J.; Batten, S. *Introduction to Turfgrass Science & Culture*; Burgess Publishing Co.: Minneapolis, MN, USA, 1979.
4. Hanna, W.; Raymer, P.; Schwartz, B. Warm-Season Grasses: Biology and Breeding. In *Turfgrass: Biology, Use, and Management, Agronomy Monograph No. 56*; Stier, J.C., Horgan, B.P., Bonos, S.A., Eds.; American Society of Agronomy: Madison, WI, USA, 2013; pp. 543–590.
5. Uddin, M.K.; Juraimi, A.S. Salinity Tolerance Turfgrass: History and Prospects. *Sci. World J.* **2013**, *2013*, 409413. [CrossRef]
6. Jespersen, D.; Leclerc, M.; Zhang, G.; Raymer, P. Drought performance and physiological responses of bermudagrass and seashore paspalum. *Crop Sci.* **2019**, *59*, 778–786. [CrossRef]
7. Martiniello, P. Effect of traffic stress on cool-season turfgrass under a Mediterranean climate. *Agron. Sustain. Dev.* **2007**, *27*, 293–301. [CrossRef]
8. Głąb, T.; Szewczyk, T.; Gondekc, W.; Knagad, K.; Tomasikd., J.; Kowalik, M.K. Effect of plant growth regulators on visual quality of turfgrass. *Sci. Hort.* **2020**, *267*, 109314. [CrossRef]
9. Fan, J.; Zhang, W.; Amombo, E.; Hu, L.; Kjorven, J.O.; Chen, L. Mechanisms of Environmental Stress Tolerance in Turfgrass. *Agronomy* **2020**, *10*, 522. [CrossRef]
10. Katuwal, K.B.; Xiao, B.; Jespersen, D. Physiological Responses and Tolerance Mechanisms of Seashore Paspalum and Centipedegrass Exposed to Osmotic and Iso-osmotic Salt stresses. *J. Plant Physiol.* **2020**, *248*, 153154. [CrossRef]
11. Torun, H. Combined efficiency of salicylic acid and calcium on the antioxidative defense system in two different carbon-fixative turfgrasses under combined drought and salinity. *S. Afr. J. Bot.* **2022**, *144*, 72–82. [CrossRef]
12. Kozłowska, M.; Bandurska, H.; Bres, W. Response of Lawn Grasses to Salinity Stress and Protective Potassium Effect. *Agronomy* **2021**, *11*, 843. [CrossRef]
13. Shi, H.; Ye, T.; Zhong, B.; Liu, X.; Chan, Z. Comparative proteomic and metabolomic analyses reveal mechanisms of improved cold stress tolerance in bermudagrass (*Cynodon dactylon* (L.) Pers.) by exogenous calcium. *J. Integr. Plant Biol.* **2014**, *56*, 1064–1079. [CrossRef]
14. Augustyniak, A.; Perlikowski, D.; Rapacz, M. Insight into cellular proteome of *Lolium multiflorum/Festuca arundinacea* introgression forms to decipher crucial mechanisms of cold acclimation in forage grasses. *Plant Sci.* **2018**, *272*, 22–31. [CrossRef] [PubMed]
15. Liu, M.; Sun, T.; Liu, C.; Zhang, H.; Wang, W.; Wang, Y.; Xiang, L.; Chan, Z. Integrated physiological and transcriptomic analyses of two warm- and cool-season turfgrass species in response to heat stress. *Plant Physiol. Biochem.* **2022**, *170*, 275–286. [CrossRef] [PubMed]
16. Dionne, J.; Rochefort, S.; Huff, D.R.; Desjardins, Y.; Bertrand, A.; Castonguay, Y. Variability for freezing tolerance among 42 ecotypes of green-type annual bluegrass. *Crop Sci.* **2010**, *50*, 321–336. [CrossRef]
17. Radkowski, A.; Radkowska, I.; Bocianowski, J.; Sladkovska, T.; Wolski, K. The Effect of Foliar Application of an Amino Acid-Based Biostimulant on Lawn Functional Value. *Agronomy* **2020**, *10*, 1656. [CrossRef]
18. Feng, D.; Gao, Q.; Liu, J.; Tang, J.; Hua, Z.; Sun, X. Categories of exogenous substances and their effect on alleviation of plant salt stress. *Eur. J. Agron.* **2023**, *142*, 126656. [CrossRef]
19. Clemente, R.; Arco-Lázaro, E.; Pardo, T.; Martín, I.; Sánchez-Guerrero, A.; Sevilla, F.; Bernal, M.P. Combination of soil organic and inorganic amendments helps plants overcome trace element induced oxidative stress and allows phytostabilisation. *Chemosphere* **2019**, *223*, 223–231. [CrossRef]
20. Ahire, M.I.; Mundada, P.S.; Nikam, T.D.; Bapat, V.A.; Penna, S. Multifaceted roles of silicon in mitigating environmental stresses in plants. *Plant Physiol. Biochem.* **2021**, *169*, 291–310. [CrossRef]

21. Xu, X.; Zou, G.; Li, Y.; Sun, Y.; Liu, F. Silicon application improves strawberry plant antioxidation ability and fruit nutrition under both full and deficit irrigation. *Sci. Hortic.* **2023**, *309*, 111684. [CrossRef]
22. Jin, X.; u Rahman, M.K.; Ma, C.; Zheng, X.; Wu, F.; Zhou, X. Silicon modification improves biochar's ability to mitigate cadmium toxicity in tomato by enhancing root colonization of plant-beneficial bacteria. *Ecotoxicol. Environ. Saf.* **2023**, *249*, 114407. [CrossRef]
23. Zahedi, S.M.; Hosseini, M.S.; Hoveizeh, N.F.; Kadkhodaei, S.; Vaculík, M. Comparative morphological, physiological and molecular analyses of drought-stressed strawberry plants affected by SiO_2 and SiO_2-NPs foliar spray. *Sci. Hortic.* **2023**, *309*, 111686. [CrossRef]
24. Gao, A.; Chen, C.; Zhang, H.; Yang, B.; Yu, Y.; Zhang, W.; Fang-Jie Zhao, F.J. Multi-site field trials demonstrate the effectiveness of silicon fertilizer on suppressing dimethylarsenate accumulation and mitigating straighthead disease in rice. *Environ. Pollut.* **2023**, *316*, 120515. [CrossRef] [PubMed]
25. da Silva, H.F.O.; Tavares, O.C.H.; da Silva, L.S.; Zonta, E.; da Silva, E.M.R.; Júnior, O.J.S.; Nobre, C.P.; Berbara, R.L.L.; García, A.C. Arbuscular mycorrhizal fungi and humic substances increased the salinity tolerance of rice plants. *Biocatal. Agric. Biotechnol.* **2022**, *44*, 102472. [CrossRef]
26. Wang, W.; Shi, J.; Qu, K.; Zhang, X.; Jiang, W.; Huang, Z.; Guo, Z. Composite film with adjustable number of layers for slow release of humic acid and soil remediation. *Environ. Res.* **2023**, *218*, 114949. [CrossRef]
27. Alsamadany, H. Physiological, biochemical and molecular evaluation of mungbean genotypes for agronomical yield under drought and salinity stresses in the presence of humic acid. *Saudi J. Biol. Sci.* **2022**, *29*, 103385. [CrossRef] [PubMed]
28. Faccin, D.; Di Piero, R.M. Extracts and fractions of humic substances reduce bacterial spot severity in tomato plants, improve primary metabolism and activate the plant defense system. *Physiol. Molecul. Plant Pathol.* **2022**, *121*, 101877. [CrossRef]
29. Chen, L.; Li, W.; Zhao, Y.; Zhang, S.; Meng, L. Mechanism of sulfur-oxidizing inoculants and nitrate on regulating sulfur functional genes and bacterial community at the thermophilic compost stage. *J. Environ. Manag.* **2023**, *326*, 116733. [CrossRef]
30. Liu, H.; Luo, L.; Jiang, G.; Li, G.; Zhu, C.; Meng, W.; Zhang, J.; Jiao, Q.; Du, P.; Li, X.; et al. Sulfur enhances cadmium bioaccumulation in Cichorium intybus by altering soil properties, heavy metal availability and microbial community in contaminated alkaline soil. *Sci. Total Environ.* **2022**, *837*, 155879. [CrossRef]
31. Sparks, D.L.; Page, A.L.; Helmke, P.A.; Loeppert, R.H. *Methods of Soil Analysis, Part 3: Chemical Methods*; John Wiley & Sons: Hoboken, NJ, USA, 2020.
32. Campbell, D.J. Determination and use of soil bulk density in relation to soil compaction. In *Developments in Agricultural Engineering*; Elsevier: Amsterdam, The Netherlands, 1998; Volume 11, pp. 113–139.
33. Netto, A.T.; Campostrini, E.J.; Oliveira, G.; Bressan, S.R.E. Photosynthetic pigments, nitrogen, chlorophyll a fluorescence and SPAD-502 readings in coffee leaves. *Sci. Hort.* **2005**, *104*, 199–209. [CrossRef]
34. Morris, K.N. A Guide to the National Turfgrass Evaluation Program (NTEP) Turfgrass Ratings. 2022. Available online: https://www.ntep.org/reports/ratings.htm (accessed on 25 August 2022).
35. Duncan, D.B. Multiple range and multiple F tests. *Biometrics* **1995**, *11*, 1–42. [CrossRef]
36. Wu, P.; Cogill, S.; Qiu, Y.; Li, Z.; Zhou, M.; Hu, Q.; Chang, Z.; Noorai, R.E.; Xia, X.; Saski, C.; et al. Comparative transcriptome profiling provides insights into plant salt tolerance in seashore paspalum (*Paspalum vaginatum*). *BMC Genom.* **2020**, *21*, 131. [CrossRef]
37. Huang, S.; Jiang, S.; Liang, J.; Chen, M.; Shi, Y. Current knowledge of bermudagrass responses to abiotic stresses. *Breed. Sci.* **2019**, *69*, 215–226. [CrossRef] [PubMed]
38. Gopinath, L.; Moss, J.Q.; Wu, Y. Evaluating the freeze tolerance of bermudagrass genotypes. *Agrosyst. Geosci. Environ.* **2021**, *4*, e20170. [CrossRef]
39. Schmidt, R.E. Iron for turfgrass nutrition. *Golf Course Manegm.* **2004**, 113–116.
40. Bello, S.K.; Alayafi, A.H.; AL-Solaimani, S.G.; Abo-Elyousr, K.A.M. Mitigating Soil Salinity Stress with Gypsum and Bio-Organic Amendments: A Review. *Agronomy* **2021**, *11*, 1735. [CrossRef]
41. Marschner, M. *Mineral Nutrition of Higher Plants*, 2nd ed.; Academic Press: London, UK, 1995; pp. 200–255.
42. Liu, A.; Hu, Z.; Bi, A.; Fan, J.; Gitau, M.M.; Amombo, E.; Chen, L.; Fu, J. Photosynthesis, antioxidant system and gene expression of bermudagrass in response to low temperature and salt stress. *Ecotoxicology* **2016**, *25*, 1445–1457. [CrossRef]
43. Chen, Y.; Clapp, C.E.; Magen, H. Mechanisms of plant growth stimulation by humic substances: The role of organo-iron complexes. *Soil Sci. Plant Nutr.* **2004**, *50*, 1089–1095. [CrossRef]
44. Hunter, A.; Butler, T. Effect of humic acid on growth and development of Agrostis stolonifera grass in a sand- based root zone. *Inter. Turfgrass Soci. Res. J.* **2005**, *10*, 937–943.
45. Shalaby, T.A.; Abd-Alkarim, E.; El-Aidy, F.; Hamed, E.; Sharaf-Eldin, M.; Taha, N.; El-Ramady, H.; Bayoumi, Y.; dos Reis, A.R. Nano-selenium, silicon and H_2O_2 boost growth and productivity of cucumber under combined salinity and heat stress. *Ecotoxicol. Environ. Saf.* **2021**, *212*, 111962. [CrossRef]
46. Nadeem, M.; ul Haq, M.A.; Saqib, M.; Maqsood, M.; Iftikhar, I.; Ali, T.; Awais, M.; Ullah, R.; He, Z. Nutrients, Osmotic and Oxidative Stress Management in Bread Wheat (*Triticum aestivum* L.) by Exogenously Applied Silicon Fertilization Under Water Deficit Natural Saline Conditions. *Silicon* **2022**, *14*, 11869–11880. [CrossRef]
47. Salem, E.M.M.; ·Kenawey, M.K.M.; Saudy, H.S.; Mubarak, M. Influence of Silicon Forms on Nutrients Accumulation and Grain Yield of Wheat Under Water Deficit Conditions. *Gesunde Pflanz.* **2022**, *74*, 539–548. [CrossRef]

48. Yin, J.; Jia, J.; Lian, Z.; Hu, Y.; Guo, J.; Huo, H.; Zhu, Y.; Gong, H. Silicon enhances the salt tolerance of cucumber through increasing polyamine accumulation and decreasing oxidative damage. *Ecotoxicol. Environ. Saf.* **2019**, *169*, 8–17. [CrossRef] [PubMed]
49. Chavarria, M.R.; Wherley, B.; Jessup, R.; Chandra, A. Leaf anatomical responses and chemical composition of warm-season turfgrasses to increasing salinity. *Curr. Plant Biol.* **2020**, *22*, 100147. [CrossRef]
50. Abdel-Fatah, G.H.; El-Sayed, B.A.; Shahin, S.M. The role of humic acid in reducing the harmful effect of irrigation with saline water on tifway turf. *J. Boil. Chem. Environ. Sci.* **2008**, *3*, 75–89.
51. Valizadeh-rad, K.; Motesharezadeh, B.; Alikhani, H.A.; Jalali, M.; Etesami, H.; Javadzarin, I. Morphophysiological and Nutritional Responses of Canola and Wheat to Water Deficit Stress by the Application of Plant Growth-Promoting Bacteria, Nano-Silicon, and Silicon. *J. Plant Growth Regul.* **2022**, 1–17. [CrossRef]
52. Valizadeh-rad, K.; Motesharezadeh, B.; Alikhani, H.A.; Jalali, M. Direct and Residual Effects of Water Deficit Stress, Different Sources of Silicon and Plant-Growth Promoting Bacteria on Silicon Fractions in the Soil. *Silicon* **2022**, *14*, 3403–3415. [CrossRef]
53. Esmaili, S.; Tavallali, V.; Amiri, B.; Bazrafshan, F.; Sharafzadeh, S. Foliar Application of Nano-Silicon Complexes on Growth, Oxidative Damage and Bioactive Compounds of Feverfew Under Drought Stress. *Silicon* **2022**, *14*, 10245–10256. [CrossRef]
54. Van Dyke, A.; Johnson, P.G.; Grossl, P.R. Influence of humic acid on water retention and nutrient acquisition in simulated golf putting greens. *Soil Use Manag.* **2009**, *25*, 255–261. [CrossRef]
55. Rizwan, M.; Ali, S.; Ibrahim, M.; Farid, M.; Adrees, M.; Bharwanaand, S.A.; Abbas, F. Mechanisms of silicon-mediated alleviation of drought and salt stress in plants: A review. *Environ. Sci. Poll. Res.* **2015**, *22*, 15416–15431. [CrossRef]
56. Pompeiano, A.; Giannini, V.; Gaetani, M.; Vita, F.; Guglielminetti, L.; Bonari, E.; Volterrani, M. Response of warm–season grasses to N fertilization and salinity. *Sci Hort.* **2014**, *177*, 92–98. [CrossRef]
57. Prokopiuk, K.; Żurek, G.; Rybka, K. Turf covering for sport season elongation cause no stress for grass species as detected by Chl a fluorescence. *Urban For. Urban Green.* **2019**, *41*, 14–22. [CrossRef]
58. Sharma, P.; Mayur Mukut Murlidhar Sharma, M.M.M.; Patra, A.; Vashisth, M.; Mehta, S.; Singh, B.; Tiwari, M.; Pandey, V. The Role of Key Transcription Factors for Cold Tolerance in Plants. In *Transcription Factors for Abiotic Stress Tolerance in Plants*; Wani, H.S., Ed.; Elsevier Inc.: Amsterdam, The Netherlands, 2020. [CrossRef]
59. Hajiboland, R. Silicon-Mediated Cold Stress Tolerance in Plants. In *Silicon and Nano-Silicon in Environmental Stress Management and Crop Quality Improvement*; Etesami, H., Al Saeedi, A.H., El-Ramady, H., Fujita, M., Pessarakli, M., Hossain, M.A., Eds.; Academic Press: Cambridge, MA, USA, 2022; pp. 161–180. [CrossRef]
60. Nguyen, P.N.; Do, P.T.; Pham, Y.B.; Doan, T.O.; Nguyen, X.C.; Lee, W.K.; Nguyen, D.D.; Vadiveloo, A.; Um, M.-J.; Ngo, H.-H. Roles, mechanism of action, and potential applications of sulfur-oxidizing bacteria for environmental bioremediation. *Sci. Total Environ.* **2022**, *852*, 158203. [CrossRef] [PubMed]

horticulturae

MDPI

Review

Melatonin-Induced Detoxification of Organic Pollutants and Alleviation of Phytotoxicity in Selected Horticultural Crops

Golam Jalal Ahammed [1,*] and Xin Li [2,*]

1 College of Horticulture and Plant Protection, Henan University of Science and Technology, Luoyang 471023, China
2 Key Laboratory of Tea Quality and Safety Control, Ministry of Agriculture and Rural Affairs, Tea Research Institute, Chinese Academy of Agricultural Sciences, Hangzhou 310008, China
* Correspondence: ahammed@haust.edu.cn (G.J.A.); lixin@tricaas.com (X.L.)

Abstract: Environmental pollution with organic pollutants has increased drastically in recent decades. Despite the importance of minimizing organic pollutant content such as pesticide residue in edible crops, our understanding of induced xenobiotic metabolism in plants is poor. Melatonin is a potent stress-relieving biomolecule, which exerts beneficial effects on xenobiotic metabolism in plants. Exogenous melatonin treatment not only improves photosynthesis, antioxidant defense, and plant growth but also reduces pollutant residue and xenobiotic uptake. The overexpression of melatonin biosynthetic genes enhances organic pollutant metabolism, while the suppression of endogenous melatonin biosynthesis increases organic pollutant residue in horticultural products. Studies have revealed that the glutathione-dependent detoxification pathway plays a critical role in the melatonin-induced enhanced detoxification of xenobiotics. Moreover, a role for *RESPIRATORY BURST HOMOLOG (RBOH)*-derived reactive oxygen species signaling has been revealed which potentially acts upstream of glutathione-dependent xenobiotic metabolism. Based on the literature, here, we reviewed the effects of organic pollutants on plants and how melatonin aids plants in enduring the effects of organic pollutant-induced stress. We also discussed the potential melatonin signaling mechanism in enhanced pesticide metabolism. Our assessment suggests that melatonin has positive impacts on plant tolerance to organic pollution, which can be used to improve the food safety of edible horticultural crops.

Keywords: melatonin; pesticide degradation; food safety; glutathione; xenobiotic; detoxification

1. Introduction

Thousands of organic synthetic compounds are extensively exploited in a range of industries, including agrochemicals, pharmaceuticals, food processing, toiletries, printing, textiles, petrochemicals, steel manufacturing, and so on [1]. Additionally, new synthetic chemicals are being introduced nearly every day around the world. As a result of extensive production, usage, and frequent release, environmental pollution with organic pollutants has become a serious environmental concern [2]. Organic pollutants can extensively disperse, and many organic pollutants have a long half-life, and thus they continue to pollute the environment [3]. Because of the acute and chronic impacts of toxic organic pollutants on all living organisms, the bioaccumulation of such substances has considerably increased the burden and potential threats to the environment and human health [3,4]. Alarmingly, certain organic pollutants are known to cause cancer, genetic mutations, and birth defects [4,5]. In addition, consuming organic pollutant-contaminated crops for a long time may result in serious illnesses [6]. Nonetheless, individual susceptibility, the duration and mode of exposure, and the kind of organic pollutants play a role in determining the health effects.

Due to the scarcity of freshwater resources, reclaimed water is widely used in agriculture, despite it possibly being a significant contributor of organic contaminants to edible crops [7]. When plants are grown in contaminated soils or irrigated with polluted water, organic pollutants are accumulated in the leaves, fruits, and stems of many crops that are often consumed by humans [6,8]. Nonetheless, the accumulation of organic pollutants in the above-ground sections of plants may vary greatly depending on factors including the hydrophobicity, lipophilicity, and chemical structure of the pollutants as well as the plant species/genotype and absorption mechanism [6,9]. In particular, Cucurbitaceae family members including cucumber, melon, pumpkin, squash, and zucchini have been shown to have elevated levels of organic pollutants in their above-ground sections [6]. Moreover, pesticide residues have been found in a wide variety of ready-to-eat foods and drinks such as vegetables, fruits, and fruit juices, and they are notoriously difficult to remove using normal preparation methods such as washing and peeling [10,11]. If residue levels in crops are too high, farmers have to abandon everything they grow on that polluted land, causing a total financial loss. Therefore, it is imperative to reduce the residue of organic pollutants by establishing growing strategies for safer crop production [12]. Plants can degrade or detoxify organic pollutants [13]. Thus, taking advantage of the in planta detoxification of organic pollutants is critical for ensuring their absence in the human diet [14]. However, the capacity of plants to detoxify organic pollutants is often limited by the high phytotoxicity of the xenobiotic substance at high concentrations [15]. Therefore, cultivating safer horticultural crops requires an in-depth knowledge of how plants efficiently detoxify organic pollutants.

Melatonin, also known as *N*-acetyl-5-methoxytryptamine, is an endogenous signaling molecule found in eukaryotic organisms [16]. It plays a significant role in a variety of biological processes in plants [17]. Since the discovery of phytomelatonin in 1995, numerous studies have investigated its effects on plants over the years. Phytomelatonin is gaining recognition as the plant hormone upon the recently identified first melatonin receptor PMTR1 [18]. Melatonin promotes seed germination, increases the production of lateral roots, delays leaf senescence, and modulates the blooming time in plants grown in unfavorable conditions [16,19–21]. Melatonin is a key antioxidant that removes reactive oxygen species (ROS) and reactive nitrogen species (RNS) [17]. Additionally, it regulates gene expression indirectly by activating or inhibiting stress-responsive transcription factors [22,23]. Exogenous melatonin application improves plant tolerance to a variety of stresses such as drought, salt, heat, cold, waterlogging, heavy metals, and organic pollutants via regulating endogenous melatonin production and the activities of antioxidant enzymes [24–29]. Since climate change and environmental pollution are increasingly threatening agricultural production, crop yields, and food security, melatonin has been the subject of increased study due to its stress ameliorative properties.

Recent literature has focused on the remarkable benefits of melatonin in enhancing plant adaptation to unfavorable conditions as well as the unique tolerance mechanisms and the network of regulation in plant defense via melatonin [30–36]. The tremendous potential of melatonin in modulating plant tolerance to organic pollutant-induced stress has been revealed in specific research [7,10,37,38]. Phytomelatonin not only plays a crucial role in alleviating phytotoxicity induced by organic pollutants such as different types of pesticides, polycyclic aromatic hydrocarbons (PAHs), and endocrine disruptor bisphenol-A (BPA) [7,37,39], but it also reduces pollutant concentrations in plant tissue, possibly by decreasing their uptake and/or promoting in vivo degradation [7,10,37,40]. This article reviews the current state of knowledge regarding the role of melatonin in plant tolerance to organic pollutants and associated food safety, with the goal of serving as a reference for future studies of phytomelatonin and pointing researchers in novel directions regarding its applications, particularly with regards to enhancing food safety.

2. Organic Pollutants and Phytotoxicity

Organic pollutants are carbon (C)-based anthropogenic compounds that cause adverse effects on the environment and human health [3]. As a special group of chemical pollutants, organic pollutants are different from inorganic (mostly metals) pollutants. In recent decades, many different types of organic pollutants have been released in large quantities into the environment as a consequence of massive anthropogenic activities [1]. Organic pollutants can be classified in several ways. Based on the degradability of organic pollutants, the pollutants can be divided into two categories: labile organic pollutants and recalcitrant organic pollutants [3]. Again, according to the boiling point, organic pollutants can also be divided into two categories: volatile organic compounds (240 °C~260 °C) and semivolatile organic compounds (250 °C~400 °C) [41]. Despite this classification, the margin between volatile organic compounds and semivolatile organic compounds is to a certain extent unclear, and many pollutants fall into both classes. Many industrial chemicals, pesticides, phenols, ethers, ketones, phthalate esters, pyridines, and anilines belong to semivolatile organic compounds and they are typically more resistant to environmental degradation than volatile organic compounds [3,41]. Organic pollutants such as persistent organic pollutants (POPs), polyaromatic hydrocarbons (PAHs), and polychlorinated biphenyls (PCBs) have a long half-life [39,42,43]. In contaminated soils, organic pollutants are absorbed by the roots and then transported to the shoots, where they exert a devastating effect on photosynthesis and other crucial physiological processes [6,15].

Additionally, the shoot can also accumulate lipophilic organic pollutants directly from the atmosphere [43]. Moreover, many pesticides are applied to the foliage, causing their accumulation in the shoot [37,38]. Many hazardous chemicals, including organic pollutants, exert their toxicity primarily via inhibiting photosynthetic processes (Figure 1). Chloroplasts are particularly vulnerable to organic pollutants [44]. Organic pollutants accumulated in chloroplast thylakoids and microsomal compartments interfere with fundamental photosynthetic processes [45]. Both the intact and photo-modified forms of organic pollutants stifle photosynthesis by affecting primary photochemical processes [15]. Organic pollutants block electron transport by obstructing either photosystem II (PSII) or the connection between PSII and PSI at the cytochrome b/f [46]. Moreover, variations in the amounts of photosynthetic pigments (Chl a, Chl b, and carotenoids) due to organic pollutant-induced stress eventually alter the photosynthesis process [47].

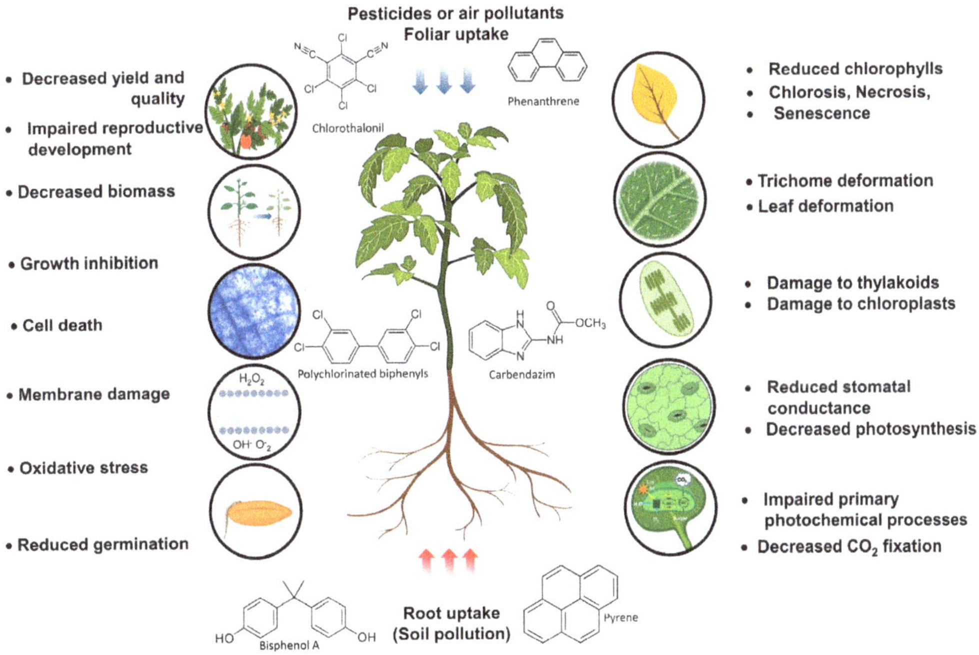

Figure 1. Deleterious effects of organic pollutants on plants and associated phytotoxicity.

Plants exposed to organic pollutants, such as pesticides and PAHs showed visible symptoms including white spots on leaves, trichome and leaf deformations, chlorosis, and necrosis as well as a decrease in biomass accumulation [15,48]. Moreover, oxidative stress, ultrastructural abnormalities, cell death, modifications to antioxidant systems, and reduced plant growth are critical signs of organic pollutant-induced negative consequences [44,49]. Bisphenol A is a xenoestrogen that can cause serious health problems in humans. BPA has been shown to be hazardous to plants as well [50]. Reduced seed germination, decreased photosynthesis, stunted growth, and delayed reproductive development are common effects of BPA on plants [51,52]. BPA treatment also reduced the quantum yield of photosystem II (Fv/Fm) and increased ROS accumulation, lipid peroxidation, and BPA accumulation [7]. Likewise, synthetic pesticides are also phytotoxic [8,15]. Most pesticides suppress PSII activity, cause photoinhibition, inhibit the electron transport chain in the thylakoid, degrade chlorophylls, inhibit photosynthesis, and reduce plant growth [15,38,40].

Organic pollutants cause phytotoxicity by triggering excessive ROS production, which eventually induces oxidative stress [49]. In particular, the oxidation of certain organic pollutants such as phenanthrene leads to ROS production within cellular compartments [53]. Despite being one of the most critical signaling molecules in plant biology, excessive ROS produced under stress as byproducts of aerobic metabolism can be seriously harmful [54,55]. The peroxidation of cell membranes caused by ROS is a critical sign of oxidative stress, and highly bioactive ROS can damage lipids, nucleic acids, and proteins [56]. Plants have evolved a robust antioxidative defense mechanism, comprising both enzymatic and non-enzymatic antioxidants, to remove ROS from different cellular compartments [57]. However, antioxidant-based ROS scavenging is largely rate-limiting. Chlorophyll degradation, decreased photosynthesis, and reduced protein and RNA levels are commonly the results of excessive ROS accumulation in plant cells [35,58,59].

3. Mechanisms of Pollutant Detoxification

To counteract the harmful effects of organic pollutants, plants use several detoxification methods [8]. In the classical detoxification mechanism (Figure 2), the three main steps in xenobiotic metabolism in higher plants are: phase I: conversion or transformation; phase II: conjugation; and phase III: compartmentalization (transport and sequestration) [13]. Typically, organic pollutants are initially hydroxylated by cytochrome P450 family enzymes, and then the modified organic pollutants are conjugated with glutathione (GSH), followed by transportation and sequestration in the vacuole [14]. Glutathione, which is synthesized from cysteine, is an important thiol in plant xenobiotic detoxification. The processes of GSH production in cells are enzyme-catalyzed and ATP-dependent [60,61]. The initial synthesis of γ-glutamylcysteine from γ-glutamate and $\dot{\alpha}$-cysteine occurs through the rate-limiting enzymatic action of the γ-glutamylcysteine synthetase enzyme (γ-ECS) encoded by *GSH1* [7]. Afterward, the *GSH2*-encoded glutathione synthetase enzyme (GS) adds glycine to the dipeptide (γ-glutamyl-$\dot{\alpha}$-cysteine). In plants, glutathione is found in both its reduced and oxidized forms. The enzyme glutathione reductase (GR), which is encoded by the *GR1* gene, catalyzes the conversion of oxidized glutathione disulfide (GSSG) back into reduced glutathione (GSH) [61,62]. Notably, the detoxification of xenobiotics in plants often involves glutathione S-transferases (GSTs), a well-known detoxifying enzyme, catalyzing the conjugation process between organic pollutants and GSH [7,50]. To neutralize the electronegative sites of xenobiotics, GSTs promote the nucleophilic conjugation of GSH (at the thiol group). Finally, transformed organic pollutants are contained inside vacuoles or the cellular walls [13] When key detoxification genes such as *GSH1*, *GR1*, and *GST1* are silenced in tomato plants, silenced plants show impaired detoxification potential characterized by increased ROS accumulation, lipid peroxidation, and organic pollutant accumulation as well as decreased GST activity [7]. Even while plants have their inherent detoxifying systems, they are not particularly efficient at breaking down stubborn xenobiotics [6,8]. As various plant growth regulators can promote xenobiotic metabolism in plants, the use of growth regulators is considered a useful strategy for increasing plant tolerance to organic pollutants and xenobiotic degradation in vivo [14,40,63].

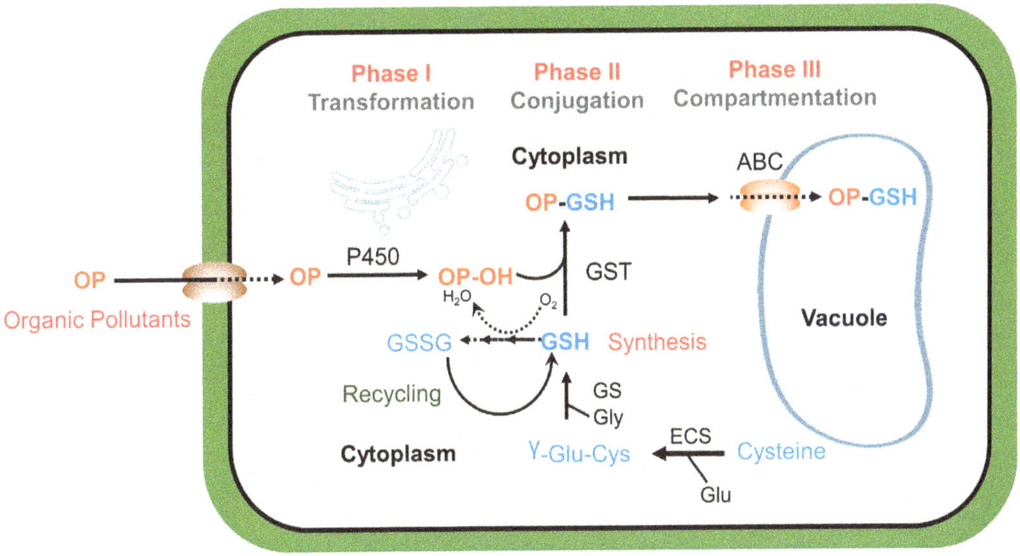

Figure 2. Mechanisms of organic pollutant detoxification in plants.

4. Melatonin: A Master Growth Regulator of Plant Stress Tolerance

4.1. Melatonin Synthesis and Sources

Researchers have uncovered the essential steps of melatonin synthesis in plants [64]. Although melatonin is synthesized in both chloroplasts and mitochondria, the chloroplastic pathway is thought to be the major route of melatonin synthesis [23]. Typically, melatonin is synthesized from tryptophan through the enzymes tryptophan decarboxylase (TDC), tryptamine 5-hydroxylase (T5H), serotonin *N*-acetyltransferase (SNAT), and *N*-acetylserotonin *O*-methyltransferase (ASMT) [59]. To be more specific, TDC converts tryptophan into tryptamine, and T5H hydroxylates tryptamine to generate 5-hydroxytryptamine (serotonin) (Figure 3). Afterward, serotonin is transformed to *N*-acetyl serotonin by SNAT, and melatonin is produced from *N*-acetyl serotonin by ASMT [65]. However, plants also use a catalytic enzyme called caffeic acid *O*-methyltransferase (COMT, involved in phenylpropane metabolism) to convert serotonin to melatonin [66,67]. COMT can substitute for ASMT to catalyze the production of melatonin from *N*-acetyl serotonin, and it can catalyze the transformation of serotonin to 5-methoxytryptamine as well [16]. The recruitment of COMT makes plant melatonin synthesis more versatile than animal synthesis [65]. Although tryptophan is required for the production of melatonin in all organisms, animals can not synthesize tryptophan and thus have to obtain it from plant-derived food [64].

There are essentially two sources of natural melatonin: 'melatonin' from animal origin and melatonin from plant origin, with the latter also being known as 'phytomelatonin' [68]. As for animal sources, melatonin was previously isolated from the pineal glands of cows; however, the risks of viral infection have led to synthetic melatonin production being the preferred option [68,69]. Despite the high yield of synthetic melatonin, the occurrence of unwanted compounds with chemically synthesized melatonin results potential health risks [68]. Notably, significant progress has been achieved in synthetic melatonin production through the use of greener protocols, resulting in the production of new melatonin derivatives with lower cytotoxicity and higher water solubility, such as sodium 4-(3-(2-acetamidoethyl)-5-methoxy-1H-indol-1-yl) butane-1-sulfonate [69,70]. As opposed to chemically synthesized melatonin, phytomelatonin derived from different plant parts such as fruit usually does not contain contaminants that are commonly found in chemical synthesis [71]. Rather, compounds associated with phytomelatonin extracts such as flavonoids, vitamins, phenols, tocopherols, and carotenoids have beneficial health effects on humans [68]. Phytomelatonin is abundant in several families of plants such as Rosaceae, Poaceae, Vitaceae, Apiaceae, and Brassicaceae [72]. In particular, fruits including cherries, grapes, apples, tomatoes, bananas, and pineapples have been reported as important sources of phytomelatonin [33]. Concentrations of phytomelatonin in different plant parts in a range of plant species were listed in our recent review [16]. However, the efficient isolation of phytomelatonin and the development of phytomelatonin-rich extracts still remain challenging tasks, which warrant further intensive studies [68,71].

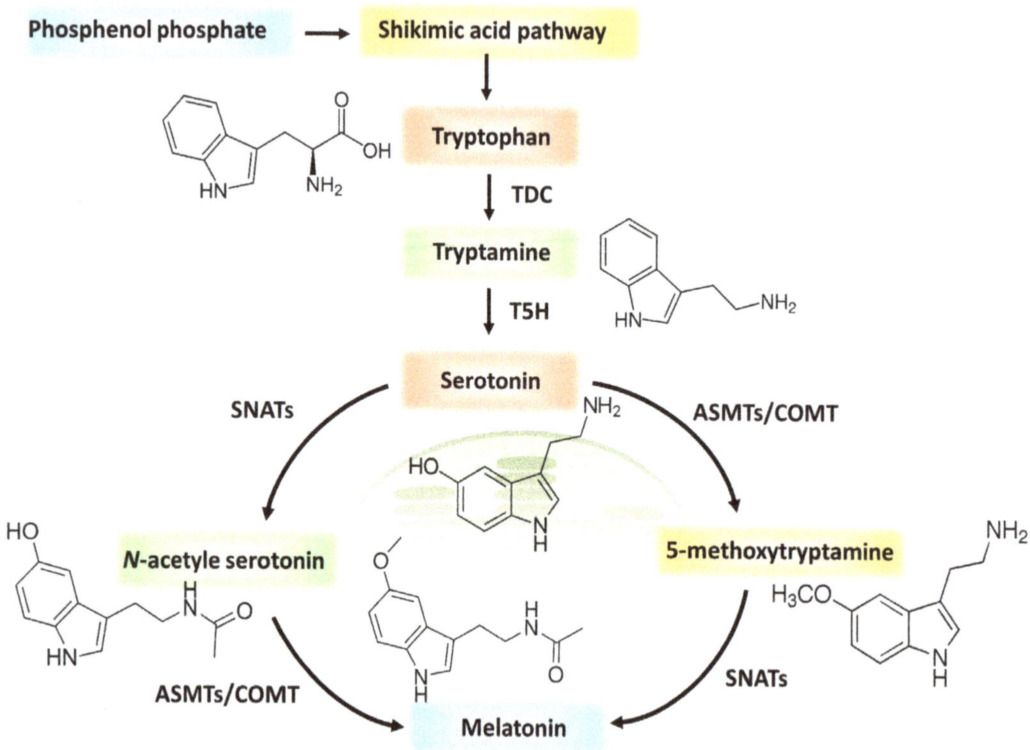

Figure 3. Melatonin biosynthetic pathway in plants.

4.2. Melatonin in Plant Physiology, Metabolism, and Abiotic Stress Tolerance

Due to the widespread effect of melatonin on gene transcription in plants, melatonin likely has a pleiotropic function in a wide variety of cellular processes [16,18,73]. Exogenous melatonin application or endogenous melatonin over-production has been shown to promote plant growth, development, and a variety of metabolic and physiological processes, including photosynthesis, carbohydrate metabolism, and nitrogen assimilation, hormone homeostasis, and so on [19,25,35,36,74]. Melatonin presumably delays postharvest fruit senescence [75–77]. The principal function of melatonin against stresses is attributed to efficient ROS scavenging [28,35]. Melatonin not only plays a role in direct ROS scavenging but also significantly improves the antioxidant defense, which includes both enzymatic and nonenzymatic antioxidants [56]. Melatonin protects plant cells and tissues from oxidative stress by increasing antioxidant gene expression and encoded enzyme activity, thus allowing plants to efficiently scavenge a wide range of ROS and RNS [17]. Recent research has shown that melatonin not only promotes primary metabolism but also stimulates secondary metabolism in plants, leading to the increased synthesis of a wide range of secondary metabolites such as polyphenols, glucosinolate, terpenoids, and alkaloid contents [23,24]. Notably, polyphenols such as flavonoids play an important role in ROS scavenging [78]. Melatonin improves the cellular redox state by maintaining the stability of GSH levels [10,40,79].

In recent years, numerous studies have revealed that melatonin can increase plant tolerance to a wide variety of biotic and abiotic stresses, including drought, salinity, heat, cold, water logging, heavy metal toxicity, and organic pollutant stress [7,80–82]. It is now well-established that melatonin has a critical role in regulating responses to abiotic stress (Figure 4). Melatonin interacts/crosstalk with hormones and signaling molecules to systematically regulate plant resistance [17,31,83,84]. Notably, increased resistance to photo-oxidative stress is mediated by melatonin-induced GSH homeostasis in cucumber [85]. Moreover, melatonin participates in xenobiotic detoxification by modulating the ascorbate (ASA)-GSH cycle and GST activity [10,38,40]. There are several lines of evidence to infer that the use of melatonin to reduce organic pollutant phytotoxicity and pollutant residue could be feasible for edible horticultural crop production.

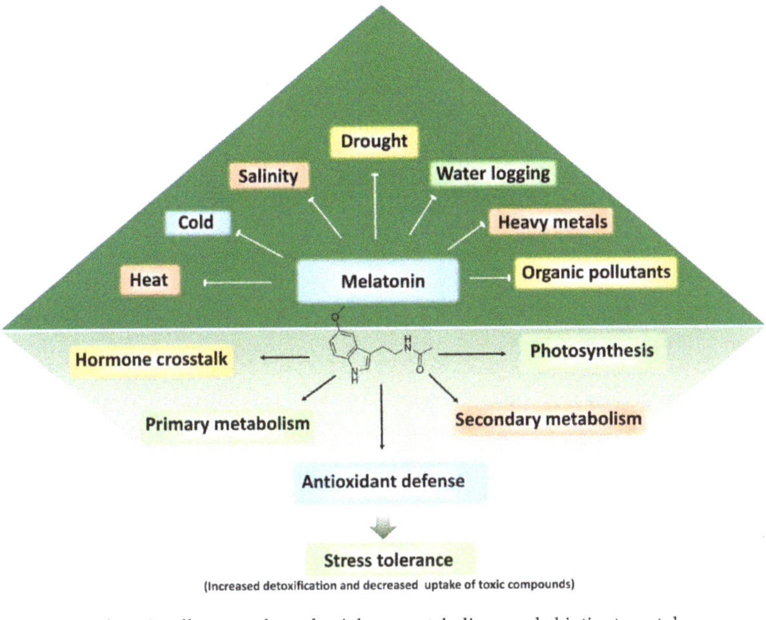

Figure 4. Melatonin effects on plant physiology, metabolism, and abiotic stress tolerance.

5. Melatonin-Induced Detoxification and Alleviation of Phytotoxicity

5.1. Exogenous Melatonin Alleviates Organic Pollutant-Induced Stress

Melatonin regulates a variety of physiological and biochemical processes in plants under stress [25,35]. It has been proposed as a possible natural safener that can protect plants from organic pollutants such as pesticide- and herbicide-induced phytotoxicity [86]. Previous studies have shown that residues of pesticides such as carbendazim, chlorothalonil, and imidacloprid in tomato and cucumber plants can be significantly decreased with the administration of exogenous melatonin [37,38,40]. However, not much is known about the detoxification mechanism triggered by melatonin in response to organic pollutants. Current knowledge of melatonin-induced detoxification is largely based on the exogenous application of melatonin and/or endogenous suppression of melatonin accumulation by using melatonin biosynthetic inhibitor p-chlorophenylalanine (CPA) [10,37]. The effects of exogenous melatonin on the detoxification of xenobiotics and alleviation of phytotoxicity are listed in Table 1. Additionally, there are a small number of pieces of genetic evidence that further strengthen the proposition that melatonin is involved in plant responses to organic pollutant-induced stress [7,38].

Table 1. Effects of exogenous melatonin on xenobiotic detoxification and alleviation of phytotoxicity.

Plant Species	Melatonin Concentrations *	Treatment Methods	Organic Pollutants	Melatonin Effects	References
Tomato (*Solanum lycopersicum* L.)	20 μM	Foliar application	Bisphenol A (BPA, 10 mg L^{-1})-root treatment	• Increased transcripts of *TDC*, *T5H*, *SNAT*, *GSH1*, *GST1* and *GR1* • Decreased ROS accumulation and lipid peroxidation • Increased *Fv/Fm*, GSH biosynthesis and regeneration • Increased BPA glutathionylation by GSH • Decreased BPA uptake	[7]
Tomato (*S. lycopersicum* L.)	100 μM	Foliar application	Chlorothalonil, 11.2 mM-foliar treatment	• Increased photosynthesis and Fv/Fm • Increased detoxification enzyme activity and gene expression • Decreased pesticide residue via H_2O_2 signaling	[37]
Tomato (*S. lycopersicum* L.)	0.5 μM	Foliar application	Carbendazim (MBC, 1 mM)-foliar treatment	• Increased chlorophyll content, Fv/Fm, photosynthesis • Decreased MDA content, decreased MBC residues in leaves (48–73%)	[38]
Lettuce (*Lactuca sativa* L.)	0.5 μM	Foliar application	Carbendazim (MBC, 1 mM)-foliar treatment	• Significantly decreased MBC residues in leaves	[38]
Chinese cabbage (*Brassica campestris* L.)	0.5 μM	Foliar application	Carbendazim (MBC, 1 mM)-foliar treatment	• Significantly decreased MBC residues in leaves	[38]
Spinach (*Spinacia oleracea* L.),	0.5 μM	Foliar application	Carbendazim (MBC, 1 mM)-foliar treatment	• Significantly decreased MBC residues in leaves	[38]
Celery (*Apium graveolens* L.)	0.5 μM	Foliar application	Carbendazim (MBC, 1 mM)-foliar treatment	• Significantly decreased MBC residues in leaves	[38]
Cucumber (*Cucumis sativus* L.)	0.5 μM	Foliar application	Carbendazim (MBC, 1 mM)-foliar treatment	• Significantly decreased MBC residues in leaves	[38]
Cucumber (*C. sativus* L.)	50 μM	Root pretreatment	Imidacloprid (IMD, 2.75 mM)-foliar treatment	• Increased Fv/Fm, chlorophyll contents, photosynthesis, improved redox state, increased antioxidant enzyme activity, GST activity, and its transcripts • Decreased H_2O_2, $O_2\cdot-$, and MDA content, decreased IMD residues in leaves	[40]
Jujube (*Ziziphus jujuba* Mill. cv. Dongzao)	0.1 mM	Mature jujube fruits (post-harvest spraying)	Fruits treated (immersed) with chlorothalonil (CHT, 10 mM), glyphosate (Gly, 2 mM), and malathion (Mal, 3 mM) solution for 2 h	• Improved firmness, reduced fruit weight loss, and decay index • Increased GSH content • Enhanced activity of GR and GST, increased antioxidants and phenolics, promoted pesticide degradation	[10]

* Only the most effective concentrations of exogenous melatonin which alleviated organic pollutant-induced phytotoxicity and/or improved the degradation of organic pollutants are presented.

5.2. Potential Mechanisms of Melatonin-Induced Xenobiotic Detoxification

Various modes of application with respect to melatonin can stimulate plant detoxification potential. The foliar spraying of melatonin is a common and practically feasible mode of application that was found to be effective for the detoxification of both shoot-sourced pesticides and root-absorbed organic pollutants such as BPA. The negative effects of BPA as manifested by decreased photochemical efficiency and increased lipid peroxidation, ROS generation, and BPA uptake were mitigated by the addition of exogenous melatonin [7]. Melatonin is a redox network modulator that promotes the detoxification of xenobiotics via the modulation of the AsA-GSH cycle, GST activity, and vacuolar sequestration [10,40]. The expression levels of melatonin biosynthesis genes such as *COMT*, *T5H*, and *SNAT* were upregulated in response to the imposition of BPA stress [7]. These transcriptional changes were accompanied by the elevated expression of *GSH1*, *GR1*, and *GST1* and the activity of GST and GR upon melatonin treatment in BPA-treated plants. Functional genetics research highlights the cooperation between melatonin and GSH in xenobiotic detoxification in plants [38]. The manipulation of GSH metabolism and the expression of associated genes, such as *GSH1*, *GR1*, and *GST1*, by virus-induced gene silencing impairs the melatonin-controlled uptake, transport, and degradation of BPA in tomato plants, indicating the mechanistic involvement of melatonin in BPA detoxification [7].

Moreover, the overexpression of *COMT1* in tomato plants promotes pesticide metabolism, which was associated with increased endogenous melatonin levels in tomato plants [38]. *COMT1* overexpression enhances antioxidant capacity and the detoxification process, leading to the alleviation of oxidative stress and a reduction in carbendazim residue in tomato leaves. Similarly, melatonin can significantly decrease chlorothalonil residue in tomato leaves along with increasing photosynthetic efficiency and antioxidant capacity [37]. Notably, the *RESPIRATORY BURST HOMOLOG (RBOH)*-dependent H_2O_2 signaling-mediated differential expression of detoxification-related genes, GSH production and/or regeneration, and GST activity, appear to play a significant role in the reduction of pesticide residue in tomato plants [87]. Similarly, endogenous H_2O_2 signaling is crucial for facilitating the melatonin-mediated detoxifying response to pesticides (Figure 5). When endogenous H_2O_2 signaling was suppressed, either by limiting NADPH oxidase-dependent H_2O_2 generation or H_2O_2 elimination by ROS scavengers, the potential of exogenous melatonin to confer a detoxifying response to pesticides was reduced, further confirming the involvement of H_2O_2 signaling in melatonin-induced xenobiotic metabolism in plants [37].

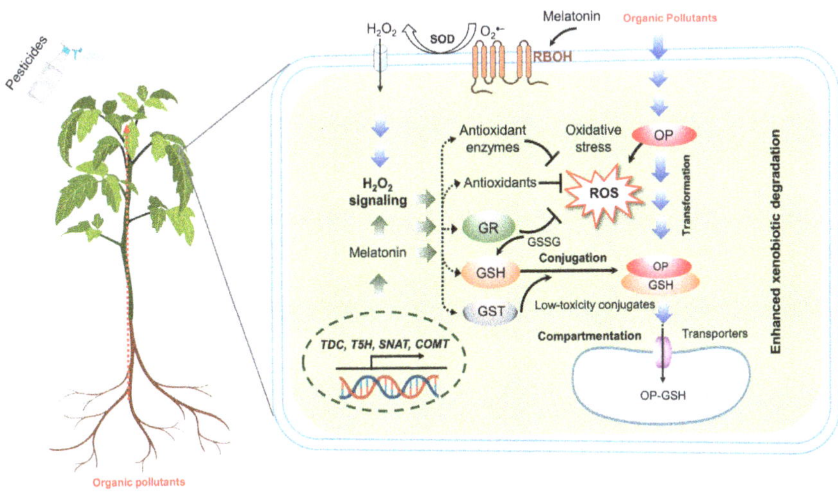

Figure 5. Mechanisms of melatonin–induced organic pollutant detoxification.

5.3. Melatonin-Induced Reduction in Pesticide Residue in Postharvest Horticultural Management

The malpractice of treating harvested fruit with pesticides to prevent fungal diseases is common in postharvest horticultural management. Although this practice can increase the shelf life of fruit, pesticide residue can harm human health [10]. Interestingly, pesticides in postharvest fruit can be degraded by melatonin treatment [75]. For instance, exogenous melatonin application can accelerate the degradation of chlorothalonil, malathion, and glyphosate in postharvest jujube fruit; however, the efficacy of melatonin-promoted pesticide degradation was significantly blunted by the administration of CPA and GSH biosynthesis inhibitor L-buthionine-sulfoximine [10]. This implies that melatonin enhances GSH-dependent detoxification, hence promoting xenobiotic metabolism in plant organs [37]. Melatonin also prolonged pesticide-delayed fruit senescence, as evidenced by increased fruit firmness and decreased weight loss and decay incidence [10].

Similar to foliar treatment, root-sourced melatonin promotes pesticide detoxification in leaves [38]. Melatonin administration increased the activity of the enzyme GST and transcripts of *GST1*, *GST2*, and *GST3*, leading to the accelerated degradation of imidacloprid [40]. Moreover, melatonin treatment improved the AsA/DHA and GSH/GSSG ratios, as well as the activity of AsA-GSH cycle enzymes, showing that melatonin might reduce imidacloprid-induced oxidative stress in cucumber via modulating the AsA-GSH cycle. In addition to in vivo detoxification, melatonin promotes soil bacterial population and the activity of dehydrogenase and peroxidase in soil polluted with PAHs, which potentially resulted in the maximum PAH removal rate, suggesting that melatonin played a beneficial role in increasing plant biomass and elevating the soil bacterial population that favored the degradation of the selected PAHs (phenanthrene and pyrene) [39].

6. Conclusions and Future Perspectives

Despite the innate ability of plants to take in and detoxify organic pollutants from environments, the accumulation of organic pollutants in plant tissue has been shown to affect plant growth and development. Most mechanistic investigations supporting organic pollutant degradation have been conducted in vitro in a chemical rather than physiological context, thus limiting our ability to comprehend the mechanisms by which plants actually degrade organic pollutants in vivo. Previous research revealed that melatonin acts as a superb biostimulator, helping in the degradation of different types of organic pollutants such as pesticides, herbicides, and BPA. Moreover, there is a close relationship between endogenous melatonin levels and organic pollutant metabolism in plants. Melatonin triggers apoplastic ROS signaling, which eventually activates antioxidant and detoxification systems to mitigate oxidative stress and pollutant metabolism in plants (Figure 5). Among different kinds of stress, organic pollutant categories are the least investigated with regard to the melatonin effect, and thus additional research into melatonin function in the stimulant category is warranted for future consideration. As melatonin has great potential for the detoxification of a broad variety of organic pollutants, future remediation technology is expected to benefit from the ongoing effort to maximize the effectiveness of melatonin in xenobiotic metabolism.

Most studies concerning organic pollutant stress have primarily investigated organic pollutant accumulation in plant tissue and subsequent phytotoxicity, wherein less attention has been paid to the molecular mechanism underlying plant tolerance to organic pollutant stress. To comprehend plant uptake, storage, and transport of organic pollutants, however, relevant knowledge is necessary. Moreover, studies revealing the melatonin effects on plant tolerance to organic pollutants have been based on exogenous application or the chemical genetic approach. To elucidate the metabolism of organic pollutants in plants and ensure food safety, functional genomic approaches have to be used. The safe cultivation of horticultural plants in areas polluted by organic pollutants is an issue that calls for researchers from the disciplines of plant physiology, molecular biology, and environmental science to work together.

The degradation of organic pollutants is closely associated with the environmental factors and activity of living organisms including plants and microbes. Moreover, endogenous melatonin biosynthesis and exogenous melatonin actions are affected by abiotic factors such as temperature and light conditions. Thus, environmental factors should be taken into consideration when exploring the role of melatonin in organic pollutant detoxification. The putative ability of melatonin to increase plant resistance to organic pollutants and decrease organic pollutant residue might provide a novel strategy to secure horticultural production. However, further research employing cutting-edge molecular techniques and mutant plants is necessary to fully comprehend the mechanisms of melatonin-induced resistance to organic pollutants.

Author Contributions: G.J.A.: conceptualization, writing—original draft, writing—review, and editing, funding acquisition, project administration. X.L.: writing—review, and editing, funding acquisition. All authors have read and agreed to the published version of the manuscript.

Funding: Research in the authors' laboratories was supported by the Ministry of Science and Technology of the People's Republic of China (DL2022026004L, QNJ2021026001, QNJ20200226001); the Zhejiang Provincial Natural Science Foundation of China (LR22C160002); the Innovation Project of the Chinese Academy of Agricultural Sciences (CAAS-ASTIP-2019-TRICAAS); the National Natural Science Foundation of China (31950410555); and the Henan University of Science and Technology Research Start-up Funds for New Faculty (13480058, 13480070).

Data Availability Statement: Not applicable.

Acknowledgments: The authors apologize to the many authors whose important work could not be referred to.

Conflicts of Interest: The authors declare no conflict of interest. The funders had no role in the design of the study; in the collection, analyses, or interpretation of data; in the writing of the manuscript; or in the decision to publish the results.

References

1. Abdul Mutalib, A.A.; Jaafar, N.F. ZnO photocatalysts applications in abating the organic pollutant contamination: A mini review. *Total Environ. Res. Themes* **2022**, *3–4*, 100013. [CrossRef]
2. Zhao, W.; Teng, M.; Zhang, J.; Wang, K.; Zhang, J.; Xu, Y.; Wang, C. Insights into the mechanisms of organic pollutant toxicity to earthworms: Advances and perspectives. *Environ. Pollut.* **2022**, *303*, 119120. [CrossRef] [PubMed]
3. Wang, S.; Wang, Y.; He, X.; Lu, Q. Degradation or humification: Rethinking strategies to attenuate organic pollutants. *Trends Biotechnol.* **2022**, *40*, 1061–1072. [CrossRef]
4. Pathak, V.M.; Verma, V.K.; Rawat, B.S.; Kaur, B.; Babu, N.; Sharma, A.; Dewali, S.; Yadav, M.; Kumari, R.; Singh, S.; et al. Current status of pesticide effects on environment, human health and it's eco-friendly management as bioremediation: A comprehensive review. *Front. Microbiol.* **2022**, *13*, 962619. [CrossRef] [PubMed]
5. Tounsadi, H.; Metarfi, Y.; Taleb, M.; El Rhazi, K.; Rais, Z. Impact of chemical substances used in textile industry on the employee's health: Epidemiological study. *Ecotoxicol. Environ. Saf.* **2020**, *197*, 110594. [CrossRef]
6. Fujita, K.; Inui, H. How does the Cucurbitaceae family take up organic pollutants (POPs, PAHs, and PPCPs)? *Rev. Environ. Sci. Bio/Technol.* **2021**, *20*, 751–779. [CrossRef]
7. Kanwar, M.K.; Xie, D.; Yang, C.; Ahammed, G.J.; Qi, Z.; Hasan, M.K.; Reiter, R.J.; Yu, J.Q.; Zhou, J. Melatonin promotes metabolism of bisphenol A by enhancing glutathione-dependent detoxification in *Solanum lycopersicum* L. *J. Hazard. Mater.* **2020**, *388*, 121727. [CrossRef]
8. Zhang, J.J.; Yang, H. Metabolism and detoxification of pesticides in plants. *Sci. Total Environ.* **2021**, *790*, 148034. [CrossRef]
9. Zhu, M.; Tang, J.; Shi, T.; Ma, X.; Wang, Y.; Wu, X.; Li, H.; Hua, R. Uptake, translocation and metabolism of imidacloprid loaded within fluorescent mesoporous silica nanoparticles in tomato (*Solanum lycopersicum*). *Ecotoxicol. Environ. Saf.* **2022**, *232*, 113243. [CrossRef]
10. Deng, B.; Xia, C.; Tian, S.; Shi, H. Melatonin reduces pesticide residue, delays senescence, and improves antioxidant nutrient accumulation in postharvest jujube fruit. *Postharvest Biol. Technol.* **2021**, *173*, 111419. [CrossRef]
11. Zhang, J.J.; Yang, H. Advance in methodology and strategies to unveil metabolic mechanisms of pesticide residues in food crops. *J. Agric. Food Chem.* **2021**, *69*, 2658–2667. [CrossRef] [PubMed]
12. Bose, S.; Kumar, P.S.; Vo, D.N. A review on the microbial degradation of chlorpyrifos and its metabolite TCP. *Chemosphere* **2021**, *283*, 131447. [CrossRef] [PubMed]
13. Peuke, A.D.; Rennenberg, H. Phytoremediation. *EMBO Rep.* **2005**, *6*, 497–501. [CrossRef]
14. Yu, G.B.; Chen, R.N.; Chen, Q.S.; Chen, F.Q.; Liu, H.L.; Ren, C.Y.; Zhang, Y.X.; Yang, F.J.; Wei, J.P. Jasmonic acid promotes glutathione assisted degradation of chlorothalonil during tomato growth. *Ecotoxicol. Environ. Saf.* **2022**, *233*, 113296. [CrossRef] [PubMed]
15. De Freitas-Silva, L.; Araujo, H.H.; Meireles, C.S.; da Silva, L.C. Plant exposure to glyphosate-based herbicides and how this might alter plant physiological and structural processes. *Botany* **2022**, *100*, 473–480. [CrossRef]
16. Wang, K.; Xing, Q.; Ahammed, G.J.; Zhou, J. Functions and prospects of melatonin in plant growth, yield and quality. *J. Exp. Bot.* **2022**, *73*, 5928–5946. [CrossRef]
17. Wang, Y.; Cheng, P.; Zhao, G.; Li, L.; Shen, W. Phytomelatonin and gasotransmitters: A crucial combination for plant physiological functions. *J. Exp. Bot.* **2022**, *73*, 5851–5862. [CrossRef]
18. Wei, J.; Li, D.-X.; Zhang, J.-R.; Shan, C.; Rengel, Z.; Song, Z.-B.; Chen, Q. Phytomelatonin receptor PMTR1-mediated signaling regulates stomatal closure inArabidopsis thaliana. *J. Pineal Res.* **2018**, *65*, e12500. [CrossRef]
19. Huangfu, L.; Chen, R.; Lu, Y.; Zhang, E.; Miao, J.; Zuo, Z.; Zhao, Y.; Zhu, M.; Zhang, Z.; Li, P.; et al. OsCOMT, encoding a caffeic acid O-methyltransferase in melatonin biosynthesis, increases rice grain yield through dual regulation of leaf senescence and vascular development. *Plant Biotechnol. J.* **2022**, *20*, 1122–1139. [CrossRef]
20. Chen, Q.; Pu, X.; Li, X.; Li, R.; Yang, Q.; Wang, X.; Guan, M.; Rengel, Z. Secrets of phytomelatonin: Possible roles in darkness. *J. Exp. Bot.* **2022**, *73*, 5828–5839. [CrossRef]

21. García-Sánchez, S.; Cano, A.; Hernández-Ruiz, J.; Arnao, M.B. Effects of temperature and light on the germination-promoting activity by melatonin in almond seeds without stratification. *Agronomy* **2022**, *12*, 2070. [CrossRef]
22. Arnao, M.; Hernández-Ruiz, J. Melatonin and reactive oxygen and nitrogen species: A model for the plant redox network. *Melatonin Res.* **2019**, *2*, 152–168. [CrossRef]
23. Zeng, W.; Mostafa, S.; Lu, Z.; Jin, B. Melatonin-mediated abiotic stress tolerance in plants. *Front. Plant Sci.* **2022**, *13*, 847175. [CrossRef]
24. Arnao, M.B.; Cano, A.; Hernandez-Ruiz, J. Phytomelatonin: An unexpected molecule with amazing performances in plants. *J. Exp. Bot.* **2022**, *73*, 5779–5800. [CrossRef] [PubMed]
25. Cai, S.Y.; Zhang, Y.; Xu, Y.P.; Qi, Z.Y.; Li, M.Q.; Ahammed, G.J.; Xia, X.J.; Shi, K.; Zhou, Y.H.; Reiter, R.J.; et al. HsfA1a upregulates melatonin biosynthesis to confer cadmium tolerance in tomato plants. *J. Pineal Res.* **2017**, *62*, e12387. [CrossRef]
26. Zhang, T.; Wang, J.; Sun, Y.; Zhang, L.; Zheng, S. Versatile roles of melatonin in growth and stress tolerance in plants. *J. Plant Growth Regul.* **2021**, *41*, 507–523. [CrossRef]
27. Li, R.; Yang, R.; Zheng, W.; Wu, L.; Zhang, C.; Zhang, H. Melatonin promotes SGT1-involved signals to ameliorate drought stress adaption in rice. *Int. J. Mol. Sci.* **2022**, *23*, 599. [CrossRef] [PubMed]
28. Gu, Q.; Wang, C.; Xiao, Q.; Chen, Z.; Han, Y. Melatonin confers plant cadmium tolerance: An update. *Int. J. Mol. Sci.* **2021**, *22*, 11704. [CrossRef] [PubMed]
29. Moustafa-Farag, M.; Elkelish, A.; Dafea, M.; Khan, M.; Arnao, M.B.; Abdelhamid, M.T.; El-Ezz, A.A.; Almoneafy, A.; Mahmoud, A.; Awad, M.; et al. Role of melatonin in plant tolerance to soil stressors: Salinity, ph and heavy metals. *Molecules* **2020**, *25*, 5359. [CrossRef]
30. Ahammed, G.J.; Xu, W.; Liu, A.; Chen, S. Endogenous melatonin deficiency aggravates high temperature-induced oxidative stress in *Solanum lycopersicum* L. *Environ. Exp. Bot.* **2019**, *161*, 303–311. [CrossRef]
31. Li, H.; Guo, Y.; Lan, Z.; Xu, K.; Chang, J.; Ahammed, G.J.; Ma, J.; Wei, C.; Zhang, X. Methyl jasmonate mediates melatonin-induced cold tolerance of grafted watermelon plants. *Hortic. Res.* **2021**, *8*, 57. [CrossRef]
32. Liu, P.; Wu, X.; Gong, B.; Lü, G.; Li, J.; Gao, H. Review of the mechanisms by which transcription factors and exogenous substances regulate ROS metabolism under abiotic stress. *Antioxidants* **2022**, *11*, 2106. [CrossRef] [PubMed]
33. Feng, X.; Wang, M.; Zhao, Y.; Han, P.; Dai, Y. Melatonin from different fruit sources, functional roles, and analytical methods. *Trends Food Sci. Technol.* **2014**, *37*, 21–31. [CrossRef]
34. Siddiqui, M.H.; Alamri, S.; Nasir Khan, M.; Corpas, F.J.; Al-Amri, A.A.; Alsubaie, Q.D.; Ali, H.M.; Kalaji, H.M.; Ahmad, P. Melatonin and calcium function synergistically to promote the resilience through ROS metabolism under arsenic-induced stress. *J. Hazard. Mater.* **2020**, *398*, 122882. [CrossRef] [PubMed]
35. Hassan, M.U.; Mahmood, A.; Awan, M.I.; Maqbool, R.; Aamer, M.; Alhaithloul, H.A.S.; Huang, G.; Skalicky, M.; Brestic, M.; Pandey, S.; et al. Melatonin-induced protection against plant abiotic stress: Mechanisms and prospects. *Front. Plant Sci.* **2022**, *13*, 902694. [CrossRef] [PubMed]
36. Ayyaz, A.; Shahzadi, A.K.; Fatima, S.; Yasin, G.; Zafar, Z.U.; Athar, H.U.R.; Farooq, M.A. Uncovering the role of melatonin in plant stress tolerance. *Theor. Exp. Plant Physiol.* **2022**, *34*, 335–346. [CrossRef]
37. Peng, X.; Wang, N.; Sun, S.; Geng, L.; Guo, N.; Liu, A.; Chen, S.; Ahammed, G.J. Reactive oxygen species signaling is involved in melatonin-induced reduction of chlorothalonil residue in tomato leaves. *J. Hazard. Mater.* **2023**, *443*, 130212. [CrossRef]
38. Yan, Y.; Sun, S.; Zhao, N.; Yang, W.; Shi, Q.; Gong, B. *COMT1* overexpression resulting in increased melatonin biosynthesis contributes to the alleviation of carbendazim phytotoxicity and residues in tomato plants. *Environ. Pollut.* **2019**, *252*, 51–61. [CrossRef]
39. Rostami, S.; Azhdarpoor, A.; Baghapour, M.A.; Dehghani, M.; Samaei, M.R.; Jaskulak, M.; Jafarpour, S.; Samare-Najaf, M. The effects of exogenous application of melatonin on the degradation of polycyclic aromatic hydrocarbons in the rhizosphere of Festuca. *Environ. Pollut.* **2021**, *274*, 116559. [CrossRef]
40. Liu, N.; Li, J.; Lv, J.; Yu, J.; Xie, J.; Wu, Y.; Tang, Z. Melatonin alleviates imidacloprid phytotoxicity to cucumber (*Cucumis sativus* L.) through modulating redox homeostasis in plants and promoting its metabolism by enhancing glutathione dependent detoxification. *Ecotoxicol. Environ. Saf.* **2021**, *217*, 112248. [CrossRef]
41. Popek, E. Environmental Chemical Pollutants. In *Sampling and Analysis of Environmental Chemical Pollutants*; Elsevier: Amsterdam, The Netherlands, 2018; pp. 13–69.
42. Mbachu, A.; Chukwura, E.I.; Amalachukwu, M. Role of microorganisms in the degradation of organic pollutants: A review. *Energy Environ. Eng.* **2020**, *7*, 1–11.
43. Ahammed, G.J.; Ruan, Y.P.; Zhou, J.; Xia, X.J.; Shi, K.; Zhou, Y.H.; Yu, J.Q. Brassinosteroid alleviates polychlorinated biphenyls-induced oxidative stress by enhancing antioxidant enzymes activity in tomato. *Chemosphere* **2013**, *90*, 2645–2653. [CrossRef] [PubMed]
44. Ahammed, G.J.; Li, X.; Xia, X.J.; Shi, K.; Zhou, Y.H.; Yu, J.Q. Enhanced photosynthetic capacity and antioxidant potential mediate brassinosteriod-induced phenanthrene stress tolerance in tomato. *Environ. Pollut.* **2015**, *201*, 58–66. [CrossRef] [PubMed]
45. Lei, M.; Raza, I.; Deeba, F.; Jamil, M.; Naeem, R.; Azizullah, A.; Khattak, B.; Shah, A.; Ali, I.; Jin, Z.; et al. Pesticide-induced physiological, metabolic and ultramorphological alterations in leaves of young maize seedlings. *Pol. J. Environ. Stud.* **2020**, *29*, 2247–2258. [CrossRef]

46. Váňová, L.; Kummerová, M.; Klemš, M.; Zezulka, Š. Fluoranthene influences endogenous abscisic acid level and primary photosynthetic processes in pea (*Pisum sativum* L.) plants in vitro. *Plant Growth Regul.* **2008**, *57*, 39–47. [CrossRef]
47. Ahammed, G.J.; Choudhary, S.P.; Chen, S.; Xia, X.; Shi, K.; Zhou, Y.; Yu, J. Role of brassinosteroids in alleviation of phenanthrene-cadmium co-contamination-induced photosynthetic inhibition and oxidative stress in tomato. *J. Exp. Bot.* **2013**, *64*, 199–213. [CrossRef] [PubMed]
48. Alkio, M.; Tabuchi, T.M.; Wang, X.; Colon-Carmona, A. Stress responses to polycyclic aromatic hydrocarbons in Arabidopsis include growth inhibition and hypersensitive response-like symptoms. *J. Exp. Bot.* **2005**, *56*, 2983–2994. [CrossRef] [PubMed]
49. Ahammed, G.J.; He, B.B.; Qian, X.J.; Zhou, Y.H.; Shi, K.; Zhou, J.; Yu, J.Q.; Xia, X.J. 24-Epibrassinolide alleviates organic pollutants-retarded root elongation by promoting redox homeostasis and secondary metabolism in *Cucumis sativus* L. *Environ. Pollut.* **2017**, *229*, 922–931. [CrossRef] [PubMed]
50. Ahammed, G.J.; Wang, Y.; Mao, Q.; Wu, M.; Yan, Y.; Ren, J.; Wang, X.; Liu, A.; Chen, S. Dopamine alleviates bisphenol A-induced phytotoxicity by enhancing antioxidant and detoxification potential in cucumber. *Environ. Pollut.* **2020**, *259*, 113957. [CrossRef]
51. Jiao, L.; Ding, H.; Wang, L.; Zhou, Q.; Huang, X. Bisphenol A effects on the chlorophyll contents in soybean at different growth stages. *Environ. Pollut.* **2017**, *223*, 426–434. [CrossRef]
52. Ali, I.; Liu, B.; Farooq, M.A.; Islam, F.; Azizullah, A.; Yu, C.; Su, W.; Gan, Y. Toxicological effects of bisphenol A on growth and antioxidant defense system in *Oryza sativa* as revealed by ultrastructure analysis. *Ecotoxicol. Environ. Saf.* **2016**, *124*, 277–284. [CrossRef] [PubMed]
53. Korte, F.; Kvesitadze, G.; Ugrekhelidze, D.; Gordeziani, M.; Khatisashvili, G.; Buadze, O.; Zaalishvili, G.; Coulston, F. Organic toxicants and plants. *Ecotoxicol. Environ. Saf.* **2000**, *47*, 1–26. [CrossRef]
54. Pelaez-Vico, M.A.; Fichman, Y.; Zandalinas, S.I.; Van Breusegem, F.; Karpinski, S.M.; Mittler, R. ROS and redox regulation of cell-to-cell and systemic signaling in plants during stress. *Free Radic. Biol. Med.* **2022**, *193*, 354–362. [CrossRef] [PubMed]
55. Mittler, R.; Zandalinas, S.I.; Fichman, Y.; Van Breusegem, F. Reactive oxygen species signalling in plant stress responses. *Nat. Rev. Mol. Cell Biol.* **2022**, *23*, 663–679. [CrossRef]
56. Yang, S.J.; Huang, B.; Zhao, Y.Q.; Hu, D.; Chen, T.; Ding, C.B.; Chen, Y.E.; Yuan, S.; Yuan, M. Melatonin enhanced the tolerance of *Arabidopsis thaliana* to high light through improving anti-oxidative system and photosynthesis. *Front. Plant Sci.* **2021**, *12*, 752584. [CrossRef]
57. Hasanuzzaman, M.; Bhuyan, M.; Zulfiqar, F.; Raza, A.; Mohsin, S.M.; Mahmud, J.A.; Fujita, M.; Fotopoulos, V. Reactive oxygen species and antioxidant defense in plants under abiotic stress: Revisiting the crucial role of a universal defense regulator. *Antioxidants* **2020**, *9*, 681. [CrossRef] [PubMed]
58. Altaf, M.A.; Shahid, R.; Altaf, M.M.; Kumar, R.; Naz, S.; Kumar, A.; Alam, P.; Tiwari, R.K.; Lal, M.K.; Ahmad, P. Melatonin: First-line soldier in tomato under abiotic stress current and future perspective. *Plant Physiol. Biochem.* **2022**, *185*, 188–197. [CrossRef] [PubMed]
59. Zhao, C.; Nawaz, G.; Cao, Q.; Xu, T. Melatonin is a potential target for improving horticultural crop resistance to abiotic stress. *Sci. Hortic.* **2022**, *291*, 110560. [CrossRef]
60. Hasan, M.K.; Ahammed, G.J.; Sun, S.; Li, M.; Yin, H.; Zhou, J. Melatonin inhibits cadmium translocation and enhances plant tolerance by regulating sulfur uptake and assimilation in *Solanum lycopersicum* L. *J. Agric. Food Chem.* **2019**, *67*, 10563–10576. [CrossRef]
61. Hasan, M.K.; Liu, C.; Wang, F.; Ahammed, G.J.; Zhou, J.; Xu, M.X.; Yu, J.Q.; Xia, X.J. Glutathione-mediated regulation of nitric oxide, S-nitrosothiol and redox homeostasis confers cadmium tolerance by inducing transcription factors and stress response genes in tomato. *Chemosphere* **2016**, *161*, 536–545. [CrossRef]
62. Yu, G.B.; Zhang, Y.; Ahammed, G.J.; Xia, X.J.; Mao, W.H.; Shi, K.; Zhou, Y.H.; Yu, J.Q. Glutathione biosynthesis and regeneration play an important role in the metabolism of chlorothalonil in tomato. *Chemosphere* **2013**, *90*, 2563–2570. [CrossRef]
63. Sharma, A.; Kumar, V.; Kumar, R.; Shahzad, B.; Thukral, A.K.; Bhardwaj, R.; Tejada Moral, M. Brassinosteroid-mediated pesticide detoxification in plants: A mini-review. *Cogent Food Agric.* **2018**, *4*, 1436212. [CrossRef]
64. Tan, D.X.; Reiter, R.J. An evolutionary view of melatonin synthesis and metabolism related to its biological functions in plants. *J. Exp. Bot.* **2020**, *71*, 4677–4689. [CrossRef] [PubMed]
65. Mannino, G.; Pernici, C.; Serio, G.; Gentile, C.; Bertea, C.M. Melatonin and phytomelatonin: Chemistry, biosynthesis, metabolism, distribution and bioactivity in plants and animals-An overview. *Int. J. Mol. Sci.* **2021**, *22*, 9996. [CrossRef] [PubMed]
66. Chang, J.; Guo, Y.; Yan, J.; Zhang, Z.; Yuan, L.; Wei, C.; Zhang, Y.; Ma, J.; Yang, J.; Zhang, X.; et al. The role of watermelon caffeic acid O-methyltransferase (ClCOMT1) in melatonin biosynthesis and abiotic stress tolerance. *Hortic. Res.* **2021**, *8*, 210. [CrossRef] [PubMed]
67. Liu, D.D.; Sun, X.S.; Liu, L.; Shi, H.D.; Chen, S.Y.; Zhao, D.K. Overexpression of the melatonin synthesis-related gene *SlCOMT1* improves the resistance of tomato to salt Stress. *Molecules* **2019**, *24*, 1514. [CrossRef]
68. Arnao, M.B.; Hernández-Ruiz, J. Phytomelatonin versus synthetic melatonin in cancer treatments. *Biomed. Res. Clin. Pract.* **2018**, *3*, 1–6. [CrossRef]
69. Contente, M.L.; Farris, S.; Tamborini, L.; Molinari, F.; Paradisi, F. Flow-based enzymatic synthesis of melatonin and other high value tryptamine derivatives: A five-minute intensified process. *Green Chem.* **2019**, *21*, 3263–3266. [CrossRef]
70. Zhang, J.; Yan, X.; Tian, Y.; Li, W.; Wang, H.; Li, Q.; Li, Y.; Li, Z.; Wu, T. Synthesis of a new water-soluble melatonin derivative with low toxicity and a strong effect on sleep aid. *ACS Omega* **2020**, *5*, 6494–6499. [CrossRef]

71. Perez-Llamas, F.; Hernandez-Ruiz, J.; Cuesta, A.; Zamora, S.; Arnao, M.B. Development of a phytomelatonin-rich extract from cultured plants with excellent biochemical and functional properties as an alternative to synthetic melatonin. *Antioxidants* **2020**, *9*, 158. [CrossRef] [PubMed]
72. Tijero, V.; Munoz, P.; Munne-Bosch, S. Melatonin as an inhibitor of sweet cherries ripening in orchard trees. *Plant Physiol. Biochem.* **2019**, *140*, 88–95. [CrossRef] [PubMed]
73. Sun, H.; Cao, X.; Wang, X.; Zhang, W.; Li, W.; Wang, X.; Liu, S.; Lyu, D. RBOH-dependent hydrogen peroxide signaling mediatesmelatonin-induced anthocyanin biosynthesis in red pear fruit. *Plant Sci.* **2021**, *313*, 111093. [CrossRef]
74. Moreno, J.E.; Campos, M.L. Waking up for defense! Melatonin as a regulator of stomatal immunity in plants. *Plant Physiol.* **2022**, *188*, 14–15. [CrossRef] [PubMed]
75. Aghdam, M.S.; Mukherjee, S.; Flores, F.B.; Arnao, M.B.; Luo, Z.; Corpas, F.J. Functions of melatonin during postharvest of horticultural crops. *Plant Cell Physiol.* **2021**. [CrossRef] [PubMed]
76. Arnao, M.B.; Hernandez-Ruiz, J. Melatonin as a plant biostimulant in crops and during post-harvest: A new approach is needed. *J. Sci. Food Agric.* **2021**, *101*, 5297–5304. [CrossRef]
77. Xu, T.; Chen, Y.; Kang, H. Melatonin is a potential target for improving post-harvest preservation of fruits and vegetables. *Front. Plant Sci.* **2019**, *10*, 1388. [CrossRef]
78. Ahanger, M.A.; Bhat, J.A.; Siddiqui, M.H.; Rinklebe, J.; Ahmad, P. Integration of silicon and secondary metabolites in plants: A significant association in stress tolerance. *J. Exp. Bot.* **2020**, *71*, 6758–6774. [CrossRef]
79. Gu, Q.; Chen, Z.; Yu, X.; Cui, W.; Pan, J.; Zhao, G.; Xu, S.; Wang, R.; Shen, W. Melatonin confers plant tolerance against cadmium stress via the decrease of cadmium accumulation and reestablishment of microRNA-mediated redox homeostasis. *Plant Sci.* **2017**, *261*, 28–37. [CrossRef]
80. Debnath, B.; Islam, W.; Li, M.; Sun, Y.; Lu, X.; Mitra, S.; Hussain, M.; Liu, S.; Qiu, D. Melatonin mediates enhancement of stress tolerance in plants. *Int. J. Mol. Sci.* **2019**, *20*, 1040. [CrossRef]
81. Hoque, M.N.; Tahjib-Ul-Arif, M.; Hannan, A.; Sultana, N.; Akhter, S.; Hasanuzzaman, M.; Akter, F.; Hossain, M.S.; Sayed, M.A.; Hasan, M.T.; et al. Melatonin modulates plant tolerance to heavy metal stress: Morphological responses to molecular mechanisms. *Int. J. Mol. Sci.* **2021**, *22*, 11445. [CrossRef]
82. Nawaz, K.; Chaudhary, R.; Sarwar, A.; Ahmad, B.; Gul, A.; Hano, C.; Abbasi, B.H.; Anjum, S. Melatonin as master regulator in plant growth, development and stress alleviator for sustainable agricultural production: Current status and future perspectives. *Sustainability* **2020**, *13*, 294. [CrossRef]
83. Mukherjee, S. Insights into nitric oxide-melatonin crosstalk and N-nitrosomelatonin functioning in plants. *J. Exp. Bot.* **2019**, *70*, 6035–6047. [CrossRef] [PubMed]
84. Guo, Y.; Yan, J.; Su, Z.; Chang, J.; Yang, J.; Wei, C.; Zhang, Y.; Ma, J.; Zhang, X.; Li, H. Abscisic acid mediates grafting-induced cold tolerance of watermelon via interaction with melatonin and methyl jasmonate. *Front. Plant Sci.* **2021**, *12*, 785317. [CrossRef] [PubMed]
85. Li, H.; He, J.; Yang, X.; Li, X.; Luo, D.; Wei, C.; Ma, J.; Zhang, Y.; Yang, J.; Zhang, X. Glutathione-dependent induction of local and systemic defense against oxidative stress by exogenous melatonin in cucumber (*Cucumis sativus* L.). *J. Pineal Res.* **2016**, *60*, 206–216. [CrossRef] [PubMed]
86. Acosta, M.G.; Cano, A.; Hernández-Ruiz, J.; Arnao, M.B. Melatonin as a possible natural safener in crops. *Plants* **2022**, *11*, 890. [CrossRef]
87. Zhou, Y.; Xia, X.; Yu, G.; Wang, J.; Wu, J.; Wang, M.; Yang, Y.; Shi, K.; Yu, Y.; Chen, Z.; et al. Brassinosteroids play a critical role in the regulation of pesticide metabolism in crop plants. *Sci. Rep.* **2015**, *5*, 9018. [CrossRef] [PubMed]

horticulturae

Article

The Effect of Salinity and Drought on the Essential Oil Yield and Quality of Various Plant Species of the Lamiaceae Family (*Mentha spicata* L., *Origanum dictamnus* L., *Origanum onites* L.)

Michalis K. Stefanakis [1,2], Anastasia E. Giannakoula [1,*], Georgia Ouzounidou [3], Charikleia Papaioannou [4], Vaia Lianopoulou [1] and Eleni Philotheou-Panou [1]

[1] Laboratory of Plant Physiology, Department of Agriculture, International Hellenic University, 54700 Sindos, Greece; michstefanakis@yahoo.gr (M.K.S.); vaia_l@hotmail.com (V.L.); epanoufilotheou@gmail.com (E.P.-P.)
[2] Department of Chemistry, University of Crete, Voutes, 71003 Heraklion, Greece
[3] Hellenic Agricultural Organization-Demeter, Institute of Food Technology, 1 S. Venizelou Str., 14123 Lycovrissi, Greece; geouz@nagref.gr
[4] Laboratory of Genetics, Department of Biology, University of Patras, 26504 Patras, Greece; xpapaioannou@upatras.gr
* Correspondence: agianna@ihu.gr; Tel.: +30-6937322511

Abstract: *Mentha spicata* L., *Origanum dictamnus* L., and *Origanum onites* L. are aromatic plants that produce very important essential oils. They are considered model plants with beneficial health properties due to their antioxidant content. Enhancing the yield while maintaining the quality of essential oil is of significant commercial importance. Salinization and drought cause various effects on the yield and quality of the bioactive constituents in essential oil. By assessing the response of these plants and their secondary metabolites accumulation to different salt stress and irrigation levels, this study aims to gain insights into how plants adapt to and cope with salinity and drought. A pot experiment was conducted in the spring of 2020 to assess the effect of salinity and drought stress on the growth and essential oils content of the three aromatic plant species mentioned above. The soil mixture used was perlite and peat in a ratio of 1:1:6, while four salinity treatments (25, 50, 100, and 150 mM NaCl) and two levels of irrigation were applied (100% and 50%). Salinity significantly affects total chlorophyll concentration especially in higher concentrations (100 and 150 mM) in *M. spicata* plants, especially under 50% soil water irrigation. Under the same conditions, *M. spicata* contained the higher proline concentration, which was significantly greater than that in *O. dictamnus* and *O. onites*. Similar variations of malondialdehyde and hydrogen hyperoxide were revealed among the three species, with significantly higher values in *M. spicata* when subjected to both excess salinity and drought conditions. The major compounds identified in *M. spicata* were carvone, in *O. dictamnus* carvacrol, and *p*-cymene and in *O. onites* carvacrol. It is important to highlight that *O. onites* had the highest concentration of essential oil, and that the concentration increased with the increase of NaCl. This suggests that the presence of NaCl in the soil may have a stimulating effect on the production of essential oil in *O. onites*. However, it is plausible that the stress caused by NaCl triggers a physiological response in *O. onites*, leading to increased production of essential oil. This could be a protective mechanism to enhance the plant's resistance to the stressor. Overall, *O. onites* and *O. dictamnus* appeared to be more resistant to these stress conditions than *M. spicata*, since they maintained their growth and essential oil quality indicators at higher levels. These two species possess mechanisms that prevent or minimize lipid peroxidation, thus protecting their cell membranes and maintaining their ultrastructure integrity.

Keywords: proline; secondary metabolites; lipid peroxidation; essential oils; aromatic plants

1. Introduction

The cultivation of aromatic plants aims to exploit the parts that contain active compounds that are used in various ways, especially for medicinal purposes [1]. The essential oils (EOs) that act as antiparasitics or disinfectants display a broad range of antioxidant, antimicrobial, antiseptic, antifungal, antibacterial, cytotoxic/anticancer, antigenotoxic, and antiviral activities [1]. The morphological and chemical diversity of the genus *Origanum* is extraordinary. Out of 67 taxa, nine species thrive in Greece, including 18 hybrids, the majority of which are found throughout the Mediterranean region [2]. The classification of *Origanum* is quite intricate. It is difficult, for many species, to define sectional boundaries, making their morphological markers inefficient for differentiating them [3]. Genetic, environmental, physiological, and agronomic variables affect the volatile oils' chemical compositions [4,5]. Carvacrol is the main essential oil component in oregano herbs, responsible for the characteristic "oregano" scent. Other volatile compounds that dominate the essential oils of oregano species are thymol, *p*-cymene, and γ-terpinene, whereas in *O. onites* the dominant compound is borneol [1,6–9]. The genus *Mentha* includes more than 30 species of herbaceous perennial plants. *Mentha spicata* (*M. spicata*) spreads mainly in the temperate and subtemperate zones of the world. It is considered a rich source of essential oils, which are widely used in the pharmaceutical industry and in food production [10]. The major compounds found in the essential oil of *M. spicata* are carvone, limonene, and 1,8-cineole. These findings are consistent with literature data [1,11].

Changes in climatic conditions lead to increased biotic and abiotic stress for the plants. The productivity of aromatic plants was affected globally due to these stresses [12]. Plant growth and essential oil production are influenced by various environmental factors, drought and salinity stresses being two of them. Other factors such as geographical area of cultivation, cultivation cycle, cultivation conditions, harvest year, cultivars or varieties, and age, can also influence essential oil production and its composition [13].

Saline soil has a negative impact on plants, leading to physiological and metabolic disturbances. This affects various aspects of plant development, growth, yield, and the quality of essential oils, especially in aromatic plants. Salinity can hinder seed germination, reduce survival percentage, alter morphological characteristics, and decrease overall EO yield and its components [14]. The discovery of high-yielding genotypes of these plants is particularly promising, as numerous studies have documented the response of aromatic plants to salinity stress [15–18]. The tolerance to salinity depends on the interaction between salinity and other environmental factors (such as drought stress) [19]. Salt stress is a significant challenge for aromatic plants and can have various negative effects on their growth and productivity. One of the primary impacts of salt stress is a decrease in osmotic potential, leading to reduced availability of water for plants. This can result in water stress and negatively affect plant growth. In addition to water stress, salt stress also affects the physical structure of the soil. It diminishes water permeability and soil aeration, further hindering plant growth.

Drought stress significantly reduces plant yield and modifies the polyphenol content as well as the antioxidant capacities of plants [20]. Its impact on the amount of essential oils, antioxidant properties, and polyphenol content, however, is also a subject of debate. Despite reports of increased essential oil content and improved antioxidant activity during drought treatment, the scientific data were insufficient to make any broader conclusions, according to Kren et al. (2012) and Szabo et al. (2020) [21,22]. Thus, further studies to determine the optimal water supply for different species are paramount.

Overall, salt stress poses a significant obstacle to agricultural production. The aim of this study is to understand the adverse effects of salt stress in combination with drought in specific Lamiaceae species (*M. spicata, O. dictamnus, O. onites*), in order to develop strategies to mitigate its impact and improve plant productivity in these challenging environments. We chose these three species of the Lamiaceae family as they are endemic to the collection areas while these areas are at risk of salinity and drought due to the intensive use of fertilizers and climate change.

2. Materials and Methods

2.1. Plant Material and Experimental Design

Origanum dictamus (dittany) plants were transferred from the primary material of the area of Messara (Coordinates; 34°57′57″ N, 24°50′26″ E). *Origanum onites* (oregano) plants were collected from the seed of a mother plantation in Agiasos of Lesvos, (Coordinates; 39°04′77″ N, 26°22′06″ E). *Mentha spicata* (spearmint) was transferred as rooted cuttings from plants of the Stylia area in Corinthia (Coordinates; 37°59′54″ N, 22°33′37″ E).

The experiment was established in spring of 2020 with the above plants of three-year growth set up in pots in a greenhouse. The soil substrate consisted of a mixture of soil, perlite, and peat in a ratio of 1:1:6. Fertilization was applied to the plants on 10 November 2020 with 8 units of N, 8 units of P, and 13 units of K in the following forms: 65 g NH_4NO_3, 60 g KH_2PO_4, and 70 g KNO_3. On 22 March 2021, four salinity treatments were chosen after preliminary experiments (25 mM NaCl, 50 mM NaCl, 100 mM NaCl, and 150 mM NaCl); per treatment, 1.462 g NaCl (*w*/*v*), 2.926 NaCl, 5.85 NaCl, and 8.78 g NaCl were dissolved and each pot was irrigated with 0.39 g NaCl per 0.5 l water. The experimental design was completely randomized with four replicates.

Every two days, the soil moisture was measured using a Theta probe ML2, and watering schedules were adjusted to keep the soil moisture levels at 50% of the water capacity. Water was applied at a 50% capacity to all plants with the exception of the control plants. A tape measure was used to measure the height of the plants. Harvesting was performed at 8 cm height using a sickle. The plants were then dried in an airy shed.

2.2. Chlorophyll Estimation

After sixty days from the beginning of the treatment with salt, a sample of a third of the top completely grown leaf was taken for chlorophyll analysis. Fresh leaf blade material (0.1 g) was put in 25 mL glass test tubes, and 15 mL of 96% (*v*/*v*) ethanol was poured into each tube in order to estimate the amount of chlorophylls. The plant material-filled tubes were kept in an incubator set at 79.8 °C for three to four hours, or until the samples had completely discolored. The measurement of chlorophyll a and b absorption was conducted at 665 and 649 nm, correspondingly. The total chlorophyll calculation method was applied according to Wintermans and Mots (1988) [23].

2.3. Determination of Proline

Each plot's fully grown leaf samples were sliced into small pieces. Roughly 0.3 g of these samples was weighed and then individually added to glass vials holding 10 mL of 80% (*v*/*v*) ethanol. After 30 min of heating at 60 °C, the extracts were filtered and diluted with 80% (*v*/*v*) ethanol to a volume of 20 milliliters. The acid ninhydrin reagent method was used to quantify the free Pro content in various plant extracts [24]. Then, 500 mM of thick H_2SO_4 was mixed with around one gram of ninhydrin. Next, two mM each of the aqueous alcohol extract and acid ninhydrin were put into test tubes. To reduce evaporation, glass marbles were placed over the test tubes, which were kept at 95 °C for 60 min. They were then allowed to cool to room temperature. Lastly, 4 mL of toluene was added to each sample replicate. After the solution layers were separated, the toluene layer was carefully decanted, put in glass corvettes, and its absorbance at 518 nm was measured.

2.4. Estimation of Lipid Peroxidation and Hydrogen Hyperoxide

At the end of the experimental period (see above), the amount of lipid peroxidation of Lamiaceae leaves was quantified, and the concentration of malondialdehyde (MDA) was determined through reaction with 2-thiobarbituric acid (TBA) [25]. To a sample of 0.1 g of fresh leaf blade tissue, 0.5 mL of 0.1% (*w*/*v*) trichloroacetic acid (TCA) was added to homogenize the samples. For ten minutes, the homogenate was centrifuged at 15,000× *g* and 4 °C. A mixture of 0.5 mL of the supernatant and 1.5 mL of 0.5% TBA diluted in 20% TCA was prepared. Incubation lasted 25 min at 95 °C. Further incubation in an ice bath brought the process to a halt. Following a 10 min centrifugation at 10,000× *g* and

4 °C, the absorbance of the supernatant was measured at 532 and 600 nm. The value at 532 nm was used as the substrate for the non-specific absorption value at 600 nm. Using the MDA extinction coefficient of 155 mM^{-1} cm^{-1} and Lambert–Beer's equation, the concentration of MDA was determined [26]. The findings are displayed as μmol MDA g^{-1} FW. Following the literature procedure, the H_2O_2 concentration of leaves was determined [27,28]. Then, 500 mg of fresh leaf material was homogenized in 3 mL of 1% TCA. Following centrifugation, 1.5 mL of 1 M KI and 10 mM K buffer (0.75 mL) were added to 0.75 mL of the filtrate, and the absorbance of each sample was measured at 390 nm.

2.5. Isolation of Essential Oils

For 10 days, the collected plant material was allowed to air dry at room temperature in the shade and darkness. In order to minimize hydrodistillation overheating artifacts, each sample was subjected to hydrodistillation three times for two hours using a modified Clevenger-type apparatus with a water-cooled oil receiver. Following normal protocols, the volatiles were trapped in 5 mL of gas chromatography-grade n-pentane [29]. They were then dried over anhydrous sodium sulfate and stored in closed, airtight Pyrex containers at −4 °C. The amount of essential oil is stated as mL per 100 g^{-1} d.w.

2.6. GC-MS Analysis of Essential Oils

The GC-MS analysis of the extracted volatile oils was carried out using a Shimadzu GC-17A gas chromatograph (Shimadzu Corporation, Kyoto, Japan) coupled with a Shimadzu GCMS-QP 5050 mass-selective quadrupole mass spectrometer as the detector and the appropriate data system. The fused silica capillary column (Supelco, Merck, München, Germany, SBP-5, with 0.25 μm film thickness, 30 m × 0.25 mm i.d.) was directly linked to the ion source, and the GC was outfitted with a Grob-type split-splitless injector. A carrier gas of helium with a back pressure of 0.8 Atm was employed. The injector temperature was 250 °C and the oven temperature program was started at 50 °C for 5 min and then increased at a rate of 5 °C/min up to 150 °C, was retained at this temperature for 10 min and increased again at a rate of 5 °C/min up to 280 °C, where it remained for 20 min. The range of the scan was 30–700 m/z. With an ionization energy of 70 eV, a GS-MS detection electron ionization system was employed.

2.7. Identification and Quantification of Essential Oils Components

Based on the mass spectrometer's detection of the total number of fragments (total ion count) of the metabolites, the components were quantified. The chemical components were identified by analyzing their mass spectra using the NIST21, NIST107, and PMW_TOX2 mass spectra libraries, and by comparing each component's retention time (Rt) with those of commercially accessible compounds [30]. Additionally, components were identified by contrasting the data with those from the literature [31]. Retention indices were computed using the standard hydrocarbon retention durations (C_9–C_{25}) as a guide, based on the work of Van den Dool and Kratz (1963) [32]. Co-injection with standard substances was also performed when needed.

2.8. Statistical Analysis

The experimental layout included three species and four salt concentrations, and four replicates (plants) per treatment. Data were subjected to analysis of variance (ANOVA). For comparison of means, the Duncan multiple range test was used ($p \leq 0.05$) using the SPSS 24.0 statistical package (SPSS, Inc., Chicago, IL, USA).

3. Results

3.1. Plant Growth

In dittany and oregano, a 100% watering regime and higher NaCl concentrations caused a trend toward an increase in plant height at the end of bloom that was roughly 17% higher than that of the control (Table 1). At the same conditions, no changes in spearmint height were revealed. The application of 50% irrigation and high NaCl concentration induced similar results in plant height in all of the plants. The height of the plants did not differ between salinity treatments, but the dry weight of dittany and oregano increased significantly. The ratio of dry weight to fresh weight did not seem to be affected between salinity treatments.

Table 1. Effect of NaCl in height (cm) and dry weight/fresh weight of different Lamiaceae species *Mentha spicata* L., *Origanum dictamnus* L., and *Origanum onites* L. Each value is the mean of 4 replications ± standard error.

Treatnents	First Month of Treatment 100% Irrigation	At the End of Blooming 100% Irrigation	First Month of Treatment 50% Irrigation	At the End of Blooming 50% Irrigation	DW/FW 100% Irrigation	DW/FW 50% Irrigation
0 mM NaCl						
Spearmint	16.27 ± 1.72	32.00 ± 2.45	16.00 ± 1.30	31.00 ± 2.30	0.68 ± 0.07	0.64 ± 0.07
Dittany	17.20 ± 2.39	23.60 ± 1.14	17.00 ± 2.10	23.00 ± 1.40	0.17 ± 0.05	0.16 ± 0.02
Oregano	14.60 ± 1.82	25.10 ± 0.90	14.50 ± 1.82	24.60 ± 0.89	0.63 ± 0.03	0.70 ± 0.06
25 mM NaCl						
Spearmint	16.17 ± 1.47	33.67 ± 1.86	16.00 ± 1.40	33.00 ± 1.60	0.67 ± 0.02	0.60 ± 0.03
Dittany	15.20 ± 1.31	26.60 ± 3.91	15.00 ± 1.10	26.00 ± 2.10	0.21 ± 0.03	0.20 ± 0.04
Oregano	13.00 ± 0.71	27.80 ± 3.96	12.60 ± 0.60	26.90 ± 2.40	0.64 ± 0.06	0.67 ± 0.07
50 mM NaCl						
Spearmint	15.40 ± 1.79	31.67 ± 3.37	15.00 ± 1.50	32.50 ± 3.20	0.72 ± 0.07	0.67 ± 0.06
Dittany	17.20 ± 2.77	26.20 ± 1.92	17.00 ± 1.60	26.00 ± 1.92	0.22 ± 0.02	0.20 ± 0.03
Oregano	14.60 ± 2.30	27.60 ± 2.07	14.00 ± 1.30	26.40 ± 1.87	0.61 ± 0.09	0.57 ± 0.08
100 mM NaCl						
Spearmint	15.17 ± 0.75	31.17 ± 1.94	15.00 ± 0.50	31.50 ± 1.40	0.70 ± 0.11	0.68 ± 0.09
Dittany	19.60 ± 2.68	26.00 ± 2.60	19.00 ± 1.80	25.40 ± 2.30	0.23 ± 0.04	0.21 ± 0.03
Oregano	16.40 ± 2.05	27.60 ± 2.80	16.20 ± 2.50	26.60 ± 2.10	0.69 ± 0.02	0.60 ± 0.04
150 mM NaCl						
Spearmint	15.67 ± 1.21	32.17 ± 2.9	15.50 ± 1.10	32.00 ± 2.20	0.68 ± 0.17	0.66 ± 0.10
Dittany	17.80 ± 2.68	27.80 ± 2.24	16.00 ± 2.68	27.00 ± 2.35	0.20 ± 0.01	0.18 ± 0.02
Oregano	15.20 ± 2.05	28.80 ± 2.78	14.70 ± 1.50	27.40 ± 2.60	0.70 ± 0.05	0.63 ± 0.08

3.2. *Secondary Metabolites*

3.2.1. Proline and Chlorophyll Content

At the highest concentration (150 mM) of NaCl and simultaneous 50% irrigation, spearmint contained the highest proline concentrations, which were significantly greater compared to those of oregano and dittany species in the same conditions (five times higher than the control), ($p \leq 0.05$). A progressive increase in proline content was revealed in all studied species with increasing dose of NaCl. Spearmint seems to be more susceptible to salinity and drought as it showed a drastic increase even at 50 mM NaCl in both irrigation treatments (by four and four-and-a-half times, respectively). Salinity or drought affected the proline concentration less in dittany and oregano showing an increase two-and-a-half and three times higher under 50% irrigation and 150 mM NaCl, respectively (Figure 1).

The highest chlorophyll contents were found in oregano that was well irrigated without salinity treatment. Salinity significantly affects total chlorophyll concentration in higher concentrations (100 and 150 mM) in spearmint plants, especially under 50% soil water irrigation (Figure 2). In addition, salinity and drought negatively affect the total chlorophyll's concentration in dittany and oregano species but to a lower degree (Figure 2). At the higher concentrations (150 mM) of NaCl and 50% irrigation, spearmint contained the lowest chlorophyll concentrations, which were significantly lower than those of oregano and dittany species ($p \leq 0.05$).

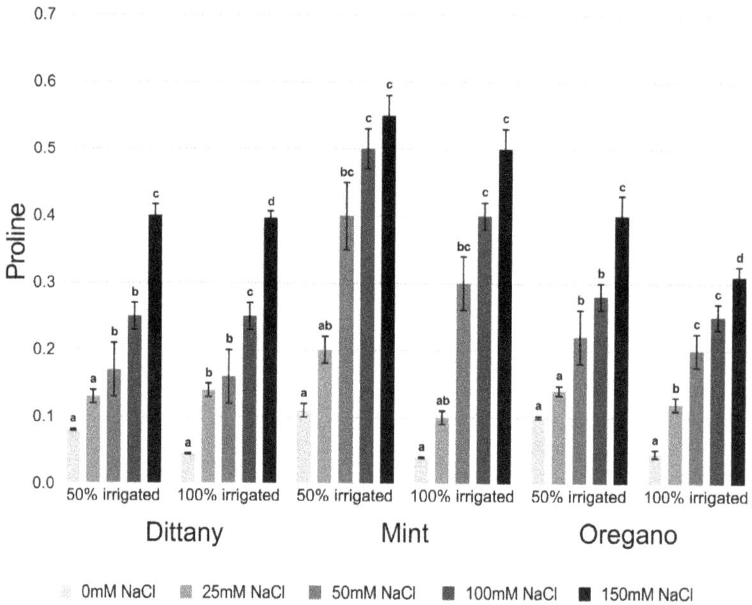

Figure 1. Effect of NaCl on proline (mg g^{-1} FW) content in leaves of different Lamiaceae species *Origanum dictamnus* L., *Mentha spicata* L., and *Origanum onites* L. Each value is the mean of 4 replications ± standard error. Means not sharing the same letter are significantly different at $p \leq 0.05$.

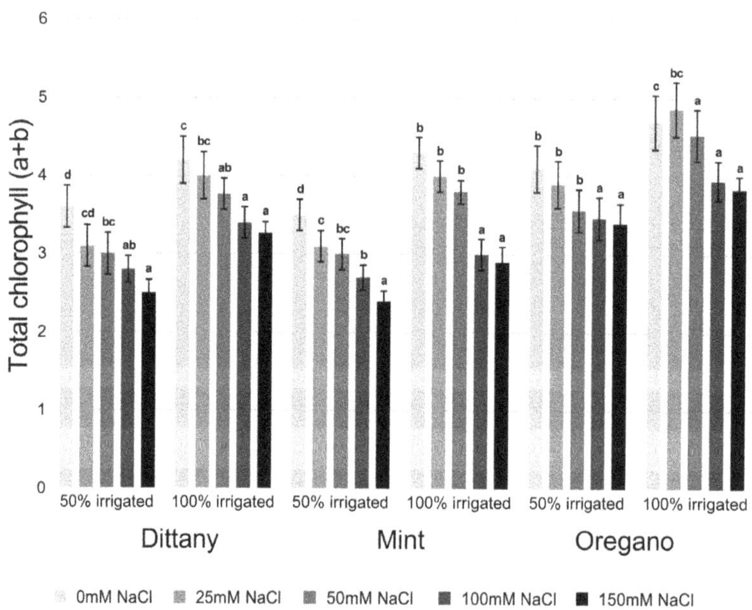

Figure 2. Effect of NaCl on total chlorophyll (a + b) (mg g^{-1} FW) content in leaves of different Lamiaceae species *Origanum dictamnus* L., *Mentha spicata* L., and *Origanum onites* L. Each value is the mean of 4 replications ± standard error. Means not sharing the same letter are significantly different at $p \leq 0.05$.

3.2.2. Lipid Peroxidation and Hydrogen Peroxide Assays

Salinity conditions combined with drought significantly increased the concentration of MDA in all species of the Lamiaceae family, as observed in Figure 3. Concentrations of MDA in salinity treatments, especially in higher concentrations of spearmint plants, were about ten times higher than those in the control plants (Figure 3). Moreover, it was noticed that MDA negatively contributed to spearmint salinity tolerance.

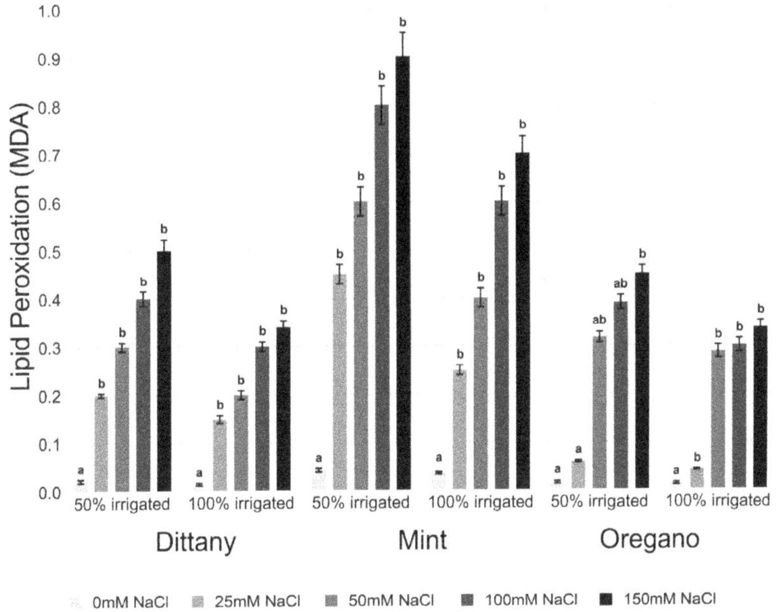

Figure 3. Effect of NaCl on MDA (μmol g^{-1} FW) content in leaves of different Lamiaceae species *Origanum dictamnus* L., *Mentha spicata* L., and *Origanum onites* L. Each value is the mean of 4 replications ± standard error. Means not sharing the same letter are significantly different at $p \leq 0.05$.

A similar trend to that of MDA was observed for H_2O_2 in all species of the Lamiaceae family. The H_2O_2 values in spearmint plants after salinity and 50% irrigation treatments were up to two times higher than those in the dittany and oregano species (Figure 4). This study demonstrated that both MDA and H_2O_2 increased in Lamiaceae plants but, in spearmint plants, after simultaneous salinity treatment and drought stress, the increase was about two times higher than in oregano plants.

3.3. *Essential Oil Content and Main Constituents*

Table 2 presents the content of the volatile fractions of the plants analyzed. It is obvious that oregano had the highest concentration of essential oil, which increased with the increase of NaCl in the soil substrate (by 25% of the control at 150 mM NaCl). In dittany, exposure to 25 mM NaCl caused a significant reduction of oil production by 45% ($p < 0.05$); then, increasing the salinity resulted in an augmentation of the essential oil that was always lower than the control by almost 19% at the higher salt level. In spearmint, it seems that the essential oil level was less affected by the presence of NaCl.

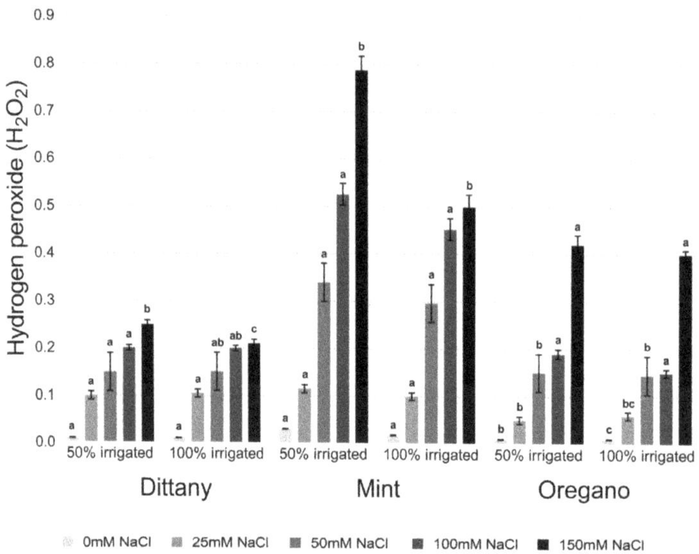

Figure 4. Effect of NaCl on H_2O_2 (μmol g^{-1} FW) content in leaves of different Lamiaceae species *Origanum dictamnus* L., *Mentha spicata* L., and *Origanum onites* L. Each value is the mean of 4 replications ± standard error. Means not sharing the same letter are significantly different at $p \leq 0.05$.

Table 2. Content of essential oils of the examined plant species under salt stress (mL 100 g^{-1} D.W.). Mean values (n = 4 ± se) not sharing the same letter are significantly different at $p \leq 0.05$.

Treatments	*M. spicata*	*O. dictamnus*	*O. onites*
0 mM NaCl	3.24 ± 0.08 b	3.61 ± 0.18 d	4.86 ± 0.09 b
25 mM NaCl	3.60 ± 0.05 c	2.00 ± 0.06 a	4.53 ± 0.08 a
50 mM NaCl	2.94 ± 0.07 a	2.11 ± 0.03 a	5.54 ± 0.15 c
100 mM NaCl	3.06 ± 0.13 ab	2.60 ± 0.12 b	6.57 ± 0.11 e
150 mM NaCl	3.15 ± 0.11 ab	2.93 ± 0.13 c	6.07 ± 0.16 d
p-value	<0.001	<0.001	<0.001

The composition of essential oils was not affected in terms of the main constituents in spearmint and dittany, while in oregano, at 25 and 50 mM NaCl treatments there was a variation in the ratio of carvacrol to its precursors. Moreover, in the treatments with 25 and 50 mM NaCl, there is a clear differentiation in the ratio of carvacrol to precursors of 0.68 and 0.65, respectively. The highest salinity procedures, such as 100 and 150 mM NaCl with a ratio of carvacrol to precursors of 0.99 and 0.98, respectively, are the same as those of the control (1.03) (Tables 3–5). Since, in salinity interventions, the plants preceded the growth stages, this difference cannot be due to a more advanced stage of ripening.

Table 3. Composition of the essential oils of *Mentha spicata* L.

Compounds	tR [1]	RI [2]	RI [3]	0 mM NaCl	25 mM NaCl	50 mM NaCl	100 mM NaCl	150 mM NaCl	*p*-Value
Limonene	8.402	1026	1029	21.69 ± 0.01 a	23.22 ± 0.05 b	21.81 ± 0.23 a	23.16 ± 0.10 b	22.88 ± 0.25 b	<0.001
1,8-Cineole	8.463	1029	1031	9.81 ± 0.03 a	9.54 ± 0.05 a	11.46 ± 0.18 b	10.40 ± 0.08 b	10.34 ± 0.11 b	<0.001
Carvone	11.912	1242	1243	53.11 ± 0.12 a	51.32 ± 0.02 a	52.70 ± 0.07 a	52.60 ± 0.18 a	50.91 ± 0.15 a	>0.05
β-Caryophyllene	14.661	1418	1419	1.81 ± 0.02 e	1.36 0.01 c	0.87 ± 0.01 a	1.11 ± 0.04 b	1.61 ± 0.04 d	<0.001
% Total				85.80	85.44	86.84	87.26	85.29	

[1] tR: Retention time (min); [2] RI: Retention Indices from experimental using a SBP-5 column using a homologous series of n-alkanes (C_9–C_{25}); [3] RI: Retention indices according to literature. Alphabetic characters (a–e) in the table demonstrate the significant differences among the various NaCl treatments, as they resulted from the post hoc analyses, based on Tukey's Honest Significant Difference (HSD) test.

Table 4. Composition of the essential oils of *Origanum dictamnus* L.

Compounds	tR [1]	RI [2]	RI [3]	0 mM NaCl	25 mM NaCl	50 mM NaCl	100 mM NaCl	150 mM NaCl	*p*-Value
α-Thujene	6.493	930	932	1.24 ± 0.02 e	1.05 ± 0.01 b	0.77 ± 0.01 a	1.14 ± 0.00 c	1.18 ± 0.01 d	<0.001
α-Terpinene	8.185	1012	1014	2.05 ± 0.00 b	1.91 ± 0.01 a	1.91 ± 0.03 a	2.05 ± 0.01 b	2.07 ± 0.00 b	<0.001
p-Cymene	8.314	1022	1024	34.81 ± 0.07 b	43.25 ± 0.04 c	33.31 ± 0.62 a	35.16 ± 0.10 b	33.57 ± 0.29 a	<0.001
γ-Terpinene	8.924	1052	1054	13.54 ± 0.07 c	10.16 ± 0.03 a	13.37 ± 0.10 c	12.41 ± 0.04 b	12.50 ± 0.12 b	<0.001
Linalool	9.602	1092	1095	2.02 ± 0.04 ab	3.05 ± 0.03 d	1.95 ± 0.05 a	2.07 ± 0.01 b	2.34 ± 0.02 c	<0.001
Thymol	12.514	1290	1290	0.12 ± 0.01 a	0.15 ± 0.01 b	0.16 ± 0.01 b	0.12 ± 0.00 a	0.16 ± 0.00 b	<0.001
Carvacrol	12.669	1299	1298	34.77 ± 0.05 b	25.25 ± 0.27 a	36.13 ± 0.37 c	34.41 ± 0.20 b	35.84 ± 0.15 c	<0.001
% Total				88.55	84.82	87.60	87.36	87.65	

[1] tR: Retention time (min); [2] RI: Retention Indices from experimental using a SBP-5 column using a homologous series of n-alkanes (C_9–C_{25}); [3] RI: Retention indices according to literature. Alphabetic characters (a–e) in the table demonstrate the significant differences among the various NaCl treatments, as they resulted from the post hoc analyses, based on Tukey's Honest Significant Difference (HSD) test.

Table 5. Composition of the essential oils of *Origanum onites* L.

Compounds	tR [1]	RI [2]	RI [3]	0 mM NaCl	25 mM NaCl	50 mM NaCl	100 mM NaCl	150 mM NaCl	*p*-Value
α-Thujene	6.493	930	932	1.20 ± 0.01 b	1.72 ± 0.02 d	2.45 ± 0.03 e	1.00 ± 0.01 a	1.29 ± 0.02 c	<0.001
α-Pinene	6.621	937	939	0.78 ± 0.01 b	1.16 ± 0.00 d	1.62 ± 0.02 e	0.67 ± 0.00 a	0.85 ± 0.00 c	<0.001
Myrcene	7.695	986	990	1.28 ± 0.00 b	1.72 ± 0.01 c	2.47 ± 0.05 d	1.16 ± 0.01 a	1.34 ± 0.00 b	<0.001
α-Terpinene	8.176	1012	1014	3.18 ± 0.02 a	4.80 ± 0.02 c	5.24 ± 0.02 d	3.43 ± 0.01 b	3.41 ± 0.02 b	<0.001
p-Cymene	8.314	1022	1024	3.45 ± 0.01 a	4.73 ± 0.01 d	5.88 ± 0.01 e	3.51 ± 0.01 b	3.66 ± 0.01 c	<0.001
Limonene	8.400	1026	1029	1.12 ± 0.00 a	1.50 ± 0.00 d	1.96 ± 0.02 e	1.16 ± 0.01 b	1.25 ± 0.02 c	<0.001
γ-Terpinene	8.924	1052	1054	8.65 ± 0.01 a	12.55 ± 0.08 d	13.96 ± 0.03 e	9.59 ± 0.00 c	8.95 ± 0.02 b	<0.001
cis-Linalool oxide	9.439	1172	1174	0.78 ± 0.00 a	1.11 ± 0.01 d	1.03 ± 0.02 c	0.82 ± 0.01 b	0.82 ± 0.01 b	<0.001
Linalool	9.602	1092	1095	11.71 ± 0.02 d	10.77 ± 0.07 a	11.40 ± 0.07 c	10.96 ± 0.08 b	11.25 ± 0.03 c	<0.001
Borneol	10.728	1162	1165	1.52 ± 0.00	1.36 ± 0.00	1.08 ± 0.00	1.59 ± 0.00	1.45 ± 0.00	–
Terpinen-4-ol	10.891	1172	1174	6.52 ± 0.00 b	8.53 ± 0.53 d	4.56 ± 0.01 a	7.62 ± 0.01 c	6.81 ± 0.02 b	<0.001
α-Terpineol	11.097	1188	1188	3.93 ± 0.01 b	3.91 ± 0.02 b	2.97 ± 0.00 a	4.16 ± 0.02 c	3.95 ± 0.05 b	<0.001
Thymol	12.514	1290	1290	0.32 ± 0.03 c	0.20 ± 0.01 a	0.17 ± 0.00 a	0.27 ± 0.01 bc	0.26 ± 0.03 b	<0.001
Carvacrol	12.669	1299	1298	45.41 ± 0.15 d	36.46 ± 0.35 b	35.44 ± 0.24 a	45.56 ± 0.10 d	44.50 ± 0.03 c	<0.001
% Total				89.85	90.50	90.24	91.49	89.80	
Ratio of carvacrol to precursors				1.03	0.68	0.65	0.99	0.98	
Monoterpene Hydrocarbons				19.66	28.18	33.58	20.52	20.75	
Sum of precursors				44.12	53.86	54.62	45.67	45.03	

[1] tR: Retention time (min); [2] RI: Retention Indices from experimental using a SBP-5 column using a homologous series of n-alkanes (C_9–C_{25}); [3] RI: Retention indices according to literature. Alphabetic characters (a–e) in the table demonstrate the significant differences among the various NaCl treatments, as they resulted from the post hoc analyses, based on Tukey's Honest Significant Difference (HSD) test.

4. Discussion

To mitigate the adverse effects of drought and salinity on aromatic plants and improve their productivity, it is crucial to understand the physiological and biochemical responses of aromatic plants under these conditions. Generally, the addition of salts to the soil solution limits the ability of the plant to absorb water, resulting in delayed growth and development [33]. The growth, yield, and quality of aromatic Lamiaceae plants can be adversely affected by salinity [34].

We focused on three species of aromatic plants in order to investigate their tolerance to saline and drought conditions because they are widely recognized for their commercial importance. By studying the physiological and biochemical responses of these aromatic species to salt stress, we aimed to gain a better knowledge of their tolerance mechanisms. This observation can be valuable for breeding and selecting salt-tolerant cultivars, as well as for implementing effective strategies to mitigate the negative effects of salt stress on the cultivation of aromatic plants.

Among the three aromatic species, dittany was revealed to be more resistant to salt and drought stress compared to oregano and spearmint, as there was a significant increase in its height and the ratio DW/FW (Table 1). The increase in salinity accelerated the entry of plants into flowering, especially for oregano. The vegetative growth of plants lasts throughout spring for spearmint, whereas a maximum growth at the end of April and early May was observed for oregano. As a result, the growth and the height of the three aromatic plants under salinity treatments, especially in the higher concentrations and a simultaneous 50% of irrigation, was poor.

Salt stress is known to have detrimental effects on plant growth, development, and antioxidant capacity. One of the reasons for the reduced plant growth under salt stress is the accumulation of excess salts around the root zone, which interferes with water uptake. The increase in salinity accelerated the entry of plants into flowering, especially for island oregano. Similar results that show the negative relationship between the salinity stress and the plant growth parameters were also reported previously [35], although the response was different in different genotypes. Drought stress leads to turgor loss, trim down in photoassimilation, and metabolites that are required for cell division. As a consequence, impaired mitosis and cell elongation and expansion result in reduced growth [13,35,36]. Likewise, we observed a significant and uniform reduction in all measured production parameters (plant height, canopy diameter, leaf area, fresh root weight, and biomass production) under drought stress, across all the tested species.

The concentration of certain ions increases under salt stress, which can have inhibitory effects on plant metabolism. This specific ion toxicity can disrupt plant functions and lead to mineral nutrient imbalances and deficiencies [13,35].

The decreased chlorophyll content under water stress and salinity might be a result of the ion uptake disturbances with concomitant reduced photosynthesis and respiration of the aromatic plants. Among the three species, spearmint revealed that it lost a remarkable portion of the photosynthetic pigment resulting in growth inhibition.

Plant grown under salt and water stress may exhibit significantly higher levels of certain secondary plant metabolites compared to controls cultivated in normal conditions [13]. Additionally, those stresses adversely affect the content of total carbohydrates, fatty acids, and proteins, while increasing the level of amino acids, particularly proline. The distinctive cyclic structure of proline's side chain gives proline an exceptional conformational rigidity compared to other amino acids. It is essential in the primary metabolism and is the most extensively studied osmolyte due to its great importance for stress tolerance [18]. For osmotic adjustment, proline contributes to the stabilization of subcellular structures (e.g., membranes and proteins), scavenging free radicals, and buffering cellular redox potential under stress conditions. The rapid breakdown of proline upon stress relief may provide sufficient reducing agents that support mitochondrial oxidative phosphorylation and the generation of ATP for recovery from stress, as well as repairing stress-induced damage [37].

Proline is considered to be a protective osmolyte associated with the degree of stress tolerance; its accumulation indicates that the plants are under stress [38]. There is evidence that the synthesis and accumulation of proteins is related to the resistance of plants to water stress [39,40]. However, a correlation between proline accumulation and stress tolerance is not always apparent [41], and other authors have suggested that it is an indicator of some type of damage [42]. The influence of drought and salt stress on secondary compound accumulation is highly debatable and a lot of contradictory results are reported [24,26,34,43,44]. In this experiment, we observed an augmentation of proline biosynthesis in the more sensitive spearmint compared to the more resistant dittany and oregano. Similar results in proline content have also been observed as a result of water stress in oregano plants [45].

The findings of this study highlight the importance of lipid peroxidation and TBARs content as indicators of salt stress response in aromatic plants. Understanding the biochemical role of MDA is important in assessing the extent of lipid peroxidation and its impact on cellular structures. The increase in MDA levels indicates higher levels of lipid peroxidation and suggests the presence of cellular membrane damage. This can have detrimental effects on various cellular processes and overall plant performance. The contrasting responses observed in spearmint (sensitive), oregano, and dittany (tolerant) species provide insights into the mechanisms underlying salt and drought tolerance and susceptibility. Our results indicated that spearmint is more susceptible to salinity and drought-induced lipid peroxidation, leading to membrane damage and ultrastructure changes. On the other hand, the tolerant oregano and dittany have mechanisms that prevent or minimize lipid peroxidation, thus protecting their cell membranes and maintaining their ultrastructure integrity. Similar results were reported by Krause et al. [46], who found that salt treatment caused an increase in lipid peroxidation, as revealed by high TBA content in sensitive barley genotypes, but no significant effect was shown in the most salt-tolerant genotypes. According to Chiappero et al. [47], drought effects were ameliorated by the reduction of MDA (malonyldialdehyde) levels, thus avoiding the accumulation of ROS and increasing antioxidant enzyme activities and the antioxidant level, specifically in relation to the total phenolic compounds. In the current experiment, stress treatments induced the enhancement of hydrogen peroxide, especially in spearmint plants compared to oregano plants. On the contrary, this enhancement of H_2O_2 indicates the prevalence of oxidative stress and this may be one of the possible mechanisms by which the toxicity of salinity and drought could be manifested in the plant tissues [24,48]. Nevertheless, H_2O_2 accumulation is another ROS that is implicated in enhanced lipid peroxidation and membrane damage, causing cell death [49,50].

Regarding the quantitative and qualitative composition of spearmint plants, the essential oil was not affected when referring to carbon as the most predominant component in all treatments. This result is in accordance with a previous study performed by Ounoki [51] and reveals that the essential oil composition of spearmint was unaffected by the treatments applied to the plant material [51]. In dittany plants, no statistically significant difference was observed in the quantitative and qualitative composition of dew essential oil. Salinity did not seem to affect the concentration of carvacrol, which is the main component. It should be noted that the quantities of *p*-cymene and γ-terpinene in the treatment with 25 mM NaCl seem to increase, whereas carvacrol decreases compared to controls.

In pennyroyal plant grown under water stress, a 40% increase in EO content was also shown [52], with comparable responses being observed for other Lamiaceae species grown under water stress, such as *Lavandula angustifolia* [12], *Ocimum basilicum* [12], and *Thymus vulgaris* [46,53]. However, in a one-year field trial, the researchers [19] observed that the Lamiaceae subjected to drought stress did not present variations in the amount of EOs, whereas the EO content actually decreased in *Lavandula latifolia* and *Salvia sclarea*. Other authors have also indicated negative effects on EO content due to water stress in *Mentha arvensis* [47,51,54] *Hyssopus officinalis* [55], *Rosmarinus officinalis* [56], *Salvia officinalis* [57], and *Thymus vulgaris* [58].

Finally, in oregano plants the quantitative and qualitative composition of the dew essential oil was affected after salinity treatments and drought. The percentage of the main component carvacrol did not significantly differ in treatments with both 100 and 150 mM NaCl. On the other hand, the essential oil extracted by plants treated with 25 and 50 mM NaCl were almost 10% poorer in carvacrol, with a decrease in the amount of monoterpenoids at the same time (Table 5). The monoterpene alcohol carvacrol is a compound with the characteristic flavor of oregano also found in other Lamiaceae species. This phenolic monoterpene, among others, is widely used for its pharmaceutical, cosmetic-, and food-related properties. γ-terpinene and *p*-cymene are considered the main precursors of carvacrol [51]. It has been suggested that the induction of EO yield under drought stress may be because when plants grow under stress conditions, they only allocate low amounts of carbohydrates from photosynthesis to plant development, and instead use these for the synthesis of secondary and reserve metabolites, thus generating a balance between growth and defense [12,24,46]. The increase in EO in this case could be working as a mechanism by which to dissipate "extra" energy [12,28]. Indeed, in favorable conditions, most of the plant's energy is directed towards primary metabolism and only a small portion to the secondary metabolism [46].

Further research in this area can contribute to the development of improved aromatic varieties with enhanced salt and drought tolerance.

5. Conclusions

Comparing the studied species, spearmint is more susceptible to salinity and drought, which induces lipid peroxidation, leading to membrane damage and ultrastructure changes. In contrast, oregano and dittany have developed mechanisms for preventing or minimizing lipid peroxidation. This indicates their ability to withstand salinity and drought stress without significant cell membrane damage or ultrastructure changes. These mechanisms may involve the activation of antioxidant enzymes or the accumulation of specific molecules that protect against oxidative stress.

Dittany plants have a strong ability to tolerate high levels of salinity in the soil, as indicated by the increase in essential oil yield. However, the concentration of carvacrol remained constant, regardless of the salinity levels.

Future work in the field should be undertaken to discern the roles and mechanisms affected by proline and other secondary metabolites, to reduce oxidative stress, and to study the response of chemotype of aromatics plants under salt and drought conditions.

Author Contributions: M.K.S., A.E.G. and V.L. contributed to the experimental part, literature search, and drafted the paper. G.O. and C.P. contributed to data analysis. E.P.-P. contributed to the collection, identification, and cultivation of the plant material and to the critical reading of the manuscript. M.K.S. supervised the phytochemical experiments and contributed to a critical reading of the manuscript. E.P.-P., A.E.G. and V.L. designed the study and supervised the experiments. G.O., C.P. and M.K.S. contributed to drafting and critically reading the manuscript. All authors have read and agreed to the published version of the manuscript.

Funding: This research received no external funding.

Institutional Review Board Statement: Not applicable for studies not involving humans or animals.

Informed Consent Statement: Not applicable for studies not involving humans or animals.

Data Availability Statement: The original contributions presented in the study are included in the article.

Conflicts of Interest: The authors declare no conflicts of interest.

References

1. Stefanakis, M.K.; Papaioannou, C.; Lianopoulou, V.; Philotheou-Panou, E.; Giannakoula, A.E.; Lazari, D.M. Seasonal Variation of Aromatic Plants under Cultivation Conditions. *Plants* **2022**, *11*, 2083. [CrossRef]
2. Dimopoulos, P.; Raus, T.; Bergmeier, E.; Constantinidis, T.; Iatrou, G.; Kokkini, S.; Strid, A.; Tzanoudakis, D. Vascular plants of Greece: An annotated checklist. Supplement. *Willdenowia* **2016**, *46*, 301–347. [CrossRef]
3. Karaca, M.; Ince, A.G.; Aydin, A.; Ay, S.T. Cross-genera transferable e-microsatellite markers for 12 genera of the Lamiaceae family. *J. Sci. Food Agric.* **2013**, *93*, 1869–1879. [CrossRef]
4. Tibaldi, G.; Fontana, E.; Nicola, S. Growing conditions and postharvest management can affect the essential oil of *Origanum vulgare* L. ssp. hirtum (link) Ietswaart. *Ind. Crops Prod.* **2011**, *34*, 1516–1522. [CrossRef]
5. Tonk, F.; Yüce, S.; Bayram, E.; Akçali Giachino, R.R.; Sönmez, Ç.; Telci, İ.; Furan, M. Chemical and genetic variability of selected Turkish oregano (*Origanum onites* L.) clones. *Plant Syst. Evol.* **2010**, *288*, 157–165. [CrossRef]
6. Stefanakis, M.K.; Touloupakis, E.; Anastasopoulos, E.; Ghanotakis, D.; Katerinopoulos, H.E.; Makridis, P. Antibacterial activity of essential oils from plants of the genus *Origanum*. *Food Control* **2013**, *34*, 539–546. [CrossRef]
7. Papaioannou, C.; Stefanakis, M.K.; Batargias, C.; Kilias, G.; Anastasopoulos, E.; Katerinopoulos, H.E.; Papasotiropoulos, V. Genetic Profiling and Volatile Oil Content of Oregano Genotypes from Greece. *Rev. Bras. Farmacogn.* **2020**, *30*, 295–300. [CrossRef]
8. Stefanakis, M.K.; Anastasopoulos, E.; Katerinopoulos, H.E.; Makridis, P. Use of essential oils extracted from three *Origanum* species for disinfection of cultured rotifers (*Brachionus plicatilis*). *Aquac. Res.* **2014**, *45*, 1861–1866. [CrossRef]
9. Stefanaki, A.; van Andel, T. Mediterranean aromatic herbs and their culinary use. In *Aromatic Herbs in Food. Bio-Active Compounds, Processing and Applications*; Academic Press: Cambridge, MA, USA, 2021; pp. 93–121. [CrossRef]
10. Bardaweel, S.K.; Bakchiche, B.; AL Salamat, H.A.; Rezzoug, M.; Gherib, A.; Flamini, G. Chemical composition. antioxidant antimicrobial and antiproliferative activities of essential oil of *Mentha spicata* L. (Lamiaceae) from Algerian Saharan atlas. *BMC Complement. Med. Ther.* **2018**, *18*, 201. [CrossRef] [PubMed]
11. Mahboubi, M. *Mentha spicata* L. essential oil. phytochemistry and its effectiveness in flatulence. *J. Tradit. Complement. Med.* **2018**, *11*, 75–81. [CrossRef] [PubMed]
12. Mahajan, M.; Kuiry, R.; Pal, P.K. Understanding the consequence of environmental stress for accumulation of secondary metabolites in medicinal and aromatic plants. *J. Appl. Res. Med. Aromat. Plants* **2020**, *18*, 100255. [CrossRef]

13. Mulugeta, S.M.; Radácsi, P. Influence of Drought Stress on Growth and Essential Oil Yield of *Ocimum* Species. *Horticulturae* **2022**, *8*, 175. [CrossRef]
14. Shao, H.B.; Chu, L.Y.; Jaleel, C.A.; Zhao, C.X. Water-deficit stress-induced anatomical changes in higher plants. *Comptes Rendus Biol.* **2008**, *331*, 215–225. [CrossRef] [PubMed]
15. Ahl, H.A.; Omer, E.A. Medicinal and aromatic plants production under salt stress. A review. *Herba Pol.* **2011**, *57*, 72–87.
16. Khorsandi, O.; Hassani, A.; Sefidkon, F.; Shirzad, H.; Khorsandi, A. Effect of salinity (NaCl) on growth, yield, essential oil content and composition of *Agastache foeniculum* kuntz. *IJMAPR* **2010**, *26*, 438–451. [CrossRef]
17. Dehghani Bidgoli, R.; Azarnezhad, N.; Akhbari, M.; Ghorbani, M. Salinity stress and PGPR effects on essential oil changes in *Rosmarinus officinalis* L. *Agric. Food Secur.* **2019**, *8*, 2. [CrossRef]
18. Sarmoum, R.; Haid, S.; Biche, M.; Djazouli, Z.; Zebib, B.; Merah, O. Effect of Salinity and Water Stress on the Essential Oil Components of Rosemary (*Rosmarinus officinalis* L.). *Agronomy* **2019**, *9*, 214. [CrossRef]
19. García-Caparrós, P.; Romero, M.J.; Llanderal, A.; Cermeño, P.; Lao, M.T.; Segura, M.L. Effects of Drought Stress on Biomass, essential oil content, nutritional parameters and costs of production in six *Lamiaceae* Species. *Water* **2019**, *11*, 573. [CrossRef]
20. Benitez, L.; Vighi, I.; Nogueira do Amaral, M.; Auler Moraes, G.; Rodrigues, L.; Braga, A. Correlation of proline content and gene expression involved in the metabolism of this amino acid under abiotic stress. *Acta Physiol. Plant.* **2016**, *38*, 267. [CrossRef]
21. Kren, S.; Sönmez, Ç.; Özçakal, E.; Kurttas, Y.S.K.; Bayram, E.; Gürgülü, H. The effect of different irrigation water levels on yield and quality characteristics of purple basil (*Ocimum basilicum* L.). *Agric. Water Manag.* **2012**, *109*, 155–161. [CrossRef]
22. Szabó, K.; Zubay, P.; Zámboriné, N. What shapes our knowledge of the relationship between water deficiency stress and plant volatiles? *Acta Physiol. Plant.* **2020**, *42*, 130. [CrossRef]
23. Wintermans, J.F.; De Mots, A. Spectrophotometric characteristics of chlorophylls a and b and their pheophytins in ethanol. *Biochim. Biophys. Acta* **1965**, *109*, 448–453. [CrossRef]
24. Giannakoula, A.; Therios, I.; Chatzissavvidis, C. Effect of Lead and Copper on Photosynthetic Apparatus in Citrus (*Citrus aurantium* L.) Plants. The Role of Antioxidants in Oxidative Damage as a Response to Heavy Metal Stress. *Plants* **2021**, *10*, 155. [CrossRef]
25. Heath, R.L.; Packer, L. Photoperoxidation in isolated chloroplasts. *Arch. Biochem. Biophys.* **1968**, *125*, 189–198. [CrossRef]
26. Binott, J.J.; Owuoche, J.O.; Bartels, D. Physiological and molecular characterization of Kenyan barley (*Hordeum vulgare* L.) seedlings for salinity and drought tolerance. *Euphytica* **2017**, *213*, 139. [CrossRef]
27. Alexieva, V.; Sergiev, I.; Mapelli, S.; Karanov, E. The effect of drought and ultraviolet radiation on growth and stress markers in pea and wheat. *Plant Cell Environ.* **2001**, *24*, 1337–1344. [CrossRef]
28. Velikova, V.; Yordanov, I.T.; Edreva, A.M. Oxidative stress and some antioxidant systems in acid rain-treated bean plants Protective role of exogenous polyamines. *Plant Sci.* **2000**, *151*, 59–66. [CrossRef]
29. Council of Europe. *European Pharmacopoeia*, 5th ed.; Council of Europe: Strasbourg, France, 2005; pp. 217, 2567–2570.
30. National Institute of Standards and Technology (NIST). 2017. Available online: http://webbook.nist.gov/chemistry/name-ser.html (accessed on 21 December 2021).
31. Adams, R. *Identification of Essential Oil Components by Gas Chromatography/Mass Spectroscopy*, 4th ed.; Allured Publishing Corporation: Carol Stream, IL, USA, 2007.
32. Van den Dool, H.; Kratz, P.D. A generalization of the retention index system including linear temperature programmed as-liquid partition chromatography. *J. Chromatogr. A* **1963**, *11*, 463–471. [CrossRef] [PubMed]
33. Al-Fraihat, A.H.; Al-Dalain, S.Y.; Zatimeh, A.A.; Haddad, M.A. Enhancing Rosemary (*Rosmarinus officinalis*, L.) Growth and Volatile Oil Constituents Grown under Soil Salinity Stress by Some Amino Acids. *Horticulturae* **2023**, *9*, 252. [CrossRef]
34. Avasiloaiei, D.I.; Calara, M.; Brezeanu, P.M.; Murariu, O.C.; Brezeanu, C. On the Future Perspectives of Some Medicinal Plants within *Lamiaceae* Botanic Family Regarding Their Comprehensive Properties and Resistance against Biotic and Abiotic Stresses. *Genes* **2023**, *14*, 955. [CrossRef] [PubMed]
35. Estaji, A.; Roosta, H.; Rezaei, S.; Hosseini, S.; Niknam, F. Morphological, physiological and phytochemical response of different Satureja hortensis L. accessions to salinity in a greenhouse experiment. *J. Appl. Res. Med. Aromat. Plants* **2018**, *10*, 25–33. [CrossRef]
36. Olfati, J.; Moqbeli, E.; Fathollahi, S.; Estaji, A. Salinity stress effects changed during *Aloe vera* L. vegetative growth. *J. Stress Physiol. Biochem.* **2012**, *8*, 152–158.
37. Iqbal, H.; Yaning, C.; Waqas, M.; Shareef, M.; Raza, S.T. Differential response of quinoa genotypes to drought and foliage-applied H_2O_2 in relation to oxidative damage, osmotic adjustment and antioxidant capacity. *Ecotoxicol. Environ. Saf.* **2018**, *164*, 344–354. [CrossRef] [PubMed]
38. Signorelli, S. The Fermentation Analogy: A Point of View for Understanding the Intriguing Role of Proline Accumulation in Stressed Plants. *Front. Plant Sci.* **2016**, *7*, 1339. [CrossRef] [PubMed]
39. Ghorbanli, M.; Gafarabad, M.; Amirkian, T.; Mamaghani, B.A. Investigation of proline, total protein, Chlorophyll, ascorbate and dehydroascorbate changes under drought stress in Akria and Mobil tomato cultivars. *Plant Sci.* **2013**, *45*, 241–249.
40. Rezayian, M.; Niknam, V.; Ebrahimzadeh, H. Differential responses of phenolic compounds of *Brassica napus* under drought stress. *Iran. J. Plant Physiol.* **2018**, *8*, 2417–2422.
41. Szabados, L.; Savoure, A. Proline: A multifunctional amino acid. *Trends Plant Sci.* **2009**, *15*, 1360–1385. [CrossRef]

42. Hien, D.; Jacobs, M.; Angenon, G.; Hermans, C.; Thanh, T.; Le Van Son, T.; Roosens, N. Proline accumulation and D 1-pyrroline-5-carboxylate synthetase gene properties in three rice cultivars differing in salinity and drought tolerance. *Plant Sci.* **2003**, *165*, 1059–1068. [CrossRef]
43. Witzel, K.; Matrosa, A.; Strickert, M.; Kaspar, S.; Peukert, M.; Mühlinge, K.H.; Börnera, A.; Mock, H.P. Salinity Stress in Roots of Contrasting Barley Genotypes Reveals Time-Distinct and Genotype-Specific Patterns for Defined Proteins. *Mol. Plant* **2014**, *7*, 336–355. [CrossRef]
44. Uddin, M.K.; Juraimi, A.S.; Ismail, M.R.; Othman, R.; Rahim, A.A. Relative salinity tolerance of warm season turfgrass species. *J. Environ. Biol.* **2011**, *32*, 309–312.
45. Abdali, R.; Rahimi, A.; Siavash Moghaddam, S.; Heydarzadeh, S.; Arena, C.; Vitale, E.; Zamanian, M. The Role of Stress Modifier Biostimulants on Adaptive Strategy of Oregano Plant for Increasing Productivity under Water Shortage. *Plants* **2023**, *12*, 4117. [CrossRef]
46. Krause, S.T.; Liao, P.; Crocoll, C.; Boachon, B.; Förster, C.; Leidecker, F.; Wiese, N.; Zhao, D.; Wood, J.C.; Buell, C.R.; et al. The biosynthesis of thymol, Carvacrol, and thymohydroquinone in Lamiaceae proceeds via cytochrome P450s and a short-chain dehydrogenase. *Proc. Natl. Acad. Sci. USA* **2021**, *118*, e2110092118. [CrossRef]
47. Chiappero, J.; Cappellari, L.d.R.; Palermo, T.B.; Giordano, W.; Banchio, E. Influence of Drought Stress and PGPR Inoculation on Essential Oil Yield and Volatile Organic Compound Emissions in *Mentha piperita*. *Horticulturae* **2022**, *8*, 1120. [CrossRef]
48. Allel, D.; Ben-Amar, A.; Abdelly, C. Leaf photosynthesis, chlorophyll fluorescence and ion content of barley (*Hordeum vulgare*) in response to salinity. *J. Plant Nutr.* **2018**, *41*, 497–508. [CrossRef]
49. Sun, S.Q.; He, M.; Cao, T.; Yusuyin, Y.; Han, W. Antioxidative responses related to H_2O_2 depletion in *Hypnum plumaeforma* under the combined stress induced by Pb and Ni. *Environ. Monit. Assess.* **2010**, *163*, 303–312. [CrossRef]
50. Singh, S.; Eapen, S.; D'Souza, S. Cadmium accumulation and its influence on lipid peroxidation and antioxidative system in an aquatic plant, *Bacopa monnieri* L. *Chemosphere* **2006**, *62*, 233–246. [CrossRef]
51. Ounoki, R.; Ágh, F.; Hembrom, R.; Ünnep, R.; Szögi-Tatár, B.; Böszörményi, A.; Solymosi, K. Salt Stress Affects Plastid Ultrastructure and Photosynthetic Activity but Not the Essential Oil Composition in Spearspearmint (*Mentha spicata* L. var. crispa "Moroccan"). *Front. Plant Sci.* **2021**, *12*, 739467. [CrossRef] [PubMed]
52. Jamil, M.; Bae, L.D.; Yong, J.K.; Ashraf, M.W.; Chun, L.S.; Shik, R.E. Effect of salt (NaCl) stress on germination and early seedling growth of four vegetables species. *J. Cent. Eur. Agric.* **2006**, *7*, 273–282.
53. Aziz, E.A.; Hendawi, S.T.; Azza, E.E.D.; Omer, E.A. Effect of soil type and irrigation intervals on plant growth, essential oil and constituents of *Thymus vulgaris* plant. *Am.-Eurasian J. Agric. Environ. Sci.* **2008**, *4*, 443–450.
54. Misra, A.; Srivastava, N.K. Influence of water stress on Japanese mint. *J. Herbs Spices Med. Plants* **2000**, *7*, 51–58. [CrossRef]
55. Danesh-Shahraki, H.; Pirbalouti, A.G.; Rajabzadeh, F.; Kachouei, M.A. Water Deficit Stress Mitigation by the Foliar Spraying of Salicylic Acid and Proline on the Volatile Oils and Growth Features of Hyssop (*Hyssopus officinalis* L.). *J. Essent. Oil Bear. Plants.* **2023**, *26*, 115–129. [CrossRef]
56. Singh, M.; Ramesh, S. Effect of irrigation and nitrogen on herbage, oil yield and water-use efficiency in rosemary grown under semi-arid tropical conditions. *J. Med. Aromat. Plant Sci.* **2000**, *22*, 659–662.
57. Govahi, M.; Ghalavand, A.; Nadjafi, F.; Sorooshzadeh, A. Comparing different soil fertility systems in Sage (*Salvia officinalis*) under water deficiency. *Ind. Crops Prod.* **2015**, *74*, 20–27. [CrossRef]
58. Rahimi, A.; Mohammadi, M.M.; Siavash Moghaddam, S.; Heydarzadeh, S.; Gitari, H. Effects of Stress Modifier Biostimulants on Vegetative Growth, Nutrients, and Antioxidants Contents of Garden Thyme (*Thymus vulgaris* L.) Under Water Deficit Conditions. *J. Plant Growth Regul.* **2022**, *41*, 2059–2072. [CrossRef]

MDPI AG
Grosspeteranlage 5
4052 Basel
Switzerland
Tel.: +41 61 683 77 34

Horticulturae Editorial Office
E-mail: horticulturae@mdpi.com
www.mdpi.com/journal/horticulturae

www.ingramcontent.com/pod-product-compliance
Lightning Source LLC
Chambersburg PA
CBHW040315270926
51850CB00007BC/1494
9783725867523